# Advancements in Real-Time Simulation of Power and Energy Systems

# Advancements in Real-Time Simulation of Power and Energy Systems

Editors

**Panos Kotsampopoulos**
**Md Omar Faruque**

MDPI • Basel • Beijing • Wuhan • Barcelona • Belgrade • Manchester • Tokyo • Cluj • Tianjin

*Editors*
Panos Kotsampopoulos
National Technical University of Athens
Greece

Md Omar Faruque
Florida State University
USA

*Editorial Office*
MDPI
St. Alban-Anlage 66
4052 Basel, Switzerland

This is a reprint of articles from the Special Issue published online in the open access journal *Energies* (ISSN 1996-1073) (available at: https://www.mdpi.com/journal/energies/special_issues/ advancements_real_time_simulation_power_energy_systems).

For citation purposes, cite each article independently as indicated on the article page online and as indicated below:

LastName, A.A.; LastName, B.B.; LastName, C.C. Article Title. *Journal Name* **Year**, *Volume Number*, Page Range.

**ISBN 978-3-0365-1214-3 (Hbk)**
**ISBN 978-3-0365-1215-0 (PDF)**

# Contents

# About the Editors

**Panos Kotsampopoulos** received the Diploma in Electrical and Computer Engineering from the National Technical University of Athens (NTUA), Greece in 2010 and the PhD degree in 2017 from the same school. Since 2010, he has been been working on research projects in the Smart RUE research group of NTUA, where he is currently a senior researcher and responsible for the development of the laboratory infrastructure of the Electric Energy Systems Laboratory. He was a guest researcher at the Austrian Institute of Technology (AIT), Vienna in 2012 and 2013. He has participated in several European research projects as a principal investigator for ICCS-NTUA and in national industrial projects. He is chair of the IEEE PES Task Force "Innovative teaching methods for modern power and energy systems", chapter co-leader of the IEEE WG P2004 "Recommended Practice for Hardware-in-the-Loop Simulation Based Testing of Electric Power Apparatus and Controls", and a member of other task forces and working groups. He is editor of the "IEEE Open Access Journal of Power and Energy" and member of the Editorial Board of the journal "Energies". He is chair of IEEE Young Professionals Greece and co-founder of the energy community "Collective Energy". He is a senior member of IEEE.

**Md Omar Faruque** received his Ph.D. from the University of Alberta, Canada in 2008. Since then, he has been working at the Center for Advanced Power Systems at Florida State University. In 2013, he was also appointed as faculty at the FAMU-FSU College of Engineering. His research interest is in the area of modeling and simulation, smart grid and renewable energy integration, energy management and demand response, and ship power system design. He has published more than 100 publications, including 40 in peer reviewed journals. He has been serving as the chair of the IEEE Power and Energy Society (PES) Task Force on "Real-Time Simulation of Power and Energy System" since 2012. He is also a co-chair of the working group "Modeling and Analysis of System Transient using Digital Programs". Prof. Faruque has been a member of many other IEEE task forces and working groups. He also served as a guest editor of many Special Issues of journals of IEEE, IET, and Energies.

# Preface to "Advancements in Real-Time Simulation of Power and Energy Systems"

Modern power and energy systems are characterized by the wide integration of distributed generation, storage and electric vehicles, adoption of ICT solutions, and interconnection of different energy carriers and consumer engagement, posing new challenges and creating new opportunities. Advanced testing and validation methods are needed to efficiently validate power equipment and controls in the contemporary complex environment and support the transition to a cleaner and sustainable energy system. Real-time hardware-in-the-loop (HIL) simulation has proven to be an effective method for validating and de-risking power system equipment in highly realistic, flexible, and repeatable conditions. Controller hardware-in-the-loop (CHIL) and power hardware-in-the-loop (PHIL) are the two main HIL simulation methods used in industry and academia that contribute to system-level testing enhancement by exploiting the flexibility of digital simulations in testing actual controllers and power equipment. This Special Issue aims to address recent advances in real-time HIL simulation in several domains (also in new and promising areas), including technique improvements, to promote its wider use.

The Special Issue is composed of 14 papers submitted from Europe, North America, and Asia, covering the following areas:

- Advancements in real-time digital simulation of power and energy systems.

- Advancements in CHIL simulation of power and energy systems.

- Advancements in PHIL simulation of power and energy systems.

K. Sidwall et al. summarize various recent advancements in real-time digital simulation, which have been enabled by the growth in high-performance processing space and the emerging availability of high-end processors for embedded designs. These include predictive switching of power electronic converters, streaming GOOSE and Sampled Values (IEC 61850-90-5), FPGA-based data streaming, and improved models of synchronous machines, transformers, and transmission lines. The second paper by M. Mirz et al. describes recent improvements to the open-source real-time simulator DPsim to study important use cases that involve grid-connected power electronics. It is shown that dynamic phasors, which result from shifted frequency analysis, allow the user to efficiently combine the characteristics of conventional phasor and electromagnetic transient simulation. A real-time hybrid simulator for dynamic performance testing of power electronic replica controllers in large power systems was developed and tested by J. Song et al. The hybrid simulator is a co-simulation tool that synthesizes a real-time simulator with a transient stability program to perform real-time dynamic simulation of a large power system. F. D'Agostino et al. illustrate in detail the development of a multiphysics real-time simulation framework, able to mimic the behavior of a DC electric ship equipped with electric propulsion, rotating generators, and battery energy storage systems. This section is concluded by L. Estrada et al. that present a systematic methodology for HIL simulation of power converters using LabVIEW software and FPGA.

CHIL simulation is addressed in the next set of papers. D. Sodin et al. developed and tested an under-frequency load shedding (UFLS) protection scheme in a CHIL setup. The under-frequency technology was implemented in an intelligent electronic device (IED) and it was shown that locally measured RoCoF can be effectively used for bringing a high level of flexibility to a system-wide scheme. J. Song et al. present industrial experiences of the first thyristor controlled series compensator (TCSC) replica controller installed in Korea (in 2019) through a review of its configuration, test platform, and practical application. The control performance was accurately verified under the entire Korean power system. L. Pellegrino et al. introduce the Joint Test Facility for Smart Energy Networks with Distributed Energy Resources (JaNDER) that allows users to exchange data in real time between two or more laboratory infrastructures and create a virtual infrastructure combining resources installed in different locations. Remote CHIL test cases are reported.

PHIL simulation is addressed at the last set of papers. A comparison of DER voltage regulation technologies using real-time simulations is performed by A. Summers et al. A novel state estimation-based particle swarm optimization for distribution voltage regulation was developed and tested in a PHIL setup. M. Barragán-Villarejo et al. propose a power system hardware-in-the-loop (PSHIL) concept, which widens the focus from device-oriented to system-oriented testing in order to evaluate, in a holistic way, the impact of a given technology on the power system, considering all of its power and control components. Geographically distributed PHIL tests by re-purposing existing grid-forming converters as a power interface and using the internet are reported by S. Vogel et al. M. Syed et al. present a non-intrusive add-on controller for DER inverters and establish its practical real-world relevance via a rigorous performance evaluation utilizing a high-fidelity HIL test bed. B. Guo et al. developed a variable speed hydro-electric plant PHIL setup, which is dedicated to different hydraulic operation scheme testing and control validation, while using reduced-scale models.

Finally, J. Montoya and members of the Smart Grid International Research Facility Network (SIRFN) present a comprehensive review of methods, test procedures, studies, and experiences by employing advanced real-time laboratory techniques for the validation of a range of research and development prototypes and novel power system solutions.

**Panos Kotsampopoulos , Md Omar Faruque**

*Editors*

*Article*

# Advancements in Real-Time Simulation for the Validation of Grid Modernization Technologies

**Kati Sidwall * and Paul Forsyth**

RTDS Technologies Inc., 150 Innovation Dr Unit 100, Winnipeg, MB R3T 2E1, Canada; paf@rtds.com
* Correspondence: kati@rtds.com

Received: 22 June 2020; Accepted: 31 July 2020; Published: 7 August 2020

**Abstract:** Real-time simulation and hardware-in-the-loop testing have increased in popularity as grid modernization has become more widespread. As the power system has undergone an evolution in the types of generator and load deployed on the system, the penetration and capabilities of automation and monitoring systems, and the structure of the energy market, a corresponding evolution has taken place in the way we model and test power system behavior and equipment. Consequently, emerging requirements for real-time simulators are very high when it comes to simulation fidelity, interfacing options, and ease of use. Ongoing advancements from a processing hardware, graphical user interface, and power system modelling perspective have enabled utilities, manufacturers, educational and research institutions, and consultants to apply real-time simulation to grid modernization projects. This paper summarizes various recent advancements from a particular simulator manufacturer, RTDS Technologies Inc. Many of these advancements have been enabled by growth in the high-performance processing space and the emerging availability of high-end processors for embedded designs. Others have been initiated or supported by developer participation in power industry working groups and study committees.

**Keywords:** real-time simulation; hardware-in-the-loop; multi-rate simulation; HIL; CHIL; PHIL; EMT

---

## 1. Introduction

Real-time digital simulation and hardware-in-the-loop (HIL) testing has been used in the power industry for over twenty-five years [1]. Originally developed as a solution for flexibly testing the control and protection associated with high-voltage direct current (HVDC) projects, the application of the technology is now more varied. Hardware-in-the-loop testing is used by electric utilities, educational and research institutions, and consultants. Much of the recent application of real-time simulation has been related to grid modernization technologies and projects.

Real-time simulators run an electromagnetic transient (EMT) model of the power system. EMT simulation tools are well-established in the power industry. In providing a time-varying, instantaneous value output, EMT models represent the behavior of the network over a wide frequency range and allow for a greater depth of analysis than traditional phasor-based modelling approaches such as load flow or transient stability analysis. As the modern power system continues to move away from a conventional synchronous machine-based structure, EMT simulation results become increasingly relevant due to their ability to provide results over a large range of frequencies. For systems containing generation or load interfaced via power electronic converters, phasor-based modelling tools can provide optimistic results, as they lack the ability to represent low-level converter controls or capture fast network dynamics during transient conditions [2]. EMT simulation overcomes these issues and is thus increasingly used from a power systems planning and operations perspective. Many established EMT tools, including real-time simulators, are based on the Dommel Algorithm for network solution [3]. Real-time simulators are distinct from PC-based EMT software tools in that they make use of dedicated parallel processing hardware to execute the simulation in real time. Continuous real-time operation allows for external equipment to be connected to the simulated network in a closed loop—referred to as HIL testing. HIL testing offers the ability to test multiple devices simultaneously, the ability to represent and study the behavior of power electronics (and their interaction with protection and control), and the flexibility and safety that comes with testing in a controlled laboratory environment rather than on site. Testing real equipment rather than a model of the device allows for greater confidence in equipment behavior and reduces issues experienced during commissioning and field operation.

The term "timestep" is used to describe the time between consecutive instantaneous outputs of the EMT simulation. The timestep is related to the sampling frequency of the simulation and determines the frequency bandwidth of signals that can be accurately reproduced. Real-time simulators must complete all calculations related to solving the network in real world time equal to or less than the simulation timestep. For protection and control testing, a timestep in the 30 to 60 microsecond ($\mu$s) range is generally used. The required timestep should be determined by the user based on their required results and dynamics of interest and can be adjusted as necessary.

Real-time simulators capable of HIL testing have several components: the parallel processing hardware platform for running the simulation, input/output (I/O) facilities in order to create the closed-loop interface with the device(s) being tested, and the graphical user interface which runs on the user's PC and allows for dynamic interaction with the simulation in real time. The graphical user interface includes the power system modelling library, another key component of the real-time simulator.

As the technology gains popularity and widespread use, it is critical that the manufacturers of real-time simulators develop the technology on a continual basis to improve simulation capability, simulation fidelity, application diversity, interfacing considerations, and ease of use. This requires advancements from the simulation hardware, graphical user interface, and power system modelling perspectives. Development is also particularly important given the rapidly-changing state of the power system. Increasing penetration of converter-connected distributed energy resources (DERs) and loads, widespread use of communication protocols and the growing reliance on information and operational technology, and decentralization of control are just a few of the many aspects of grid modernization that are changing the power system as we know it. Providing a modelling platform that accurately represents the modernized system and allows for the testing of existing and novel devices in this environment is critical to real-time simulator manufacturers.

This paper details several recent advancements to the RTDS® Simulator (NovaCor™, RTDS Technologies Inc., Winnipeg, MB, Canada), a real-time simulator which was originally developed in the 1980s and is now commercially available and used widely [4]. The major original contributions to the field of real-time power system simulation presented in this paper are summarized as follows:

- An original predictive switching technique allowing for the real-time switched-resistance representation of power electronic converters as part of small-timestep circuits without the need for a numerical interface.
- Direct support for streaming routable (R-) versions of GOOSE and Sampled Values (compliant with IEC Technical Report 61850-90-5) in real time.
- A field-programmable gate array (FPGA)-based implementation of real-time IEC 61850-9-2LE/IEC 61869-9 data streaming that allows for output stream sampling rates as high as 250 kHz.
- An improved real-time phase domain synchronous machine (PDSM) model with a stator-ground fault option and access to all phases of the stator terminals, the machine neutral, and up to two points in each phase of the field winding.
- An improved real-time transformer model based on a terminal duality equivalent.
- A frequency-dependent transmission line model capable of running in real-time at a timestep sufficiently small for the testing of travelling wave-based protective relays sampling in the 1 MHz range.

These advancements will support the work of power systems engineers in several ways:

- Improving the accuracy and fidelity with which power electronics and power system components are represented in real-time. This improves the fidelity of the HIL testing process, increases the range of events that can be replicated, and increases the range of functionalities that can be tested in protection, automation, and control devices. For example, the enhanced PDSM model presented in this paper will allow engineers to perform HIL testing of 100% stator-ground fault protection. The predictive switching algorithm presented in this paper will allow for the controls of converters with higher switching frequencies to be tested with greater accuracy.

- Expanding the types of devices and schemes which can be tested in a closed loop via communication protocols without requiring protocol converters or compromising the validity of the test setup. For example, the implementation of routable (R-) GOOSE messages presented in this paper will allow engineers to test automation schemes requiring IEC 61850-based substation-to-substation communication.

- Allowing engineers to test more state-of-the-art device features and schemes at relatively early stages of development. For example, the small timestep frequency-dependent transmission line model presented in this paper allows for the closed-loop testing of emergent travelling wave-based protection systems for the first time.

Examples of recent user experiences are available as case studies in the Supplementary Materials section. Further documentation on the advancements presented in this paper are available in the References section or upon request.

## 2. Advancements in Real-Time Simulation

### 2.1. Expanding the Tightly-Coupled Network Solution

As digital processing technology has become more advanced and accessible, the parallel processing hardware of real-time simulators has improved in terms of the size and complexity of the networks that can be represented and the simulation timesteps that can be maintained in real time. The current generation (2017 to time of writing) of processing hardware for the RTDS Simulator is based on a multicore processor: IBM®'s POWER8™, which has ten available cores [5]. The processor has been custom-integrated at the embedded level for the real-time application. Running the simulation's executable code directly on the processor, without an intermediary operating system, allows for efficient simulation and a high level of control in implementation. This involved approach to design and development has enabled many of the advancements mentioned in this document.

One major area of importance to users of real-time simulators is the quantity of power system nodes that can be simulated with a given timestep using a given configuration of simulation hardware. Each node contributes to the size of the network admittance matrix. It is processor-intensive but advantageous to decompose the admittance matrix in each timestep in order to complete the nodal analysis of the circuit in the real time. Dynamic matrix decomposition in each timestep allows for the representation of non-linear components as variable admittance elements. It also allows for power system models to be embedded into the main network solution without requiring a numerical interface.

As the network (and, consequently, its admittance matrix) grows in size, the time required to decompose the matrix increases exponentially. Naturally then, there is a practical limit to the size of network which can be simulated in real-time with a given timestep. This limit is influenced by the available processing power of the simulation hardware. As a result of this, users with access to a limited quantity of simulation hardware are often required to consider the size of the network they are simulating, which sometimes means judiciously reducing the network so that it "fits" on the simulation hardware. Techniques or advancements rolled out by simulator developers which effectively reduce the processing requirements for a given network are therefore of interest.

The RTDS Simulator uses an "embedded node" approach, meaning that breaker nodes on transmission line, transformer, and machine models are solved by the component model rather than included as explicit nodes in the admittance matrix. This means that these "embedded nodes" do not count toward the node total for a given network (the "network solution" size). For example, refer to Figure 1, which shows how embedded nodes are identified in the IEEE 39 bus network [6]. Fully expanded, this network contains a total of 366 power system nodes (i.e., 122 three phase buses), but 276 of those nodes (92 buses) are embedded in component models—these buses are counted in red. Considering this, the circuit would only require a 90-node (30 bus) network solution containing the buses counted in green.

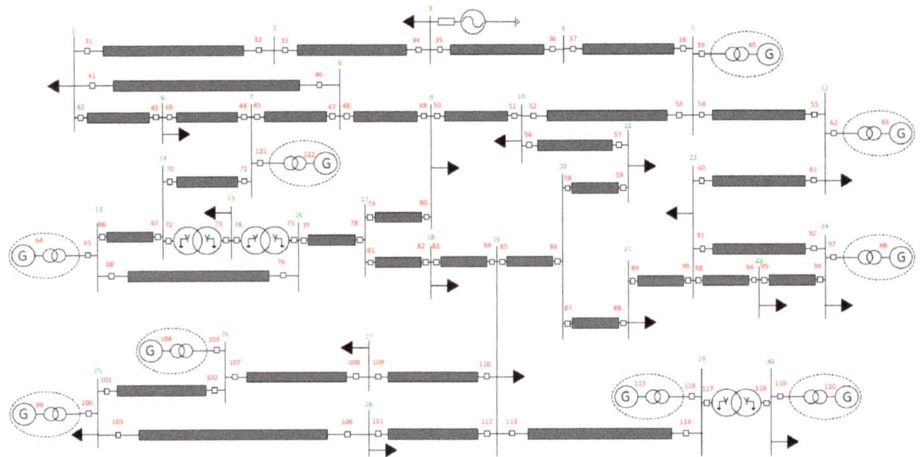

**Figure 1.** Single-line diagram showing the location of embedded power system nodes in the IEEE 39 bus system.

Typically, one core of the RTDS Simulator's processor is reserved for the network solution, completing the nodal analysis of the system, while other cores compute power system component contributions (for lines, transformers, machines, etc.) in parallel. This is one mode of scalability in the real-time simulator environment: within the multi-core processor, the quantity of licensed cores can determine the size and complexity of the network that can be represented. The second mode of scalability becomes relevant when the above-mentioned network size limit is reached. In this case, the larger network can be split into multiple subsystems which can then be solved independently and in parallel, exchanging data in real time. This decoupling is achieved via a travelling-wave transmission line model whose travel time is greater than or equal to the simulation timestep [7]. In the past, it has been possible to use this subsystem splitting technique by simulating a second subsystem either on a separate simulation hardware unit or on a second core of the original multi-core processor (i.e., within the same simulation hardware unit).

A recent advancement in this area was made possible by the higher clock rate, open design architecture, and low-latency inter-core communication of the current processor. Two cores on the same multi-core processor can now be dedicated to the network solution without decoupling into multiple subsystems. This means that the size of tightly-coupled network that can be simulated on one simulation hardware unit has doubled.

It should be noted that in many transmission networks, transmission lines of the minimum length required (i.e., with a travel time greater than or equal to the simulation timestep) are existent in the network, so no modification is required to create a subsystem split. However, in some distribution networks or other situations where longer transmission lines are not available, a line may need to be artificially lengthened in order to create a subsystem split. For high-fidelity simulation, decoupling elements such as this should be avoided if possible. Increasing the size of fully-connected network that can be represented on one unit via advancements such as the network solution using multiple cores (currently two) is helpful to users who wish to simulate larger systems where longer transmission lines may not be available.

*2.2. Processor-Based Multi-Rate Simulation*

As converter-based assets increasingly dominate many power systems worldwide, the ability to accurately and conveniently simulate power electronics—whether in the form of a kilowatt-scale renewable energy inverter or a modular multilevel converter-based HVDC station—is increasingly important. Using a real-time simulator to represent the electronics allows for HIL testing of the controls. This method has been used by manufacturers of HVDC equipment, but it is also applicable to modern controls for converter-based generation. For example, emergent aspects of grid-forming controllers such as fault current injection, harmonic sinking, and inertial characteristic can be tested in a closed loop with a simulated network prior to deployment.

There are several technical considerations when testing low-level control of power electronic schemes using a real-time simulator. In order to accurately test firing pulse control and provide information on harmonic behavior over a sufficiently high bandwidth, the simulation must proceed with an appropriately small real-time timestep. For voltage source converters, this is typically in the range of 1 to 3 μs [8]. Maintaining a stable long-term real-time simulation at these timesteps presents a challenge for off-the-shelf processing hardware. The custom integration of the RTDS Simulator's central processor at the embedded level enables the simulation of power electronics directly on the CPU. This circumvents the need for auxiliary hardware (e.g., FPGA) and any interface between hardware platforms, improving ease of use and stability.

Frequently, it is not only the behavior of the power electronic circuit but also its interactions with the AC system that are of interest to the user. While the AC system can theoretically be entirely represented with the same timestep as the power electronic circuit, this is only economically practical when the AC system of interest is relatively small. For larger networks, the processing power required to represent complex network components (including numerous complex elements such as transformers with saturation, detailed machines models, etc.) in the 1 μs range is prohibitively large. Therefore, it is advantageous to have the ability to designate a larger timestep for the AC network and run the power electronics network in parallel with a relatively small constant timestep. This is referred to as multi-rate simulation.

A multi-rate simulation environment was developed for the RTDS Simulator. This allows the user to interface multiple subnetworks running at various user-defined timesteps. These subnetworks are referred to as Substep (timestep is smaller than main timestep) and Superstep (timestep is larger than the main timestep) networks. Each subnetwork runs on a dedicated core on the central multicore processor and is interfaced with the main timestep network via a travelling wave transmission line model. Multiple Substep and Superstep networks are permitted in a given simulation case.

$$Substep\ timestep = \frac{1}{n} \times main\ timestep \tag{1}$$

where $5 \leq n \leq 64$

$$Superstep\ timestep = N \times main\ timestep \tag{2}$$

where $2 \leq N \leq 5$

The Substep and Superstep timesteps are user configurable and should be selected based on the phenomena of interest in the various parts of the network. When used judiciously, the multi-rate environment allows the user to control the level of detail provided by the simulation and the quantity of simulation hardware required to run a given network. In general, subnetworks running at a smaller timestep require a greater quantity of simulation hardware (are more "processor-intensive") and those running at a larger timestep are less processor-intensive.

For example, a user with access to a limited quantity of simulation hardware may choose to represent part of the AC network in the Superstep environment, which, due to the increased timestep, can simulate larger networks per core. In this way, the Superstep environment can act as a much better-performing alternative to a system equivalent. Superstep retains the detail of EMT simulation, providing significantly more detail than an infinite source. It also maintains the ability to represent control elements and frequency deviations. In some situations, using Superstep to increase the simulation timestep and represent a larger area may be preferable to using co-simulation with a lower-detail simulation tool. Figure 2 shows simulation results for a model which was developed to reflect the real-world network of a utility. The network includes over 200 three-phase buses, over fifty transmission lines, over twenty generators, an HVDC bipole (including DC line, converters, and switched filters) a static VAR compensator, and series capacitors. The network was first simulated using the main timestep environment to represent the entire network at a timestep of 50 µs (single-rate simulation). Then, a second simulation was run using the Superstep environment to simulate a portion of the network at two times the rate of the main timestep (multi-rate simulation). The results show the negative DC bus voltage upon energization of the first HVDC bipole.

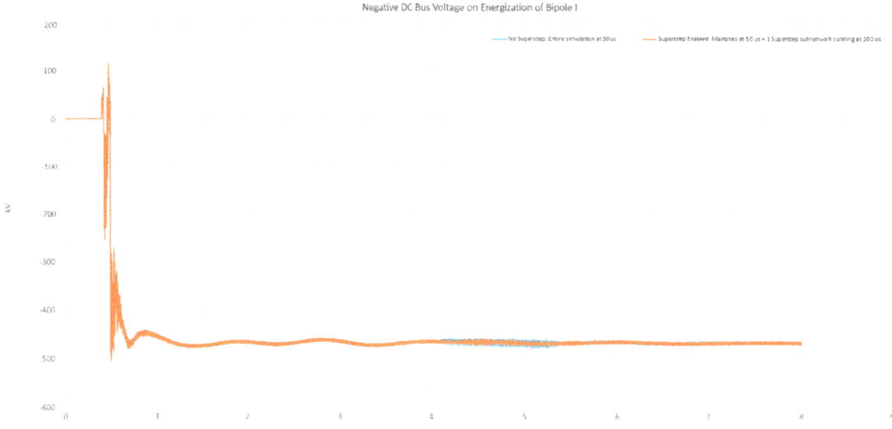


**Figure 2.** Simulation results showing negative DC bus voltage of an HVDC bipole upon energization for a single-rate simulation at 50 µs (blue curve) vs. a multi-rate simulation including a Superstep subnetwork running at 100 µs (orange curve).

The Pearson correlation coefficient for the signals shown in Figure 2 is 0.999897361872902. Table 1 below shows the Pearson correlation coefficient for several variables from the same simulation.

**Table 1.** Pearson correlation coefficients for single-rate simulation vs. multi-rate simulation involving a Superstep subnetwork.

| Variable | Pearson Correlation Coefficient |
| --- | --- |
| Negative DC Bus Voltage | 0.999897361872902 |
| Positive DC Bus Voltage | 0.999895090775645 |
| Phase A–AC Bus Voltage | 0.999982317324120 |
| Phase B–AC Bus Voltage | 0.999982328047804 |
| Phase C–AC Bus Voltage | 0.999982329635286 |
| DC Current | 0.999621462671499 |

The ability to conveniently divide the network into subnetworks running at different timesteps on the central processor for numerically stable multi-rate simulation is an advancement for this technology category.

### 2.3. Predictive Switching Algorithm for Switched-Resistance Representation of Power Electronic Converters

Maintaining a full conductance matrix decomposition for a power electronic circuit in each and every timestep (~1 to 3 μs) is challenging from a processing perspective, but doing so allows each switching instance to be represented by a change in conductance at the switching point (sometimes called switched-resistance representation or resistive switching), which is advantageous in terms of simulation accuracy and flexibility. Due to the processing challenges, an alternative modelling approach [9] was established and adopted by EMT tools in which discrete circuits using inductors and capacitors, referred to as L/C-associated discrete circuits, are used to emulate different switching states. Using L/C circuits instead of switched resistances means the network's conductance matrix does not need to be decomposed in each timestep, significantly decreasing the active processing power required to simulate a given circuit. Using this method, the matrix decomposition can instead occur prior to running the simulation, resulting in computational savings, allowing for a much more complex power electronic circuit to be simulated with the given hardware. However, there are several technical challenges associated with the L/C switching approach. When a switching event occurs, the abrupt change from one representation to the other results in an artificial energy loss. This loss increases with switching frequency and effectively limits the frequency at which a converter model can be switched if modelled using the L/C approach. Beyond a ~3 kHz switching rate, the losses become too great for results to remain valid. This method also creates some noise on the output waveforms due to current oscillations within the discrete circuits, and requires that the R, L, and C parameters are selected properly by the user to ensure a well-damped transient response and a respectively large and small impedance for the L/C circuits over the entire bandwidth of interest.

To overcome these challenges, converters of interest can typically be simulated using resistive switching, but they must be decoupled from the surrounding power electronic circuit using a traveling wave transmission line model. This can introduce significant error into the simulation. A recent advancement for the RTDS Simulator allows for the user to circumvent both L/C switching and travelling-wave-interfaced converter models when simulating power electronic networks.

In order to enable the dynamic decomposition of the network matrix for power electronic circuits, an algorithm was developed which uses predictive switching to reliably determine the ON/OFF status of a converter's switch devices before each simulation timestep [10]. The algorithm allows for several fixed-topology converter models to be simulated in the Substep environment with resistive switching, without requiring an interfacing transmission line model. The following models are available for predictive switching at the time of writing:

- Two-level converter
- Three-level T-type converter
- Three-level neutral point clamped (NPC) type converter
- Boost converter
- Buck converter

Embedding these resistively-switched models into the fully decomposed Substep network rather than using an L/C modelling approach also has the benefit of eliminating fictitious power losses and oscillations. The reduction in oscillations achieved through the use of resistively-switched models for a neutral point clamped (NPC) converter over an L/C discrete circuit approach can be seen in Figure 3. A high-frequency transient response is observed when the status of the switch is changed (and the representation is changed from an inductor to capacitor or vice versa). This response results from the energy loss created by the modelling approach. The reduction in power losses achieved through using the resistively-switched models, shown in Table 2, results in a cleaner waveform and increases the switching frequency bandwidth over which a converter can be accurately represented. If necessary, power losses of the converter can be defined by the user during the parametrization process in order to reflect those of the physical device being modelled.

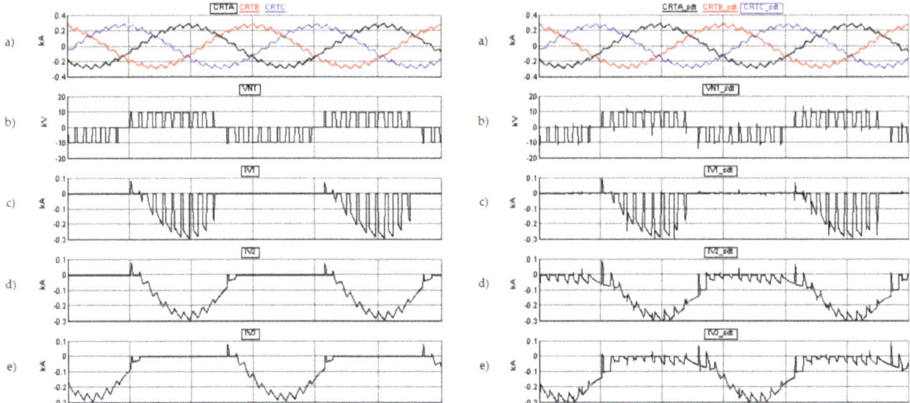

**Figure 3.** Simulation results for a neutral point clamped (NPC) converter modelled with resistive switching based on a predictive algorithm (left) and L/C associated discrete circuits (right): (**a**) AC side currents, (**b**) internal phase A voltage, (**c**) current through phase A upper valve 1, (**d**) current through phase A upper valve 2, (**e**) current through phase A lower valve 1.

**Table 2.** Power losses for resistively-switched (Substep environment) vs. L/C associated discrete circuit (ADC) models of an NPC converter.

| PWM Switching Frequency (Hz) | Power Losses for Resistively-Switched Model | Power Losses for L/C ADC Model |
|---|---|---|
| 1260 | 0.18% | 3.0% |
| 3060 | 0.22% | 6.5% |

Losses are defined as the difference in power entering the DC side vs. leaving the AC side of the converter.

### 2.4. Co-Simulation with a Real-Time Simulator and PC-Based TSA Program

As mentioned in Section 1 above, the several different power system modelling approaches that are popular today are variable in the level of detail they provide and the type of systems and dynamics they can reliably represent. Historically, phasor-based approaches such as transient stability analysis/assessment (TSA) have been widely used for system planning and operational studies. TSA simulations can run in real-time on a regular PC, as they offer phasors as an output and typically use a timestep of 1/4 or 1/2 of the system cycle (i.e., 4 to 10 ms—in the order of 100 times greater than the timestep of a typical real-time EMT simulation) [11]. The output of a TSA simulation lacks the wide frequency bandwidth offered by EMT solutions and the ability to represent subcycle phenomena associated with fast transients. However, the ability to represent larger systems with greater efficiency on a regular PC makes them ideal for many high-level power system studies.

The concept of co-simulation, in which the global representation of a given system can be achieved by composing simulations of its subsystems that are completed via different platforms or means, has been explored for some time. In a power systems context, the co-simulation of phasor-based TSA and EMT solutions could potentially be impactful for the user [12]. For users who are looking to model large-scale networks with a reasonable level of detail but who have limited access to real-time simulation hardware, EMT-TSA co-simulation could reduce the hardware requirement while maintaining a high level of detail for a portion of the network. It is important to note that the level of detail available for the portion of the network represented via TSA will be lower. Any study completed via real-time EMT-TSA co-simulation should be judiciously selected for such a study based on technical requirements and the acceptable bandwidth of results for various areas of the network.

To benefit cases where the variance in timestep and output is acceptable and co-simulation is appropriate, a TSA co-simulation platform was developed for the RTDS Simulator. Called the

TSAT-RTDS Interface (TRI), this tool provides an interface between the TSAT program (part of the DSATools™ suite of power systems analysis tools) and the real-time simulator. TSAT is familiar to many power system engineers and can represent over 5000 three-phase buses in real-time in addition to (and in parallel with) the portion of the network that is represented on the real-time simulator.

The TSAT-RTDS Interface is achieved via an FPGA that is installed in the PCIe slot of the user's PC. This creates a continuous, synchronized real-time interface, meaning that TSAT and the real-time simulator exchange outputs at the end of every TSAT timestep. The software aspect of the tool was developed with the user in mind, and is able to automatically identify potential boundaries for partitioning between the EMT and TSA portions of the simulation and provide metrics that quantify their validity. The user may also manually select boundaries and have their validity assessed by the program. While the program is capable of such calculations, boundary selection is another aspect of co-simulation that must be judiciously approached by the user in order to ensure the technical validity of the simulation.

Figure 4 shows simulation results for the Illinois 200-Bus System [13], which was divided for an EMT-TSA hybrid simulation. A 43-bus section was defined as the internal system to be modelled with the RTDS Simulator, with the remainder of the system modelled using TSAT [14]. A fault was applied at a bus within the EMT-simulated area and a nearby generator—the largest one in the system—was tripped after a short delay, causing a large disturbance to the network. The results obtained for the hybrid simulation were compared with those from a full EMT simulation and those from a full TSA simulation. The results show the active power, reactive power, and speed of the generator closest to the disturbance.

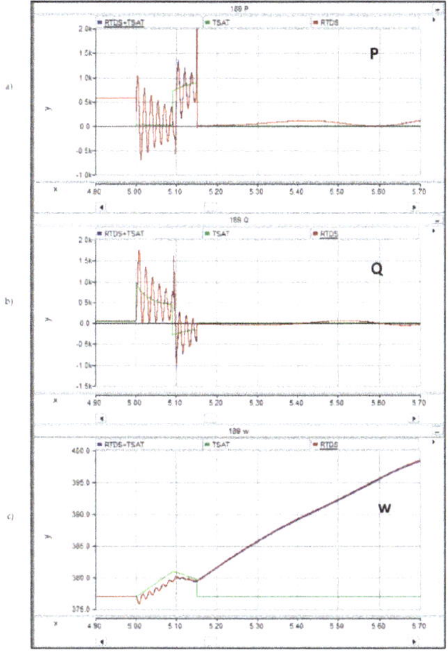

**Figure 4.** Electromagnetic transient (EMT)-only (**red**), EMT-transient stability analysis/assessment (TSA) hybrid (**blue**), and TSA-only (**green**) simulation results for Generator 189 after a fault at Bus 187: (**a**) active power, (**b**) reactive power, (**c**) speed.

In general, EMT-TSA co-simulation can be advantageous in its ability to analyze interactions between system-wide events on a large network and detailed device behavior, such as fault analysis in HVDC systems or sub-synchronous resonance studies. For users migrating to co-simulation from TSA-only simulation, another benefit is the ability to run detailed models that are not typically available in the TSA environment, including renewable energy components and HVDC or flexible alternating current transmission system (FACTS) devices. For those migrating to co-simulation from EMT-only simulation, the inclusion of a TSA-based representation of the larger surrounding system (rather than utilizing an infinite source in the real-time environment in order to reduce network matrix size) may increase the capability to accurately represent low-frequency oscillations (e.g., generator swings, inter-area dynamics, etc.).

As mentioned above in Section 2.2, in situations when an EMT-TSA co-simulation is deemed inappropriate, using the Superstep feature of the RTDS Simulator may be an alternative solution.

*2.5. Advancements in Real-Time Communication Protocol Implementation*

The use of information exchange for power systems management and control is growing. The use of Ethernet-based communication protocols to develop a hardware-in-the-loop interface is now an established practice. Software models, built-in configuration tools, and hardware for streaming standard-compliant power system data in real time are all available for the RTDS Simulator [15]. In order to allow users to test a wider variety of devices and schemes given industry advancements in automation, several recent developments have been made for the RTDS Simulator's communication protocol facilities.

Existing publish/subscribe models for IEC 61850 Sampled Values (SV) and GOOSE Messaging protocols have been expanded to include support for routable versions (R-SV and R-GOOSE), which are popular for distribution automation applications [16]. R-SV and R-GOOSE features are provided by IEC Technical Report 61850-90-5 [17] for IEC 61850-9-2 SV and IEC 61850-8-1 GOOSE packets. When using R-SV and R-GOOSE, the user specifies a destination multicast IP address. The transport protocol used in these implementations is User Datagram Protocol (UDP).

The existing processor-based model for SV was also updated to include the output sample rates specified in IEC 61869-9 and the Chinese National Standard for merging units. These are shown in Table 3.

**Table 3.** Sampling rates (IEC 61869-9) supported by GTNETx2-SV-v6 for the RTDS Simulator.

| Sampling Rate (Hz) | ASDUs per Frame | Maximum Channels per SV Stream |
|:---:|:---:|:---:|
| 4000/4800 [1] | 1 | 24 |
| 12,800/15,360 [1] | 8 | 9 |
| 4800 | 2 | 24 |
| 14,400 | 6 | 9 |
| 4800/5760 [1] | 1 | 24 |

[1] Supporting 80, 96, or 256 samples/cycle for use on a 50 or 60 Hz system, these rates have been maintained for 9-2LE backward compatibility.

Another advancement for real-time IEC 61850 Sampled Values implementation is the improvement of the existing FPGA-based model for SV streaming. In this case, FPGA-based hardware running the SV model is interfaced with the central processor via optical fiber. This model was updated for support from both the main timestep and Substep environments of the RTDS Simulator. Running the model at the Substep timestep (typically in the 1–3 µs range) allows for higher output sample rates than have previously been possible with a real-time simulator. The supported SV sample rates on the FPGA-based hardware are listed in Table 4. Sampling rates as high as 96 and 250 kHz are available, supporting users interested in DC instrument transformer applications.

**Table 4.** Sampling rates (IEC 61869-9) supported by GTFPGA-SV-v3 for the RTDS Simulator.

| Sampling Rate (Hz) | ASDUs per Frame | Publishing Rate (Frames/Second) |
| --- | --- | --- |
| 4000/4800 | 1 | 4000/4800 |
| 12,800/15,360 | 8 | 1600/1920 |
| 4800 | 2 | 2400 |
| 14,400 | 6 | 2400 |
| 4800/5760 | 1 | 4800/5760 |
| 96,000 [3] | 1 | 96,000 |
| 250,000 [3] | 1 | 250,000 |

Supporting 80, 96, or 256 samples/cycle for use on a 50 or 60 Hz system, these rates have been maintained for 9-2LE backward compatibility. [3] These rates are only supported in the Substep environment.

The implementation of Distributed Network Protocol 3 (DNP3) and IEC 60870-5-104 for the RTDS Simulator has also been improved. Rather than providing data exchange between the real-time simulator and a single external device, the user can now interface the simulator with up to four external devices simultaneously using a single instance of these protocol models. Using multiple instances of the models allows for connection to many external devices at once. The real-time simulator acts as a slave device connected to external masters. These protocols are commonly used by users for SCADA and distribution automation applications.

### 2.6. Digital Interface for Power-Hardware-in-the-Loop Simulation

Power-hardware-in-the-loop (PHIL) simulations have gained popularity due to their potential for analyzing and characterizing the response of physical converters, DERs, motors, and loads to a huge variety of system conditions in the safety of the lab [18]. However, PHIL simulations can be challenging to implement in a technically sound manner. Developments for the RTDS Simulator in this area will improve the PHIL experience for users.

Establishing a valid and stable PHIL interface is not trivial, and selecting an appropriate four quadrant amplifier is a crucial step of the process, with several technical considerations including response times, slew rate, harmonic distortion, frequency resolution, and input/output ratings and impedances. Filtering is also a major technical consideration in order to reduce the impact of noise (introduced by voltage and current sensors, physical wiring, and electromagnetic coupling between devices) on the simulation. Any error introduced by noise can be amplified in the interface via positive feedback until hardware limits are exceeded. Filters for noise reduction introduce challenges of their own, including additional delays that may be added to the interface, so filter parameters should be determined carefully.

In order to reduce noise and delay and to simplify the PHIL interface, RTDS Technologies worked with four quadrant amplifier manufacturers to develop a digital interface between the real-time simulator and the amplifier. This eliminates the need for digital to analogue conversion in the interface via analogue input and output cards.

The digital interface was developed via Aurora, an open source high-speed serial protocol developed by Xilinx. Two different interfacing methods are currently available and are selected based on the manufacturer of the amplifier being used: a direct link between the real-time simulator and amplifier using an optical cable, or intermediary FPGA-based hardware which is programmed to buffer the Aurora data at the required rate for the amplifier.

The corresponding PHIL interfacing model within the simulation software also includes embedded voltage and current sources. This reduces loop delay and improves ease of use for PHIL applications. Figure 5 includes simulation results which show the loop delay and noise associated with the improved digital interface [19]. Loop delay is case-dependent and is always less than one simulation timestep. In this example, the current is read back from the amplifier prior to the network solution calculation so that it can be used to calculate the new voltage to send out to the amplifier. The network solution therefore sees the amplifier as simply another component of the simulation. The low loop delay in the PHIL setup enables the user to mimic a fully digital simulation.

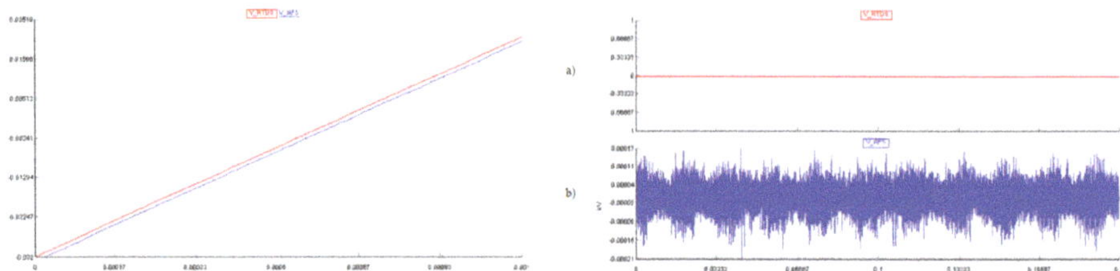

**Figure 5.** Left: Simulation results showing loop delay for a 0.1 kV-RMS sine wave; right: (**a**) a constant of 0.0 generated with the real-time simulator and (**b**) measurement from a four-quadrant amplifier receiving the signal via the Aurora protocol.

*2.7. Enhanced Power System Component Models*

2.7.1. Faulted Transformer Model Based on Terminal Duality Principle

A recent advancement in real-time power system component modelling is the development of an improved transformer model for the RTDS Simulator. This model is based on a terminal duality equivalent [20] rather than the star equivalent [21] which is frequently used in power systems simulation. The star equivalent, while providing accurate representation of the terminal conditions, is generally inaccurate in its representation of the physical transformer leakages via the inductance branches. The terminal duality model provides a more realistic representation of a transformer's magnetic circuit. It includes a leakage branch representing the gap between the first winding and the transformer limb, providing a more accurate prediction of the maximum magnetization inrush current. It also considers the mutual coupling between branches, allowing for an improved representation of terminal conditions when compared to the star equivalent.

Internal faults, tap changers, or winding offsets may create asymmetrical conditions for the model. The short circuit reactance in these situations can be evaluated with relative accuracy as the terminal duality model decomposes the leakage flux into both axial and transverse components.

This model introduces a new degree of accuracy in real-time transformer modelling, providing improved results for the transient behavior of transformers. This will in turn increase the accuracy of closed-loop tests performed on transformer protection devices via a real-time simulator.

2.7.2. Faulted Phase Domain Synchronous Machine

A phase domain synchronous machine (PDSM) model capable of modelling internal faults has also been developed for the RTDS Simulator [22]. The model allows for access (the ability to connect components) to all phases of the stator terminals, the machine neutral, and up to two points in each phase of the field winding. This allows users to apply turn-to-turn faults on the stator and on the field winding, faults between phases, and faults between phases and the field winding. The model is also appropriate for the testing of 100% stator-ground fault detection in protective devices.

With the phase domain approach, new machine inductance values are calculated every timestep in order to represent possible changes in rotor position and saturation. The self and mutual inductances of the windings are calculated as functions of rotor position and saturation to represent internal faults. The machine's differential equations are solved as an embedded part of the simulation's network solution. This embedded phase domain approach shows better performance in terms of numerical stability when compared to the approach of dq0-based models interfaced to the network (used traditionally by transient simulation programs) [23]. The interfacing delays inherent with these conventional models can cause issues with numerical stability in the presence of other interfaced components in real time.

The PDSM model allows the user to simulate the following fault scenarios and place the fault anywhere between 1% and 99% of the windings:

- Turn-to-turn faults in individual phases
- Turn-to-ground faults in individual phases
- Phase-to-phase faults
- Phase-to-ground faults
- Turn-to-turn faults in the field winding
- Turn-to-ground faults in the field winding
- Series faults in the stator and field windings
- Inherent imbalance in the windings

This model can be used to test generator protection schemes under an unprecedented range of internal fault conditions. Figure 6 shows the results of testing a commercial relay by simulating a stator-ground fault. The plot on the left shows the simulation results, including terminal voltages, machine currents, and neutral voltage and current. The plot on the right shows the relay's recorded events, indicating that the phase differential (87), neutral overvoltage (64G1), and neutral overcurrent (50N) elements were activated.

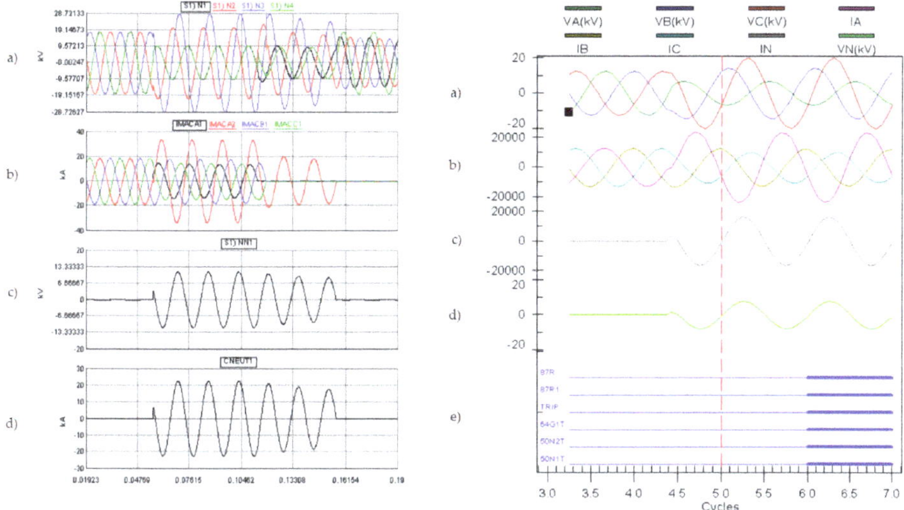

**Figure 6.** Results for testing a commercial relay for a stator-ground generator fault at 50% from the neutral. Simulation results (**left**): (**a**) stator voltages, (**b**) stator currents, (**c**) neutral voltage, (**d**) neutral current; relay recorded events (**right**): (**a**) stator voltages, (**b**) stator currents, (**c**) neutral current, (**d**) neutral voltage, (**e**) digital signals.

### 2.7.3. Substep Frequency-Dependent Transmission Line

The emergence of travelling wave-based relays is a recent advancement in the field of power systems protection. The protective devices incorporating these techniques trip securely in as fast as a few ms, record events in the MHz sampling rate range, and locate faults with relatively high accuracy [24]. In order to accurately test such a relay in a hardware-in-the-loop situation, a corresponding advancement must be made in the real-time modelling library.

A frequency-dependent phase domain (FDPD) transmission line model has long been available for the RTDS Simulator in the main timestep environment. This model uses an established method known as the universal line model [25] to represent the line and its frequency-dependent properties without the need to compute a modal transformation matrix as is necessary with a frequency-dependent modal domain model [26]. The phase domain model has also been shown to be more accurate than the modal domain model when modelling non-transposed line segments [27]. For the FDPD model, curve-fitting techniques are used to represent the frequency dependence (including damping) of the line's characteristic admittance and propagation constant. This model is more processor-intensive than both the common modal domain Bergeron transmission line model (which does not represent frequency dependence) and the frequency-dependent modal domain model.

In order to accurately test travelling-wave-based protection, the Bergeron model cannot be used, as it is a single-frequency model and cannot accurately represent travelling waves which contain a wide range of frequencies. A frequency-dependent line model must be used to properly represent the full bandwidth of behavior of the travelling wave. However, the typical main timestep (30 to 60 μs) is not sufficient for testing these protection algorithms, which are based on a relay sampling frequency in the MHz range. To overcome this challenge, a frequency-dependent transmission line model (based on the universal model) was developed for use in the Substep environment. This Substep FDPD model, capable of running in real time in the 2 μs timestep range, was used to test the first-ever travelling-wave relays used developed for primary protection of a high-voltage line [28]. It should be noted that an FDPD line model running at timesteps appropriate for travelling-wave relay testing is also available for the RTDS Simulator via FPGA. In this case, FPGA-based hardware running the frequency-dependent line model is interfaced with the central processor via optical fiber.

Figure 7 shows a simple system model that was used for testing travelling-wave relays to demonstrate the Substep frequency-dependent transmission line model [29]. A generator is connected to an infinite bus via a transformer and double-circuit transmission line. The model was run on the central processor at a 3.1 μs real-time timestep for this particular example. Timesteps as low as 1.4 μs have been used to test travelling-wave relays using the RTDS Simulator.

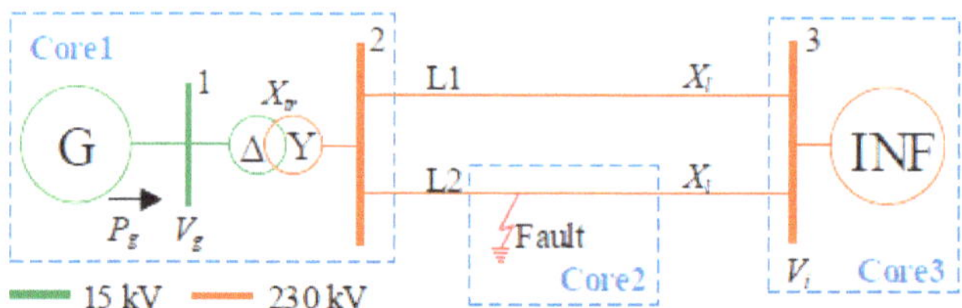


**Figure 7.** Simple system model for travelling-wave relay testing.

A permanent three-phase to ground faults was applied on L2 close to the generator side. Figure 8 includes test results from the real-time simulator (top) and relay recording (bottom). The simulation results, which show the three-phase currents on L2 at Bus 3 (from the real-time simulation running at 3.1 μs), include validation against a non-real-time simulation program in which the model was run with a 0.1 μs timestep. The relay results show the fault inception point, current variable activity, and trip signal point. The fault detection time of the relays in this test is around 1.1 ms.

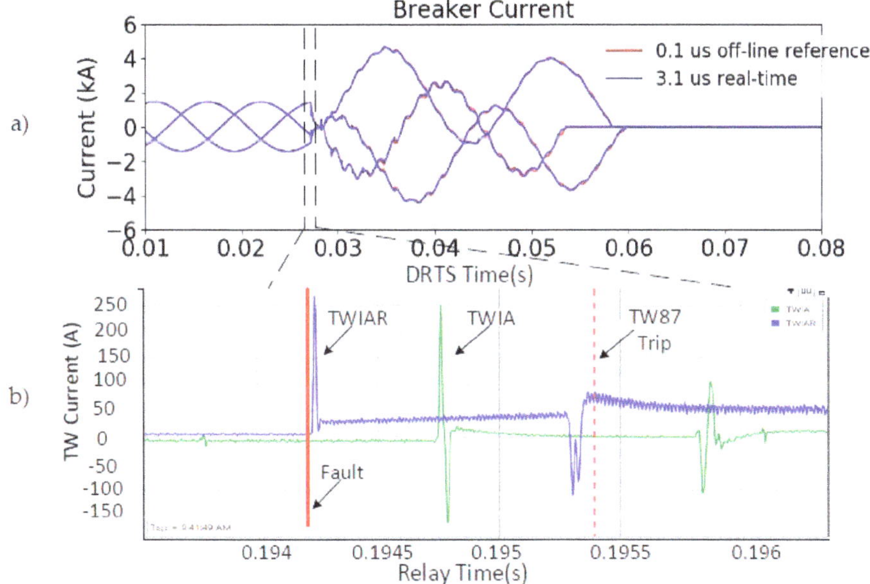



**Figure 8.** Results for testing a commercial travelling-wave relay for a three-phase to ground fault: simulation results for (**a**) three-phase current on the faulted line at the remote bus and (**b**) relay current variables.

## 3. Discussion

Real-time simulators are under continual development by their manufacturers in order to remain reflective of the modern power system and industry practices. This paper describes a non-exhaustive selection of recent developments rolled out by a particular manufacturer: RTDS Technologies Inc.

Many of the advancements detailed in this paper were enabled by growth in the high-performance processing space and the emerging availability of high-end processors with an open design architecture (e.g., the OpenPOWER™ Foundation). Others were initiated or supported by developer participation in power industry working groups and study committees.

It should also be noted that several of the advancements in this paper were motivated by needs identified by users of the real-time simulator in question. In this sense, the advancement of this technology is partially anticipatory and partially responsive. Users of real-time simulators are utilities, consulting companies, universities, research institutions, or manufacturers of power system protection and control equipment. Technical expertise combined with a high degree of control over both the hardware and the software aspects of this technology are necessary to deliver reliable and accurate solutions to these users.

More information can be found in the Supplementary Materials section or by contacting the authors.

**Supplementary Materials:** The searchable online Knowledge Base for the RTDS Simulator is available at https://knowledge.rtds.com/. A list of technical publications related to the RTDS Simulator is available at https://knowledge.rtds.com/hc/en-us/categories/360001905033-Technical-Publications. User case studies are available as follows: 1. Travelling wave relay testing: https://knowledge.rtds.com/hc/en-us/articles/360036984873-Case-Study-Travelling-wave-relay-testing 2. Distribution automation: https://knowledge.rtds.com/hc/en-us/articles/360036924154-Case-Study-Distribution-Automation 3. Large-scale simulation: https://knowledge.rtds.com/hc/en-us/articles/360042460494-Case-Study-Large-Scale-Simulation 4. HVDC and FACTS: https://knowledge.rtds.com/hc/en-us/articles/360036922834-Case-Study-De-risking-HVDC.

**Author Contributions:** K.S. acquired, analyzed, and interpreted the data, and prepared the manuscript. P.F. substantially revised and approved the manuscript. All authors have read and agreed to the published version of the manuscript.

**Funding:** This research received no external funding.

**Acknowledgments:** The authors acknowledge the contributions of the research and development teams at RTDS Technologies Inc., who were instrumental in developing the advancements discussed in this paper.

**Conflicts of Interest:** The authors declare no conflict of interest.

## References

1. Duchen, H.; Lagerkvist, M.; Lovgren, N.; Kuffel, R. Validating the Real Time Digital Simulator for HVDC Dynamic Performance Studies. In Proceedings of the ICDS, Montreal, QC, Canada, 28–30 May 1997; pp. 245–250.
2. Badrzadeh, B.; Emin, Z. The Need for Enhanced Power System Modelling Techniques and Simulation Tools. In *Reference Papers*; No. 308; CIGRE Electra: Toowong, Brisbane, Australia, 2020; pp. 54–55.
3. Dommel, H.W. Digital Computer Solution of Electromagnetic Transients in Single- and Multiphase Networks. *IEEE Trans. Power Appar. Syst.* **1969**, *88*, 388–399. [CrossRef]
4. Kuffel, R.; Giesbrecht, J.; Maguire, T.; Wierckx, R.P.; McLaren, P. RTDS—A Fully Digital Power System Simulator Operating in Real Time. In Proceedings of the International Conference on Energy Management and Power Delivery, Singapore, 21–23 November 1995.
5. Friedrich, J.; Le, H.; Starke, W.; Stuechli, J.; Sinharoy, B.; Flurh, E.; Dreps, D.; Zyuban, V.; Still, G.; Gonzalez, C.; et al. The Power8TM Processor: Designed for Big Data, Analytics, and Cloud Environments. In Proceedings of the IEEE International Conference on IC Design & Technology, Austin, TX, USA, 28–30 May 2014.
6. Athay, T.; Podmore, R.; Virmani, S. A Practical Method for the Direct Analysis of Transient Stability. *IEEE Trans. Power Appar. Syst.* **1979**, *98*, 573–584. [CrossRef]
7. Woodford, D. Validation of Digital Simulation of DC Links. *IEEE Trans. Power Appar. Syst.* **1985**, *104*, 2588–2595. [CrossRef]
8. Maguire, T.; Giesbrecht, J. Small Time-Step ($\leq 2$ µs) VSC Model for the Real Time Digital Simulator. In Proceedings of the ICPST, Montreal, QC, Canada, 19–23 June 2005; pp. 1–6.
9. Wang, K.; Xu, J.; Tai, N.; Tong, A.; Hou, J. A Generalized Associated Discrete Circuit Model of Power Converters in Real-Time Simulation. *IEEE Trans. Power Electron.* **2019**, *34*, 2220–2233. [CrossRef]
10. Maguire, T.; Elimban, S.; Tara, E.; Zhang, Y. Predicting Switch ON/OFF Statuses in Real Time Electromagnetic Transients Simulations with Voltage Source Converters. In Proceedings of the IEEE Conference on Energy Internet and Energy System Integration, Beijing, China, 20–22 October 2018.
11. Powertech Labs Inc. TSAT Transient Security Assessment Tool Brochure. Available online: https://static1.squarespace.com/static/5670ab04c647ad9f554e4d9f/t/5bb553b3eef1a13f1f4b0240/1538610102050/TSAT+rev+A.pdf (accessed on 26 May 2020).
12. Zhang, Y.; Gole, A.M.; Wu, W.; Zhang, B.; Sun, H. Development and Analysis of Applicability of a Hybrid Transient Simulation Platform Combining TSA and EMT Elements. *IEEE Trans. Power Syst.* **2013**, *28*, 357–366. [CrossRef]
13. Birchfield, A.B.; Gegner, K.M.; Xu, T.; Shetye, K.S.; Overbye, T. Statistical Considerations in the Creation of Realistic Synthetic Power Grids for Geomagnetic Disturbance Studies. *IEEE Trans. Power Syst.* **2017**, *32*, 1502–1510. [CrossRef]
14. RTDS Technologies Inc. TSAT-RTDS Interface Evaluation. 2018; Unpublished.
15. Kuffel, R.; Ouellette, D.; Forsyth, P. Real Time Simulation and Testing Using IEC 61850. In Proceedings of the Modern Electric Power System, Wroclaw, Poland, 20–22 September 2010; pp. 1–8.
16. Apostolov, A. R-GOOSE: What It Is and Its Application in Distribution Automation. In Proceedings of the International Conference & Exhibition on Electricity Distribution (CIRED), Glasgow, UK, 12–15 June 2017.
17. International Electrotechnical Commission. Communication Networks and Systems for Power Utility Automation—Part 90-5: Use of IEC 61850 to Transmit Synchrophasor Information According to IEC C37.118. IEC TR 61850-90-5:2012. Available online: https://webstore.iec.ch/publication/6026 (accessed on 26 May 2020).

18. Steurer, M.; Bogdan, F.; Ren, W.; Sloderbeck, M.; Woodruff, S. Controller and Power Hardware-In-Loop Methods for Accelerating Renewable Energy Integration. In Proceedings of the IEEE Power Engineering Society General Meeting, Tampa, FL, USA, 24–28 June 2007.

19. RTDS Technologies Knowledge Base. Power Hardware in the Loop Simulations. Available online: https://knowledge.rtds.com/hc/en-us/articles/360037124014-PHIL-Report (accessed on 26 May 2020).

20. Alvarez-Marino, C.; de Leon, F.; Lopez-Fernandez, X.M. Equivalent Circuit for the Leakage Inductance of Multiwinding Transformers: Unification of Terminal and Duality Models. *IEEE Trans. Power Deliv.* **2012**, *27*, 353–361. [CrossRef]

21. Starr, F.M. Equivalent Circuits—I. *Trans. Am. Inst. Electr. Eng.* **1932**, *57*, 287–298. [CrossRef]

22. Dehkordi, A.B.; Neti, P.; Gole, A.M.; Maguire, T.L. Development and Validation of a Comprehensive Synchronous Machine Model for a Real-Time Environment. *IEEE Trans. Energy Convers.* **2010**, *25*, 34–48. [CrossRef]

23. Gole, A.M.; Menzies, R.W.; Woodford, D.A.; Turanli, H. Improved Interfacing of Electrical Machine Models in Electromagnetic Transient Programs. *IEEE Trans. Power App. Syst.* **1984**, *9*, 2446–2451. [CrossRef]

24. Schweitzer, E.O.; Kasztenny, B.; Mynam, M.V. Performance of Time-Domain Line Protection Elements on Real-World Faults. In Proceedings of the 69th Annual Conference for Protective Relay Engineers, College Station, TX, USA, 4–7 April 2016; pp. 1–17.

25. Morched, A.; Gustavsen, B.; Tartibi, M. A Universal Model for Accurate Calculation of Electromagnetic Transients on Overhead Lines and Underground Cables. *IEEE Trans. Power Deliv.* **1999**, *14*, 1032–1038. [CrossRef]

26. Marti, J.R. Accurate Modelling of Frequency-Dependent Transmission Lines in Electromagnetic Transient Simulations. *IEEE Trans. Power Appar. Syst.* **1982**, *101*, 147–157. [CrossRef]

27. PSCAD Manual: Frequency Dependent Models. Available online: https://www.pscad.com/webhelp/EMTDC/Transmission_Lines/Frequency_Dependent_Models.htm (accessed on 26 May 2020).

28. The Modernization Story of PNM. Available online: https://selinc.com/featured-stories/pnm/ (accessed on 26 May 2020).

29. Mirzahosseini, R.; Arunprasanth, S.; Tara, E.; Samarasekera, U.; Zhang, Y. Real-Time Closed-Loop Traveling-Wave Relay Testing (TWRT) in the Environment of Multi-Machine AC Power Systems. In Proceedings of the International Conference on Power Systems Transients, Perpignan, France, 16–20 June 2019.

*Article*

# DPsim—Advancements in Power Electronics Modelling Using Shifted Frequency Analysis and in Real-Time Simulation Capability by Parallelization

**Markus Mirz** *,†, **Jan Dinkelbach** † and **Antonello Monti**

Institute for Automation of Complex Power Systems, RWTH Aachen University, 52074 Aachen, Germany; jdinkelbach@eonerc.rwth-aachen.de (J.D.); amonti@eonerc.rwth-aachen.de (A.M.)
* Correspondence: mmirz@eonerc.rwth-aachen.de
† These authors contributed equally to this work.

Received: 19 June 2020; Accepted: 21 July 2020; Published: 29 July 2020

**Abstract:** Real-time simulation is an increasingly popular tool for product development and research in power systems. However, commercial simulators are still quite exclusive due to their costs and they face problems in bridging the gap between two common types of power system simulation, conventional phasor based, and electromagnetic transient simulation. This work describes recent improvements to the open source real-time simulator DPsim to address increasingly important use cases that involve power electronics that are connected to the electrical grid and increasing grid sizes. New power electronic models have been developed and integrated into the DPsim simulator together with techniques to decouple the system solution, which facilitate parallelization. The results show that the dynamic phasors in DPsim, which result from shifted frequency analysis, allow the user to combine the characteristics of conventional phasor and electromagnetic transient simulation. Besides, simulation speed up techniques that are known from the electromagnetic domain and new techniques, specific to dynamic phasors, significantly improve the performance. It demonstrates the advantages of dynamic phasor simulation for future power systems and the applicability of this concept to large scale scenarios.

**Keywords:** real-time simulation; power electronics; shifted frequency analysis; dynamic phasors

---

## 1. Introduction

Real-time simulation is increasingly applied in power systems for testing new components and algorithms. However, power system simulators that are capable of real-time execution are still not accessible for many researchers and product developers. One reason for the high cost of real-time power system simulators is the customized hardware platform that is usually required. Early real-time simulators used special purpose processors, e.g., digital signal processors (DSP). Today, general purpose processors are commonly employed as the core computing unit but often in combination with additional hardware, such as a customized backplane to interconnect computing units or field-programmable gate arrays (FPGAs) [1]. The use of FPGAs as main computing unit is gaining traction with the advent of power electronics simulation. While this approach is only adopted by few commercial manufacturers, there are many examples of FPGA driven real-time simulators built by researchers [2–5]. An interesting alternative direction is presented in [6], where real-time simulation is conducted on a small low cost device. However, [6] is based on quasi-static assumptions, whereas the aim of this work is a more detailed represention of dynamics. DPsim (https://github.com/dpsim-simulator), the simulator employed and extended in the frame of this work, is capable of running real-time simulation on commercial off-the-shelf hardware without neglecting electromagnetic dynamics.

This is possible due to shifted frequency analysis (SFA) [7], which enables larger simulation time steps for the same simulated signal frequencies as compared to electromagnetic-transient (EMT) simulations. Larger time steps facilitate the simulation on standard hardware and in large scale because the requirements on computational speed can be relaxed. The SFA approach intends to bridge the gap between EMT and conventional phasor simulation in terms of accuracy. Applying SFA on a real signal results in one or many time-varying phasors, which are often denoted as dynamic phasors (DP).

DPsim is not the first simulator that is based on this approach. Previous developments of SFA or dynamic phasor based simulators are described in [8–10]. In contrast to DPsim, these simulators do not seem to aim at real-time execution and related topics, such as parallelization, to speed up the simulation. As described in this paper, DPsim supports the simulation of higher order harmonics to accurately represent switching power electronics. The other simulators only focus on the fundamental system frequency phasor or lower order harmonics resulting from the dq0 transform in combination with unbalanced conditions as in [9]. While previous SFA models implemented in DPsim have not required an approximation of the phasors using Fourier analysis, the detailed inverter model results presented in the following rely on it to compute the phasor values for each time step. This approach is based on the generalized state space averaging method that is explained in [11] and harmonic analysis of the inverter circuit [12].

Most importantly, DPsim is developed as open source in contrast to the aforementioned projects. It allows the user not only to compare and validate against other solutions, but also to understand the implementation in detail. Besides, DPsim supports a faster transition from offline testing to real-time execution, since the same software and hardware system is used in both cases.

The dynamic phasor approach is not only used in the development of simulators, but also for the interconnection of real-time simulators, especially if there is a considerable distance between these simulators. This idea is presented in [13], where several EMT simulators are interconnected via a dynamic phasor interface that is based on the VILLAS software project [14].

DPsim was first introduced by Mirz et al. [15], which describes the main software components and demonstrates the real-time capability of the simulator for different system sizes. It integrates the VILLAS project to facilitate the exchange of simulation data in real-time taking advantage of the vast collection of protocols supported by VILLAS. A possible simulation framework architecture that is based on DPsim and VILLAS is described in [16]. Although DPsim primarily targets the dynamic phasor domain, the solver also supports EMT and quasi-static simulations. The main input format for network description is the IEC 61970 CIM XML-RDF format, which is imported by using the CIM++ project [17]. A more detailed overview of the features of DPsim can be found in [15]. The aim of DPsim is to support real-time simulation with time steps in the range from 10 μs to several ms. Due to the frequency shift applied when computing dynamic phasors, the maximum frequency of simulated signals is not limited by the simulation time step.

Until now, DPsim only supported traditional power system components, such as synchronous generators, lines, and transformers. Here, we describe two new power electronics models and one new parallelization method, which were implemented in DPsim. First, an averaged inverter model is presented and how an associated dq0 state-space controller can be integrated into the simulation. The second model is more detailed and it extends the averaged model by harmonics resulting from switches. A new method for the DPsim solver is demonstrated that ensures real-time capability for such an inverter model with a large number of harmonics. Besides, the explanation of the subsystem decoupling using the transmission line method briefly presented in [15] is extended by a description on how the line model is adapted for SFA.

Section 2 briefly summarizes SFA and how it computes dynamic phasor variables. A detailed description of the component models and methods used to obtain the simulation results can be found in Section 3. Sections 4 and 5 present and discuss the results.

## 2. Theoretical Background

The basic concept of SFA, constructing a frequency shifted complex representation of a real valued physical signal, is also established in other domains, e.g., microwave and communications engineering [18,19]. In these domains, the concept is commonly denoted as baseband representation or complex envelope. The research reported in the power engineering domain can be roughly divided into two categories, Hilbert or Fourier based. The phasor representation can be derived from the analytical signal constructed from the Hilbert transform or by applying Fourier transform.

The Hilbert based approach is commonly applied in the power system domain, where it is referred to as envelope waveforms [8], dynamic phasors [20], or SFA [21]. The second approach relies on the Fourier transform to approximate a physical signal by a series of Fourier coefficients and it has been developed in the power electronics domain as generalized state space averaging [11]. In more recent publications, SFA and dynamic phasors have been associated to the Hilbert and Fourier based approach [7,22]. In the following, the variables that result from SFA are denoted as dynamic phasors, regardless of which one of the approaches is used.

A dynamic phasor can be considered as a projection of a signal defined as a function over time onto the time-frequency plane. Dynamic phasors are not a spectral representation over an infinite time interval, which would eliminate the time dependency as in a frequency spectrum. Instead, dynamic phasors depend on both time and frequency.

The Hilbert approach is based on the analytical signal, which can be derived from a real signal by extending it with an imaginary part that is equal to the Hilbert transform of the original signal (1).

$$\mathcal{A}\{s(t)\} = s(t) + j\mathcal{H}\{s(t)\} \tag{1}$$

Applying (1) to a bandpass signal $s_{bp}(t) = A(t)\cos(\omega_c t + \theta(t))$, results in the analytical representation corresponding to a generic bandpass signal:

$$\begin{aligned}
\mathcal{A}\{s_{bp}(t)\} &= s_{bp}(t) + j\mathcal{H}\{s_{bp}(t)\} \\
&= s_{bp,I}(t)\cos(\omega_c t) - s_{bp,Q}(t)\sin(\omega_c t) + j(s_{bp,I}(t)\sin(\omega_c t) + s_{bp,Q}(t)\cos(\omega_c t)) \\
&= (s_{bp,I}(t) + js_{bp,Q}(t))(\cos(\omega_c t) + j\sin(\omega_c t)) \\
&= \langle s_{bp}\rangle(t)e^{j\omega_c t}
\end{aligned} \tag{2}$$

with the in-phase component $s_{bp,I}(t) = A(t)\cos(\theta(t))$ and the quadrature component $s_{bp,Q}(t) = A(t)\sin(\theta(t))$. In order to calculate the dynamic phasor from the analytical signal, it is shifted by the center frequency $\omega_c$ (3).

$$\langle s\rangle(t) = \mathcal{A}\{s(t)\}e^{-j\omega_c t} \tag{3}$$

The Fourier transform is common in the power electronics domain, because it can be applied to monocomponent as well as multicomponent signals, where more than one frequency band are of interest. The disadvantage of the Fourier transform for time-frequency analysis is that it suffers from the uncertainty principle. Time and frequency resolution cannot be increased arbitrarily at the same time, which complicates the calculation of accurate dynamic phasors. Each phasor is approximated by a Fourier coefficient that is dynamically changing over time. The time dependency of the Fourier coefficients results from the sliding Fourier transform. Applying a sliding observation window of time $T$, the time dependent Fourier coefficients are expressed by the transform (4), where $\omega_c$ is the fundamental frequency.

$$\langle s\rangle_k(t) = \frac{1}{T}\int_{t-T}^{t} s(\tau)\cdot e^{-jk\omega_c \tau}d\tau \tag{4}$$

An accuracy analysis of the basic approach used in the DPsim solver, which is the combination of frequency shifted phasor variables and the trapezoidal rule as discretization method, can be found in [8].

## 3. Models and Methods

The following sections describe, in detail, the models simulated in Section 4: averaged inverter, inverter with harmonics, and transmission line. The last section explains how the simulation can be more efficiently parallelized in the case of detailed power electronics models. First, the solution process of DPsim is reviewed, including the transmission line method (TLM) for subsystem decoupling. Subsequently, the process is extended with a method to split the network not only between nodes, but also across frequencies to compensate for the increased number of equations when simulating the inverter model with harmonics.

### 3.1. Averaged Inverter Model with Controls

The averaged model with controls encompasses three kinds of components: electrical components, control components, and interface components connecting the former two. The averaged inverter model is applied in DPsim for both types of simulations DP and EMT, see Section 4.1. This section is focused on the description of how the inverter's control, which is modelled in state space, is interfaced with the inverter's and grid's electrical components in the case of DP and EMT simulation.

The inverter is modelled as Voltage Source Inverter (VSI), see Figure 1. Accordingly, a controlled voltage source represents the inverter's output based on an averaged switching model. Besides, the inverter model includes an LC filter as output filter, which is composed of two resistors, an inductor and a capacitor. The inverter model operates at low voltage (LV) level and the grid operates at medium voltage (MV) level. Therefore, the inverter includes a step-up transformer for connection to the MV grid. The chosen parameters are listed in Table 1.

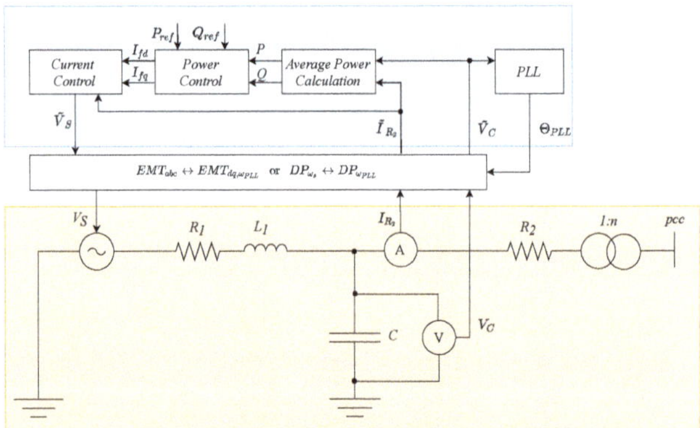

**Figure 1.** Averaged inverter model with controls.

**Table 1.** Averaged inverter example parameters.

| Attribute | Value |
| --- | --- |
| $R_1$ | $0.1\,\Omega$ |
| $L_1$ | $2\,\mathrm{mH}$ |
| $C$ | $789.3\,\mu\mathrm{F}$ |
| $R_2$ | $0.1\,\Omega$ |
| $L_{transformer}$ | $928.3\,\mathrm{mH}$ |

As mentioned in Section 3.4, DPsim follows the modified nodal analysis approach in order to solve the electrical network. Accordingly, the inverter's electrical components that are mentioned above are integrated into the system solution by stamping their contributions into system matrix and source vector. The dynamic behaviour of the components is taken into account by means of the components' DC equivalents [23]. Depending on the type of simulation, the DC equivalents represent the models either in the DP or the EMT domain.

The inverter's control is modelled in state space and it enables an operation of the inverter in grid feeding mode [24]. The control aims at injecting active and reactive power according to specified reference values. The actual power feed-in is determined through an average power calculation, including a low-pass filter. The power calculation serves as input to power and current control loop, which are both implemented as PI controllers and that finally provide the voltage set-point for the controlled voltage source. As common in the field of control engineering, the inverter's control is modelled in state space. The state space model employs quantities represented in the inverter's local dq reference frame, which rotates with the frequency measured by a phase-locked-loop (PLL). The interface connecting the electrical model with the controller's model in state space depends on the domain in which the electrical simulation is performed.

In case of an EMT simulation, the quantities are represented with respect to a stationary reference frame. The grid quantities are represented in DPsim as real-valued phase-to-ground peak quantities $x = [\hat{x}_a \ \hat{x}_b \ \hat{x}_c]$. To obtain the quantities in the inverter's local reference frame rotating with $\omega_{PLL}$, they are transformed by means of the power-invariant dq transformation, according to

$$\begin{bmatrix} x_{d,\omega_{PLL}} \\ x_{q,\omega_{PLL}} \end{bmatrix} = \sqrt{\frac{2}{3}} \underbrace{\begin{bmatrix} cos(\theta_{PLL}) & cos(\theta_{PLL} - \frac{2\pi}{3}) & cos(\theta_{PLL} + \frac{2\pi}{3}) \\ -sin(\theta_{PLL}) & -sin(\theta_{PLL} - \frac{2\pi}{3}) & -sin(\theta_{PLL} + \frac{2\pi}{3}) \end{bmatrix}}_{T_{dq,\theta_{PLL}}} \begin{bmatrix} \hat{x}_a \\ \hat{x}_b \\ \hat{x}_c \end{bmatrix}. \tag{5}$$

This means that the inverter's input variables $v_C$ and $i_{R_2}$, see Figure 1, are transformed to $\tilde{v}_C$ and $\tilde{i}_{R_2}$ according to

$$\tilde{v}_C = \begin{bmatrix} v_{C,d,\omega_{PLL}} \\ v_{C,q,\omega_{PLL}} \end{bmatrix} = T_{dq,\theta_{PLL}} \underbrace{\begin{bmatrix} \hat{v}_{C,a} \\ \hat{v}_{C,b} \\ \hat{v}_{C,c} \end{bmatrix}}_{v_C} \tag{6}$$

$$\tilde{i}_{R_2} = \begin{bmatrix} i_{R_2,d,\omega_{PLL}} \\ i_{R_2,q,\omega_{PLL}} \end{bmatrix} = T_{dq,\theta_{PLL}} \underbrace{\begin{bmatrix} \hat{i}_{R_2,a} \\ \hat{i}_{R_2,b} \\ \hat{i}_{R_2,c} \end{bmatrix}}_{i_{R_2}}. \tag{7}$$

The inverter's controllers deliver as output the quantities in the local reference frame, which are then transformed back to the grid's stationary frame by means of the inverse dq transformation

$$\begin{bmatrix} \hat{x}_a \\ \hat{x}_b \\ \hat{x}_c \end{bmatrix} = \sqrt{\frac{2}{3}} \underbrace{\begin{bmatrix} cos(\theta_{PLL}) & -sin(\theta_{PLL}) \\ cos(\theta_{PLL} - \frac{2\pi}{3}) & -sin(\theta_{PLL} - \frac{2\pi}{3}) \\ cos(\theta_{PLL} + \frac{2\pi}{3}) & -sin(\theta_{PLL} + \frac{2\pi}{3}) \end{bmatrix}}_{T^{-1}_{dq,\theta_{PLL}}} \begin{bmatrix} x_{d,\omega_{PLL}} \\ x_{q,\omega_{PLL}} \end{bmatrix}. \tag{8}$$

Correspondingly, the controller's output voltage $v_S$ is considered in the grid solution after the back transformation of $\tilde{v}_S$, according to

$$v_S = \begin{bmatrix} \hat{v}_{S,a} \\ \hat{v}_{S,b} \\ \hat{v}_{S,c} \end{bmatrix} = T^{-1}_{dq,\theta_{PLL}} \underbrace{\begin{bmatrix} v_{S,d,\omega_{PLL}} \\ v_{S,q,\omega_{PLL}} \end{bmatrix}}_{\tilde{v}_S}. \tag{9}$$

In case of a DP simulation, DPsim uses complex phase-to-phase RMS quantities denoted as $\langle x \rangle_{\omega_c} = [\langle x \rangle_{\omega_c,re} \ \langle x \rangle_{\omega_c,im}]$ following Equation (2). In contrast to the next section, we consider at this point exclusively the first order DP at center frequency $\omega_c$. The grid quantities have the grid's synchronous frequency $\omega_S$ as center frequency, which is usually equal to 50 or 60 Hz. Thus, the original EMT quantities are related to the DP grid quantities by

$$\begin{bmatrix} \hat{x}_a \\ \hat{x}_b \\ \hat{x}_c \end{bmatrix} = Re \left\{ \sqrt{\frac{2}{3}} \langle x \rangle_{\omega_S} \ e^{j\theta_S} \begin{bmatrix} 1 \\ e^{-j\frac{2\pi}{3}} \\ e^{j\frac{2\pi}{3}} \end{bmatrix} \right\}. \tag{10}$$

with $\theta_S = \omega_S t$. Instead, the inverter's input and output quantities imply the local frequency $\omega_{PLL}$ as center frequency, which is defined through the PLL's angle $\theta_{PLL}$. Accordingly, the inverter's quantities are related to the original EMT quantities by

$$\begin{bmatrix} \hat{x}_a \\ \hat{x}_b \\ \hat{x}_c \end{bmatrix} = Re \left\{ \sqrt{\frac{2}{3}} \langle x \rangle_{\omega_{PLL}} \ e^{j\theta_{PLL}} \begin{bmatrix} 1 \\ e^{-j\frac{2\pi}{3}} \\ e^{j\frac{2\pi}{3}} \end{bmatrix} \right\} \tag{11}$$

Comparing Equations (10) and (11) yields the following relationship between the DPs with two different center frequencies

$$\langle x \rangle_{\omega_{PLL}} = \langle x \rangle_{\omega_S} \ e^{j(\theta_S - \theta_{PLL})}. \tag{12}$$

Thus, the transformation from DPs with the grid's synchronous frequency $\omega_S$ as center frequency to DPs with the inverter's frequency $\omega_{PLL}$ as center frequency looks as follows

$$\begin{bmatrix} \langle x \rangle_{\omega_{PLL},re} \\ \langle x \rangle_{\omega_{PLL},im} \end{bmatrix} = \underbrace{\begin{bmatrix} cos(\theta_S - \theta_{PLL}) & -sin(\theta_S - \theta_{PLL}) \\ sin(\theta_S - \theta_{PLL}) & cos(\theta_S - \theta_{PLL}) \end{bmatrix}}_{=:T_{DP,\theta_S - \theta_{PLL}}} \begin{bmatrix} \langle x \rangle_{\omega_S,re} \\ \langle x \rangle_{\omega_S,im} \end{bmatrix}. \tag{13}$$

Accordingly, the inverter's input quantities can be obtained from

$$\tilde{v}_C = \begin{bmatrix} \langle v_C \rangle_{\omega_{PLL},re} \\ \langle v_C \rangle_{\omega_{PLL},im} \end{bmatrix} = T_{DP,\theta_S - \theta_{PLL}} \underbrace{\begin{bmatrix} \langle v_C \rangle_{\omega_S,re} \\ \langle v_C \rangle_{\omega_S,im} \end{bmatrix}}_{v_C} \tag{14}$$

$$\tilde{i}_{R_2} = \begin{bmatrix} \langle i_{R_2} \rangle_{\omega_{PLL},re} \\ \langle i_{R_2} \rangle_{\omega_{PLL},im} \end{bmatrix} = T_{DP,\theta_S - \theta_{PLL}} \underbrace{\begin{bmatrix} \langle i_{R_2} \rangle_{\omega_S,re} \\ \langle i_{R_2} \rangle_{\omega_S,im} \end{bmatrix}}_{i_{R_2}}. \tag{15}$$

Besides, the inverter's output voltage is transformed back according to

$$v_S = \begin{bmatrix} \langle v_S \rangle_{\omega_S,re} \\ \langle v_S \rangle_{\omega_S,im} \end{bmatrix} = T_{DP,\theta_{PLL}-\theta_S} \underbrace{\begin{bmatrix} \langle v_S \rangle_{\omega_{PLL},re} \\ \langle v_S \rangle_{\omega_{PLL},im} \end{bmatrix}}_{\tilde{v}_S}. \tag{16}$$

It can be noted that the transformation matrix in Equation (8), which links the inverter's dq variables $[x_{d,\omega_{PLL}}\ x_{q,\omega_{PLL}}]$ with the original EMT variables, is the same as the transformation matrix obtained when resolving Equation (11), which links the inverter's DP variables $[\langle x \rangle_{\omega_{PLL},re}\ \langle x \rangle_{\omega_{PLL},im}]$ with the original EMT variables. That is, the described interfaces ensure that, from a theoretical perspective, the inverter model in state space operates on the same quantities for EMT and DP simulation, while from a simulation perspective these quantities can significantly differ in terms of accuracy, as shown in Section 4.1.

Using these interfaces, other state space controllers can be integrated into DPsim in the same way. Control models operating on dq variables with respect to a local reference frame can be easily integrated into DPsim and applied in both EMT and DP simulations.

*3.2. Inverter Model Including Harmonics*

Here, harmonic analysis is applied to determine the dynamic phasor quantities injected into the grid by a unipolar single-phase full-bridge voltage source inverter. The harmonic frequency content of power electronics topologies could also be determined by a Fast Fourier Transform (FFT) of the electromagnetic transient waveform, which reduces the mathematical effort, but it requires extra computing capacity. Instead, the harmonic components of the PWM waveform can be computed directly, which is more precise [12].

In the following, it is assumed that the PWM signal is generated by comparing a triangular carrier signal and a reference signal of the form (17). The reference signal is defined by the nominal grid frequency $\omega_0$, the modulation index $M_r$, and the reference signal phase $\varphi$.

$$s_{ref}(t) = M_r \cos(\omega_0 t + \varphi) \tag{17}$$

Applying double Fourier integral analysis, as explained in [12], the resulting PWM waveform can be expressed in terms of Fourier coefficients (18). With respect to [12], the following equations have been reformulated in a way that allows fewer evaluations of the series term, which facilitates efficient implementation. Besides, a phase angle of the reference signal different from zero is considered. In this case, double edge natural sampling is assumed for the PWM signal. This means that the carrier signal is triangular instead of sawtooth (trailing edge), and switching is initiated whenever the carrier and reference signal cross.

$$\begin{aligned} v_{inv}(t) &= A_{nd,1} \cos(\omega_0 t + \varphi) \\ &\quad + \frac{4V_{in}}{\pi} \sum_{m=2,4,\dots}^{\infty} \sum_{n=\pm1,\pm3,\dots}^{\infty} A_{nd,2}(m,n) \cos(m\omega_{sw}t + n(\omega_0 t + \varphi)) \\ A_{nd,1} &= M_r V_{in} \\ A_{nd,2}(m,n) &= \frac{J_n(mM_r\pi/2)}{m} \sin\left(\frac{(m+n)\pi}{2}\right) \end{aligned} \tag{18}$$

$J_n$ is the Bessel function defined as:

$$J_n(x) = \sum_{k=0}^{\infty} \frac{(-1)^k}{k!\,\Gamma(k+n+1)} \left(\frac{x}{2}\right)^{2k+n}. \tag{19}$$

The switching frequency is $\omega_{sw}$, switching frequency harmonic is $m$ and the nominal grid frequency harmonic is denoted $n$.

Achieving natural sampling with a digital control system is challenging, because it is difficult to determine the exact point of crossing between reference and carrier signal. Therefore, alternatives such as regular sampling are often implemented. Regular sampling means that the reference signal is sampled and held for each carrier signal interval. Depending on the sample and hold time of the reference signal, double edge regular sampling can be further divided into symmetrical or asymmetrical sampling. In the symmetrical case, the reference is sampled once per carrier interval opposed to twice in the asymmetrical case. The harmonic analysis for a symmetrical regular sampled reference with a double edge carrier is defined by Equation (20).

$$
\begin{aligned}
v_{inv}(t) &= \frac{4V_{in}}{\pi} \sum_{n=1,3,\dots}^{\infty} A_{rd,1}(n) \cos(n(\omega_0 t + \varphi)) \\
&+ \frac{4V_{in}}{\pi} \sum_{m=1,2,\dots}^{\infty} \sum_{n=\pm 1,\pm 3,\dots}^{\infty} A_{rd,2}(n,m) \cos(m\omega_{sw}t + n(\omega_0 t + \varphi)) \\
A_{rd,1}(n) &= \frac{J_n\left(n\frac{\omega_0}{\omega_{sw}}\frac{\pi}{2}M_r\right)}{n\frac{\omega_0}{\omega_{sw}}} \sin\left(\left(n + n\frac{\omega_0}{\omega_{sw}}\right)\frac{\pi}{2}\right) \\
A_{rd,2}(n,m) &= \frac{J_n\left(\left(m + n\frac{\omega_0}{\omega_{sw}}\right)\frac{\pi}{2}M_r\right)}{m + n\frac{\omega_0}{\omega_{sw}}} \sin\left(\left(m + n + n\frac{\omega_0}{\omega_{sw}}\right)\frac{\pi}{2}\right).
\end{aligned}
$$

$$
\tag{20}
$$

If the ratio between the switching frequency of the inverter and the reference signal frequency is an integer, the harmonic analysis can be applied to calculate dynamic phasors for the fundamental and its harmonics. For consistency with the previous Equations (18) and (20), the harmonics are still denominated in terms of multiples of the switching frequency and the reference signal frequency, even though they could be expressed as multiples of $n$ only if the requirement of an integer ratio is fulfilled. Every time that the input variables $M_r$ and $\varphi$ change, the dynamic phasors must be updated according to (22) when considering natural sampling and similarly for regular sampling.

$$
\langle v \rangle_1 = A_{nd,1}\, e^{j\varphi} \tag{21}
$$

$$
\langle v \rangle_{m,n} = A_{nd,2}(m,n)\, e^{jn\varphi} \tag{22}
$$

Subsequently, the inverter is represented by voltage sources behind the LCL output filter, which is represented by basic components that are stamped into the nodal analysis system matrix. For each dynamic phasor, there is one voltage source representing one harmonic of the reference signal frequency, as depicted in Figure 2. Therefore, increasing the accuracy of the model by increasing the number of dynamic phasors reduces the simulation performance in terms of minimum computation time per simulation step, because the resulting equation set is larger. A solution to this challenge is described in Section 3.4.

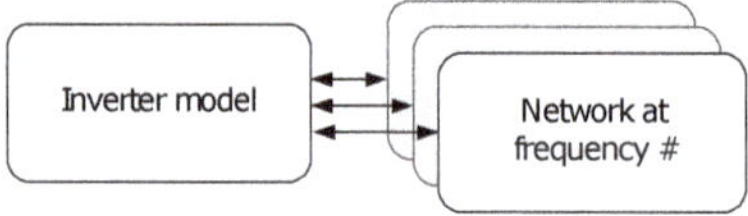


**Figure 2.** Separate computation of nonlinear inverter model from the network that is split by shifting frequency.

### 3.3. Transmission Line

The travelling wave transmission line model has two advantages over the Pi-line model. For long lines, it is more accurate than the Pi-line model and it enables the decoupling of network sections as shown in Section 3.4.

The most simple transmission line model, the Bergeron model, which is also implemented in the simulator, is based on the telegrapher's equation. This model is derived from the equations of a lossless line and losses are only approximated in a second step. The model parameters are computed for a fixed frequency, which means that they are inaccurate for deviating frequencies. Actually, the line parameters vary, especially for high frequencies, where the skin effect has a notable impact. Dynamic phasors have an advantage in this regard, because, in the best case, each phasor represents a narrow frequency band signal. Each of these phasors can be treated differently by computing the transmission line parameters for the center frequency of these bands.

The Bergeron model applied to nodal analysis can decouple the two end nodes of a transmission line. This means two network sections that are only connected by transmission lines can be solved separately and in parallel. Instead of having one large system matrix, two smaller system matrices can be computed. These two subsystems are only coupled by their source vectors, which depend on the solution of the other subsystem from the previous time step.

This transmission line model can only be used for lines, where the delay is larger than the simulation time step. Therefore, this model is usually not suitable for simulations with large time steps or networks with very short lines.

First, the commonly used EMT domain model is described. Subsequently, this model is adapted for DP variables. The Bergeron model is derived from the wave propagation Equation (24) of a lossless line with per line length inductance $L'$ and capacitance $C'$ [25].

$$-\frac{\partial v(x,t)}{\partial x} = L'\frac{\partial i(x,t)}{\partial t} \tag{23}$$

$$-\frac{\partial i(x,t)}{\partial x} = C'\frac{\partial v(x,t)}{\partial t} \tag{24}$$

The lossless line can be represented in nodal analysis by impedances and current sources, as depicted in Figure 3, with $Z_E = Z_C$ and the values of the current sources being computed from the currents at time $t - \tau$ [23].

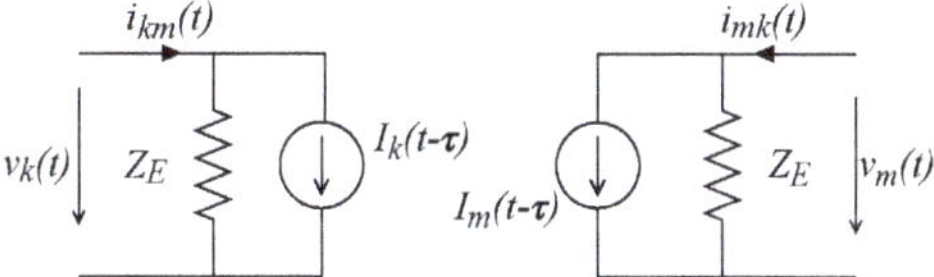


**Figure 3.** Equivalent circuit for a lossless line.

The time constant $\tau$ and the characteristic impedance $Z_C$ are calculated from the line length $d$, as follows:

$$Z_C = \sqrt{L'/C'} \quad \tau = d\sqrt{L'C'}. \tag{25}$$

Line losses can be approximated in the model by adding a resistance at the start and end of each line section and modifying the history current sources. The equivalent impedance becomes $Z_E = Z + R/4$ and the history current is calculated as:

$$
\begin{aligned}
I_k(t - \tau) = {} & \frac{-Z}{(Z + R/4)^2}(v_m(t - \tau) + (Z - R/4) \cdot i_{mk}(t - \tau)) \\
& - \frac{R/4}{(Z + R/4)^2}(v_k(t - \tau) + (Z - R/4) \cdot i_{km}(t - \tau)),
\end{aligned} \tag{26}
$$

with an analogous expression for $I_m(t - \tau)$.

It is necessary to understand how the time delay in (26) affects the DP variables in order to develop the dynamic phasor model. A bandpass signal $s(t)$ can be expressed as a baseband signal $f(t)$ shifted by $\omega_s$ (27). The Fourier transform $S(\omega)$ is given by (28).

$$s(t) = \mathbf{Re}\left\{ f(t)e^{j\omega_s t} \right\} \tag{27}$$

$$S(\omega) = \int f(t)e^{j\omega_s t}e^{-j\omega t}d\omega = F(\omega - \omega_s) \tag{28}$$

A time shift $\tau$ becomes a multiplication by $e^{-j\tau\omega}$ in the frequency domain. Hence, the time shifted signal can be computed from the Fourier transform, as follows:

$$s(t - \tau) = \mathbf{Re}\left\{ \int F(\omega - \omega_s)e^{-j\tau\omega}e^{j\omega t}d\omega \right\}. \tag{29}$$

For $y = \omega - \omega_s$, (29) becomes

$$
\begin{aligned}
s(t - \tau) &= \mathbf{Re}\left\{ \int F(y)e^{-j\tau(y+\omega_s)}e^{j(y+\omega_s)t}dy \right\} \\
&= \mathbf{Re}\left\{ e^{j(\omega_s t - \tau\omega_s)} \int F(y)e^{-j\tau y}e^{jyt}dy \right\} \\
&= \mathbf{Re}\left\{ e^{j(\omega_s t - \tau\omega_s)} f(t - \tau) \right\}.
\end{aligned}
\tag{30}
$$

This means that the time shift leads to an additional phase shift $e^{-j\tau\omega_s}$ for the DP variables. Therefore, (26) expressed in terms of dynamic phasors results in (31).

$$
\begin{aligned}
\langle I_k(t - \tau) \rangle = e^{-j\tau\omega_s} \Bigg[ &\frac{-Z}{(Z + R/4)^2}\left(\langle v_m(t - \tau) \rangle + (Z - R/4) \cdot \langle i_{mk}(t - \tau) \rangle\right) \\
&- \frac{R/4}{(Z + R/4)^2}\left(\langle v_k(t - \tau) \rangle + (Z - R/4) \cdot \langle i_{km}(t - \tau) \rangle\right) \Bigg]
\end{aligned}
\tag{31}
$$

In a dynamic phasor simulation, the carrier or shifting frequency is only considered after the actual simulation in a post processing step. Without the additional phase shift, the time delay does not affect the carrier signal and the simulation results would be incorrect.

A small example that consists of a voltage source and resistance connected by a transmission line is simulated in order to demonstrate the correctness of the DP model. The parameters of the components can be found in Table 2.

**Table 2.** Transmission line example parameters.

| Attribute | Value |
|---|---|
| $V_{sec}$ | 100 kV |
| $R$ | 5 Ω |
| $L$ | 0.16 H |
| $C$ | 1 μF |
| $R_{load}$ | 10 kΩ |

The comparison of DP to EMT simulation shows that the results are the same if the phase shift is correctly taken into account. At the beginning, transients are visible because the simulation does not start from steady state. Figure 4 compares the current through the transmission line for the DP and EMT model.

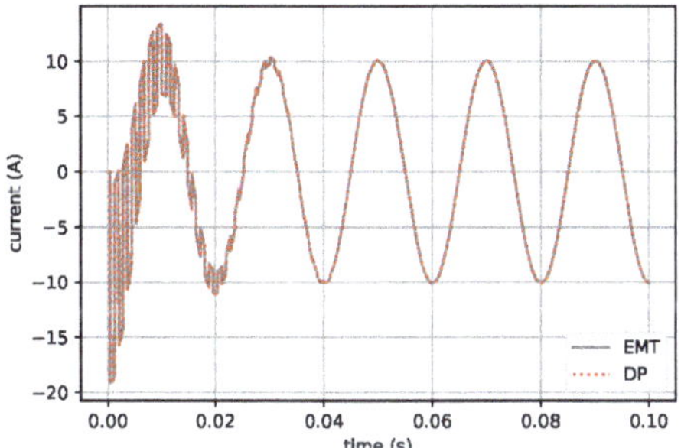

**Figure 4.** Comparison of DP and electromagnetic-transient (EMT) resistor current results for the transmission line model.

## 3.4. Parallelization of Subsystems and Frequency Bands

First, the solution process of the DPsim simulator is reviewed, including the decoupling and parallelization implemented before power electronics were added to the simulator. Subsequently, this concept is extended by a method that supports a more efficient computation of networks with many harmonics.

The main idea behind DPsim is to compute the network solution with nodal analysis [26] applied to DP variables. Therefore, the solution for one simulation step consists of three parts:

- calculating component states required for the network solution,
- solving the network,
- updating remaining component states from the network solution.

Most component related tasks can be executed in parallel as shown in Figure 5, because the components do not interact with each other directly, but only through the network.

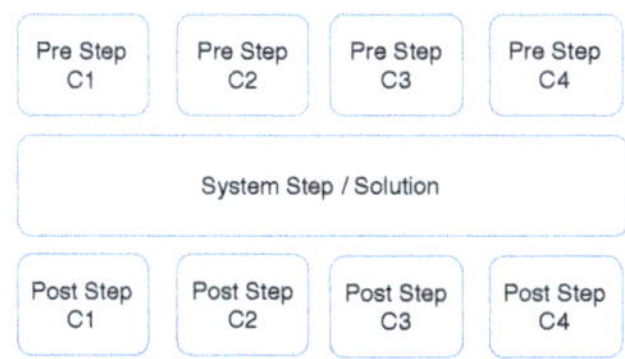

**Figure 5.** Parallelization of the component step calculations.

However, the network simulation is often the bottleneck. The TLM is commonly used in EMT real-time simulation in order to decouple the network and compute the subsystems in parallel, as shown in Figure 6. Because DPsim is computing DP variables, the EMT transmission line model has to be adapted to the DP domain, as described in Section 3.3.

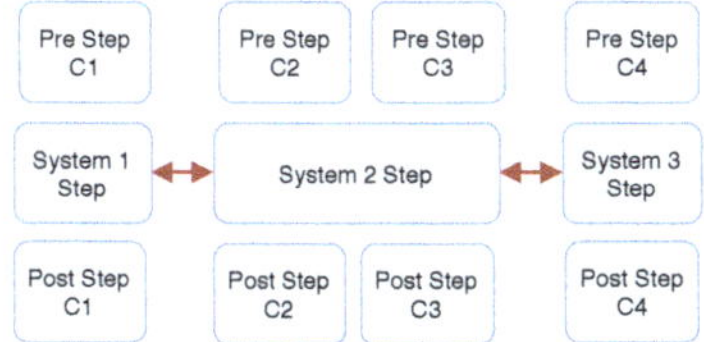

**Figure 6.** Parallel solution of decoupled subsystems.

As mentioned earlier, one of the main disadvantages of the dynamic phasor approach is that the more phasors are used to represent a time domain variable, the larger becomes the number of equations and variables to solve for. This is especially a problem for the network solution, which can be already very large because of the number of network nodes. The idea is to split the network solution into separate solutions for each frequency band, as depicted in Figure 7. The underlying assumption is that the network is usually composed of linear components. An approach on how to separate nonlinearities from the network model is proposed in [27]. Nonlinearities are implemented in the components connected to the network and they can be treated separate from the network solution. Hence, the network solution does not feature cross frequency coupling.

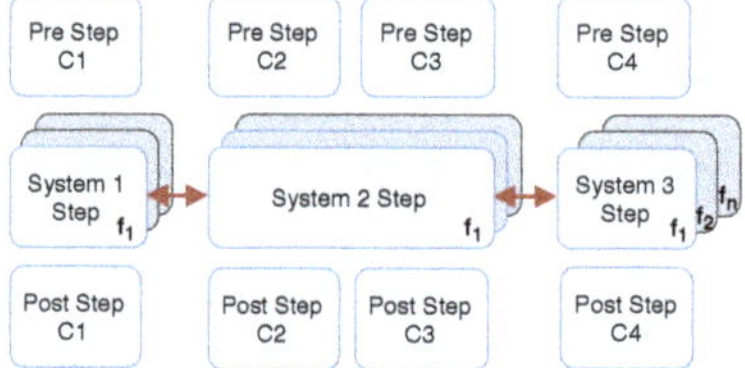


**Figure 7.** Parallel solution of decoupled subsystems for each set of dynamic phasors associated to the same shifting frequency.

## 4. Results

The first section demonstrates a scenario featuring the averaged DP inverter model integrated into DPsim in order to support the analysis of future energy systems with large shares of inverter interfaced renewable sources. The simulation scenario is a variation of the CIGRE MV network with a high share of renewable sources represented by averaged inverter models with controls, see Section 3.1. The results show that a DP simulation can be as accurate as an EMT simulation for small simulation time steps. For larger simulation steps, where the EMT simulation does not return useful results anymore, the DP results show still a good accuracy while tending towards conventional phasor results.

Section 4.2 demonstrates the extension of the inverter model by harmonics, which results in a very detailed representation of a switched inverter model. This model is compared to a switching EMT inverter model implemented in Matlab/Simulink™. The detailed dynamic phasor model, including harmonics, leads to a large set of equations, which has an impact on the computation time, especially for the network solution.

Section 4.3 presents results for the parallelization of the detailed inverter model simulation. The method is explained in Section 3.4 and it addresses the problem by splitting the solution of the network into smaller tasks, which can be executed in parallel.

### 4.1. Averaged Inverter Model with Controls

This section shows an analysis of the simulation results that were obtained for a scenario with a high penetration of renewables using two types of component models: DP and EMT models. That is, all types of grid components in the simulation (transformers, lines, loads, and renewable sources) are represented by DP or EMT models. The underlying research question is how an increase of the simulation step affects the accuracy of the simulation results when using these two modeling approaches. At this point, it should be noted that a fixed-step solver is applied, which is particularly suitable for real-time simulation, and the trapezoidal rule for numerical integration.

The two modeling approaches are investigated in the context of a grid simulation at medium voltage level, where a high penetration of renewables requires an adequate simulation of the emerging grid dynamics. The simulation scenario encompasses the first feeder of the CIGRE MV grid, see Figure 8, at a nominal voltage of 20 kV, and with a total load power of 4319 kW. In the following, a scenario is considered with 100% penetration of photovoltaic systems, i.e., the installed nominal power of the PV systems is equal to the total load power. For this, PV systems of equal size are connected to nodes N3 to N11 of the feeder. The PV systems are represented by inverters using an averaged switching model. The inverters operate in grid feeding mode targeting the injection of active and reactive power reference values. Section 3.1 outlines further details on the averaged inverter model and its control.

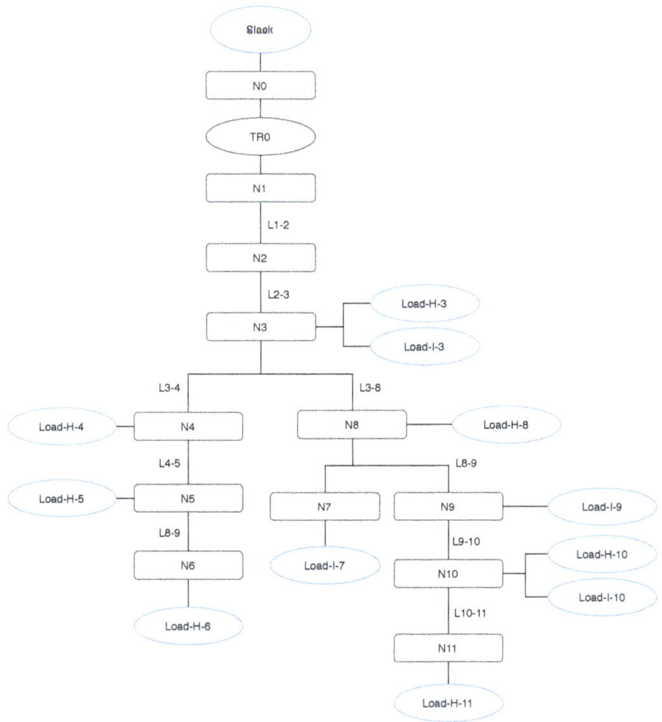

**Figure 8.** Simulation scenario considering the first feeder of CIGRE MV grid.

First, EMT and DP simulations are both executed with characteristic simulation steps. The simulations are conducted with a duration of 5 s and apply a load step event at time instant 2 s by increasing the load power about 1500 kW at node N11. Figure 9 shows the current through the line L11-10 100 ms before and 200 ms after the load step event. The DP simulation results have been analogous to the EMT results linearly interpolated to 50 µs and additionally shifted back by 50 Hz.

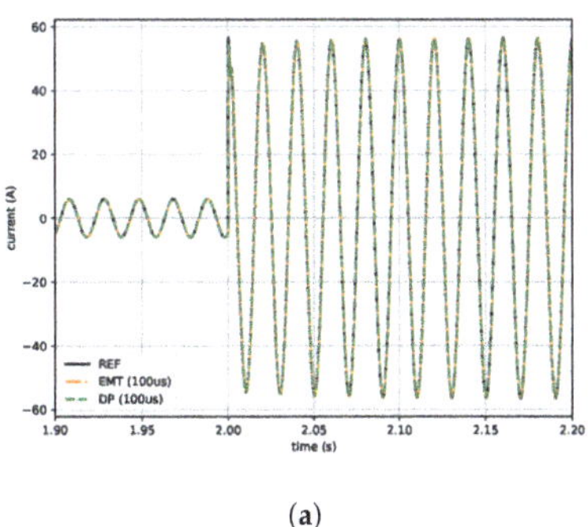

(a)

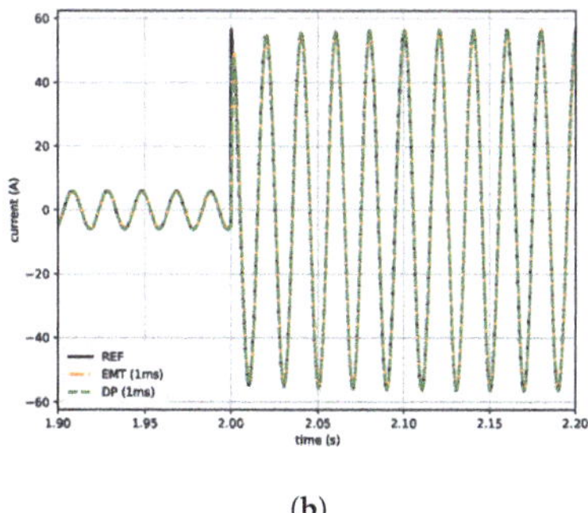

(b)

**Figure 9.** *Cont.*

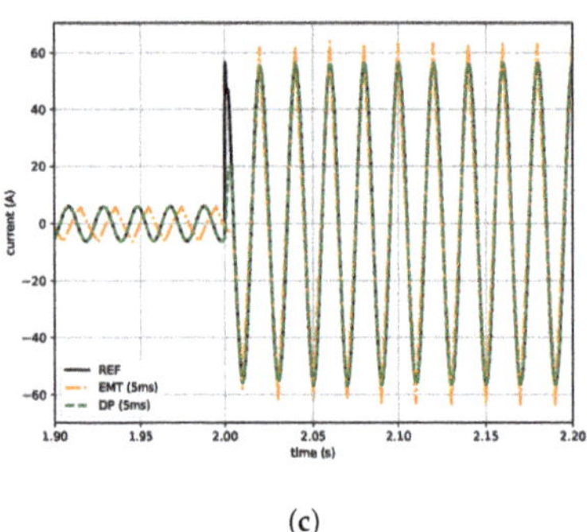 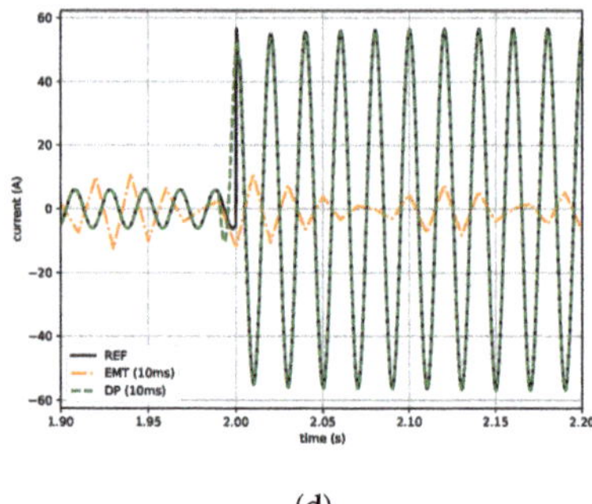

(c)             (d)

**Figure 9.** Comparison of EMT and dynamic phasor (DP) simulation results with the averaged inverter model for time steps of 100 µs (**a**), 1 ms (**b**), 5 ms (**c**) and 10 ms (**d**).

The DP and EMT simulation results are compared for characteristic simulation steps against a reference simulation, which was conducted as EMT simulation with a simulation step of 50 µs. Figure 9a demonstrates that, for a step size of 100 µs, the simulation results in the DP and EMT domain still match well with the results of the reference simulation. When increasing the simulation step to 1 ms, see Figure 9b, the results still appear very similar. However, it is noteworthy that both DP and EMT results deviate from the reference during the first oscillatory cycle. This seems to be feasible as a simulation with a step size of 1 ms is not able to capture the wide frequency spectrum occurring after the load step, for both the EMT and the 50 Hz shifted DP signal. Yet, except from this deviation during the first oscillatory cycle, the signals match well in the following oscillatory cycles after the load step event.

For larger simulation steps of 5 ms and 10 ms, see Figure 9c,d, the EMT results differ significantly from the reference results. Instead, the DP results remain quite accurate for these larger simulation steps. In particular, it is remarkable that the DP results still show the evolution of the oscillations towards the steady state behaviour well.

In the following, the difference of DP and EMT simulation results is quantatively investigated with respect to the reference results. For this, the DP and EMT simulation results obtained with steps in the range from 100 µs to 20 ms are considered. The step size is increased by 200 µs in the microseconds range and by 1 ms in the milliseconds range. For DP and EMT results of the line current L11-10, the root mean squared error (RMSE) is calculated with respect to the reference results. The evolution of the RMSE that is depicted in Figure 10 confirms that both DP and EMT results are quite accurate for simulation steps up to 1 ms.

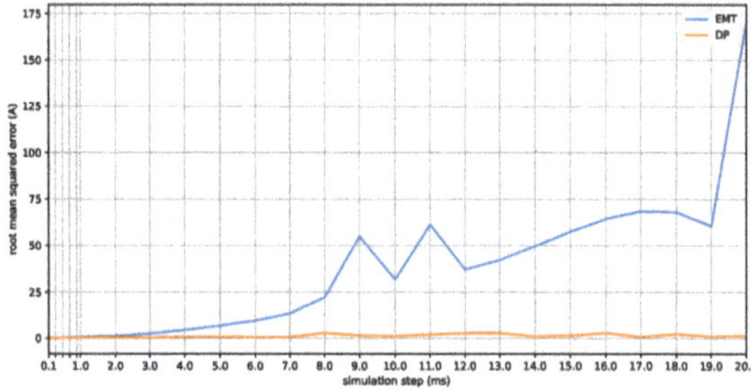

**Figure 10.** RMSE of EMT and DP simulation results with respect to the reference results.

In the range from 2 ms to 20 ms, the RMSE of the EMT results increases tremendously. This seems reasonable as in EMTP-based simulation it is well known that an appropriate accuracy is obtained only for simulation steps below 10% of the minimum cycle duration that is to be simulated [28]. This requires a simulation step below 2 ms for 50 Hz signals. As the simulated signals are shifted by 50 Hz in the DP domain, the RMSE of the DP results remains very low with respect to the EMT error.

### 4.2. Inverter Model Including Harmonics

One advantage of dynamic phasors is the ability to include selected harmonics, which is particularly interesting for power electronics modelling. The simulations presented in this section feature a DP inverter model that extends the inverter used in Section 4.1 by harmonics that are based on harmonic analysis as explained in Section 3.2. This model, which does not include switches, is validated against a switching Matlab/Simulink™ EMT model. For this comparison, the regular sampling Equation (20) are used in the harmonic analysis.

The inverter model is connected to an equivalent grid model composed of an impedance and a voltage source. Figure 11 depicts the circuit and Table 3 lists the parameters.

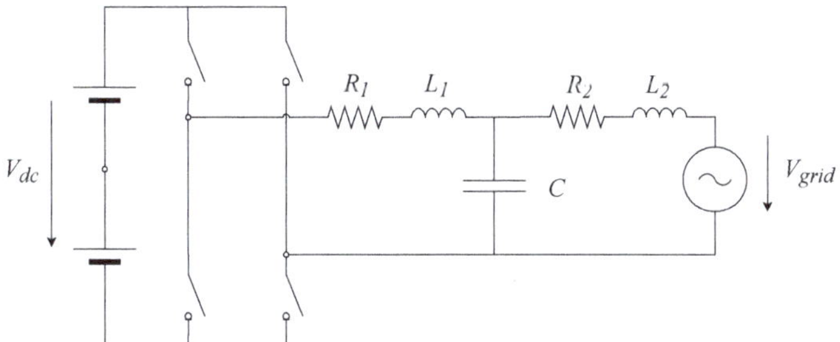


**Figure 11.** Single phase inverter with LCL output filter connected to an equivalent grid represented by a voltage source with series impedance $R_g$ and $L_g$.

**Table 3.** Inverter simulation example parameters.

| Attribute | Value |
| --- | --- |
| $R_1$ | $0.1\,\Omega$ |
| $L_1$ | $600\,\mu H$ |
| $R_2$ | $0.1\,\Omega$ |
| $L_2$ | $150\,\mu H$ |
| $R_g$ | $0.001\,\Omega$ |
| $L_g$ | $0.001\,H$ |
| $C$ | $10\,\mu F$ |
| $R_C$ | $1\,\mu\Omega$ |

The DP inverter model can be simulated with a varying number of harmonics. Here, eight and 28 additional harmonics are considered. Figure 12 shows the voltage at the output filter capacitor of the circuit depicted in Figure 11 for DP with eight additional harmonics and the Simulink EMT model. Figure 13a,b depict a smaller time interval of the simulation and demonstrate the difference in accuracy when considering different numbers of harmonics and dynamic phasors, respectively. The mean absolute error (MAE) for the results with eight harmonics is about 0.2 V, whereas the MAE for 28 harmonics is 20 % smaller.

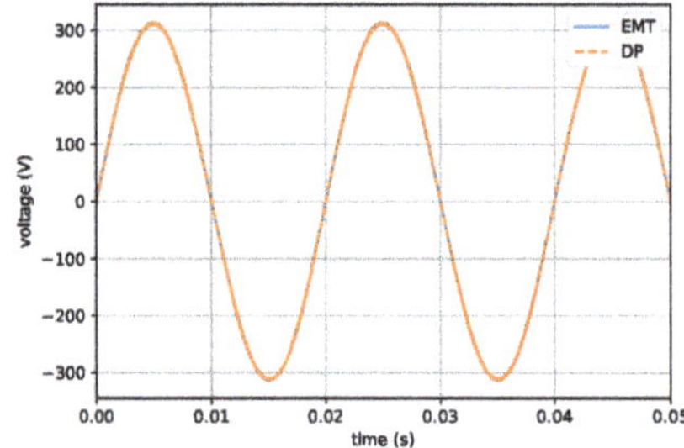

**Figure 12.** Capacitor voltage of a DP inverter model including the fundamental and eight harmonics compared to a Simulink EMT model.

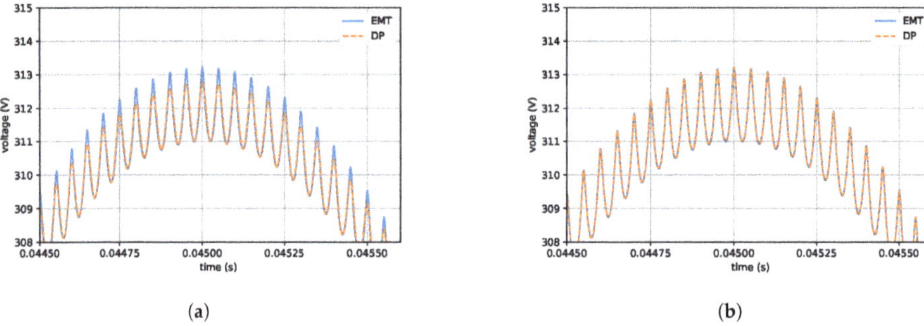

(a)  (b)

**Figure 13.** Ripple of the capacitor voltage of a DP inverter model including the fundamental and 8 harmonics (**a**) or 28 harmonics (**b**) as compared to a Simulink EMT model.

The results of the simulation with eight harmonics do not represent the ripple as accurately. It can determined from harmonic analysis that in this scenario, eight harmonics account for the phasors that are larger than 10 % of the fundamental, whereas 28 harmonics covers all harmonics larger than 1 % of the fundamental. In this example, the grid harmonics are considered up to the index $n_{max} = 7$, whereas the maximum switching harmonic is $m_{max} = 8$.

Figure 14 depicts the spectrum computed from the DP and EMT signal and shows how frequencies higher than the fundamental are accurately represented in the DP results up to a certain point, which depends on the number of dynamic phasors. Figure 15 presents the difference between the two spectra. It can be seen that high sideband harmonics are missing as well as high harmonics of the switching frequency, whereas lower harmonics are covered by the DP model.

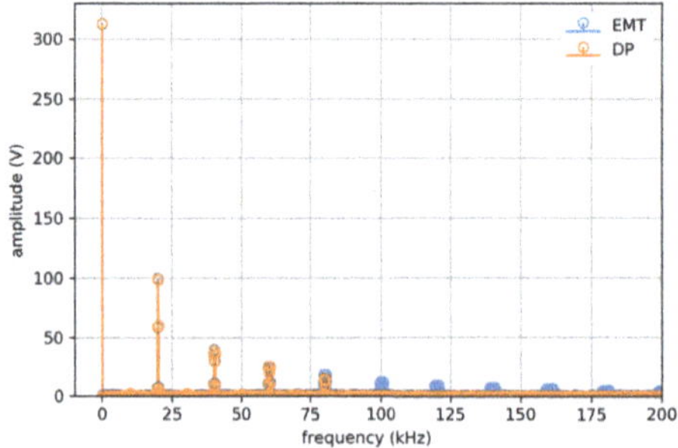

**Figure 14.** Spectrum of the DP inverter simulation with 28 harmonics as compared to Simulink EMT.

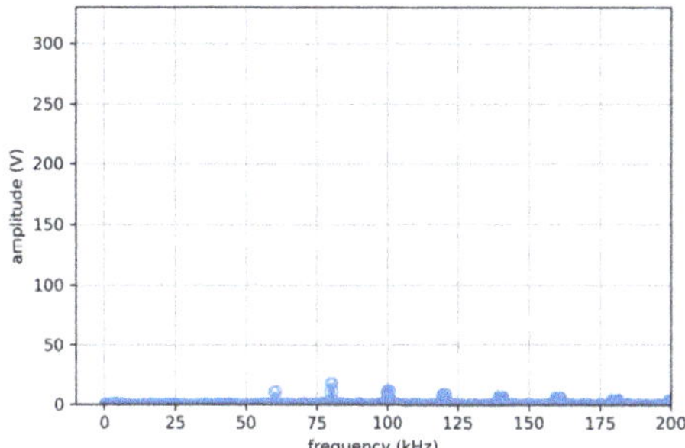

**Figure 15.** Difference of the spectrum of the inverter simulation with 28 harmonics compared to Simulink EMT.

*4.3. Parallelization for Large Systems and Power Electronics*

The previous two sections only considered the modelling accuracy, but not the simulation performance in terms of computation time, which is important for real-time execution. Especially, the DP inverter model including harmonics, described in Section 3.2, increases the number of variables drastically with considerable impact on the computation time. To improve computation time, DPsim parallelizes tasks that are not depending on each other. Larger tasks, such as the network solution, require decoupling to be able to split the task into smaller tasks that can be executed in parallel, as explained in Section 3.4.

The results for TLM decoupling in DP simulations were already presented in [15], but for consistency with the following parallelization results, the study has been repeated on different hardware, having two Intel Xeon E5-2643 v4 CPUs clocked at 3.4 GHz with 12 cores in total, running Fedora 30, and using the most recent version of DPsim. From Figure 16, it can be seen that the simulation using TLM decoupling outperforms the coupled simulation by up to about three times for 20 copies. The number of copies refers to the number of WSCC nine-bus instances that are interconnected through transmission lines for the TLM case or Pi-lines for the coupled case, respectively. All of the simulations are using eight threads in this scenario. Although it is clear that the transmission line model decoupling is better in terms of computation time, it should be noted that the transmission line model is not equivalent to the Pi-line model and it yields different simulation results.

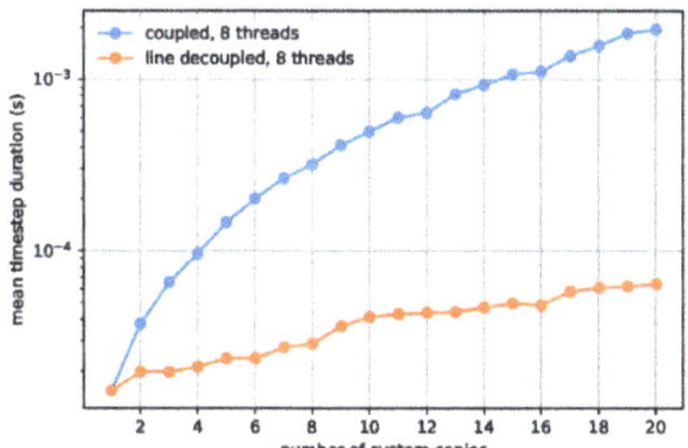

**Figure 16.** Parallelization results for eight threads comparing different decoupling methods.

The inverter model simulated in Section 4.2 and described in Section 3.2 results in many phasors of different shifting frequencies being fed into the network. At first glance, this seems to be a disadvantage of dynamic phasors, because the simulation accuracy is increased at the cost of increasing the number of equations with every phasor. As explained in Section 4.2, even considering only harmonics of at least 10% of the fundamental requires eight additional phasors. Hence, if no measures are taken, the network equation set does not only grow quadratically with the number of network nodes, but also the number of considered harmonics. To overcome this limitation of dynamic phasors, the spectral parallelization that is explained in Section 3.4 is applied to the grid connected inverter example presented in Section 4.2.

It can be seen in Figure 17 that this network split with regard to frequencies already improves the computation time significantly, even for the small example circuit that is depicted in Figure 11. The computation time is almost two times shorter if the network is solved separately for each frequency band. Besides, Figure 17 demonstrates the performance for different numbers of threads. As expected, the improvement is not noticeable when the system is not split frequency-wise. If the system is split, a higher number of threads seems to be beneficial, but the impact is not very large because the considered system is small when compared to the simulation of a complete grid.

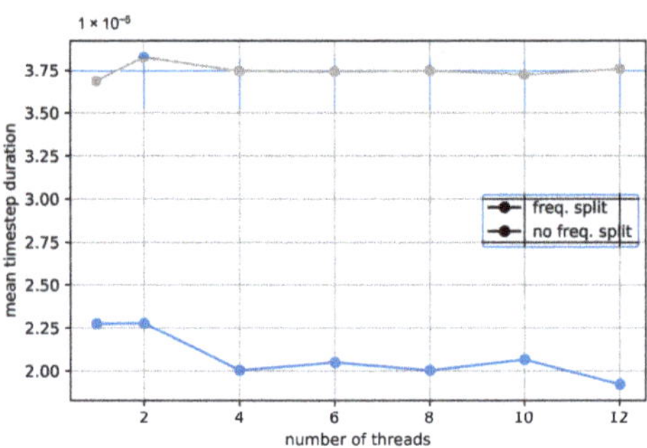


**Figure 17.** Parallelization results for single phase inverter.

The inverter model is still computed sequentially, as depicted in Figure 2 but the network solution is computed separately for each set of dynamic phasor variables that are associated to one shifting frequency. In the open loop case, the computation of harmonics inside the inverter model could be parallelized as well, but, in the general case, there could be cross frequency coupling, for example, by means of the inverter control.

## 5. Conclusions

The presented simulation results fall into two groups: power electronics modelling and real-time capability advancements by parallelization. The power electronics related results demonstrate that dynamic phasors can bridge the gap between the simulation of EMT and conventional phasor models and provide an easy method to represent either of the two modelling domains, depending on the selected simulation step. When compared to EMT models of switching power electronics, dynamic phasors allow for the user to determine and only simulate the harmonics that are relevant. However, this also requires the user to find out the relevant harmonics in advance before running the simulation. A tool to provide this information to the user automatically would be beneficial in the future. The averaged model could be used in transient stability analysis and for testing frequency control techniques. Extending the set of controllers by grid forming controllers would enable the analysis of high frequency variation conditions. The detailed inverter model featuring harmonics allows for the user to investigate the interaction of inverter dynamics at high frequencies.

The second group of results, related to advancements by parallelization, is crucial to understand what methods can be effectively used to reduce the computation time and ensure real-time execution for small time steps in the range of milli and micro seconds for real large-sized grids. Large-sized grids may be simulated in real-time to conduct controller-in-the-loop (CiL) experiments and investigate wide area oscillations. Smaller grids can be simulated at smaller time steps for hardware-in-the-loop (HiL) experiments. As demonstrated in Section 3.4, these methods can be derived from techniques employed in EMT real-time simulation or new methods specifically introduced for dynamic phasors. While the dynamic phasor based transmission line method adopts a well known method applied in EMT real-time simulation, the parallelization of network frequencies mitigates the disadvantage of dynamic phasors. A combination of subsystem and frequency wise parallelization improves the real-time capability of DPsim for large grids featuring detailed power electronics models, which is necessary for the simulation of future power systems.

**Author Contributions:** Conceptualization, writing and software: M.M. and J.D.; supervision, funding acquisition, A.M. All authors have read and agreed to the published version of the manuscript.

**Funding:** This project has received funding from the European Union's Horizon 2020 research and innovation programme under grant agreement No. 774613 and has received funding in the framework of the joint programming initiative ERA-Net Smart Energy Systems' focus initiative Integrated, Regional Energy Systems, with support from the European Union's Horizon 2020 research and innovation programme under grant agreement No. 775970.

**Conflicts of Interest:** The authors declare no conflict of interest. The funders had no role in the design of the study; in the collection, analyses, or interpretation of data; in the writing of the manuscript, or in the decision to publish the results.

## References


1. Faruque, M.O.; Strasser, T.; Lauss, G.; Jalili-Marandi, V.; Forsyth, P.; Dufour, C.; Dinavahi, V.; Monti, A.; Kotsampopoulos, P.; Martinez, J.A.; et al. Real-time simulation technologies for power systems design, testing, and analysis. *IEEE Power Energy Technol. Syst. J.* **2015**, *2*, 63–73. [CrossRef]
2. Chen, Y.; Dinavahi, V. FPGA-based real-time EMTP. *IEEE Trans. Power Deliv.* **2008**, *24*, 892–902. [CrossRef]
3. Razzaghi, R.; Paolone, M.; Rachidi, F. A general purpose FPGA-based real-time simulator for power systems applications. In Proceedings of the IEEE PES ISGT Europe 2013, Lyngby, Denmark, 6–9 October 2013; pp. 1–5.
4. Milton, M.; Benigni, A.; Bakos, J. System-level, FPGA-based, real-time simulation of ship power systems. *IEEE Trans. Energy Convers.* **2017**, *32*, 737–747. [CrossRef]
5. Li, Q.; Xiang, Y.; Mu, Q.; Zhang, X.; Li, X.; He, G. Exploration of FPGA-based electromagnetic transient real-time simulation system design using high-level synthesis. *J. Eng.* **2018**, *2019*, 1217–1220. [CrossRef]
6. Hernandez, M.E.; Ramos, G.A.; Lwin, M.; Siratarnsophon, P.; Santoso, S. Embedded real-time simulation platform for power distribution systems. *IEEE Access* **2017**, *6*, 6243–6256. [CrossRef]
7. Marti, J.R.; Dommel, H.W.; Bonatto, B.D.; Barrete, A.F. Shifted Frequency Analysis (SFA) concepts for EMTP modelling and simulation of Power System Dynamics. In Proceedings of the Power Systems Computation Conference (PSCC), Wroclaw, Poland, 18–22 August 2014; pp. 1–8.
8. Strunz, K.; Shintaku, R.; Gao, F. Frequency-adaptive network modeling for integrative simulation of natural and envelope waveforms in power systems and circuits. *IEEE Trans. Circuits Syst. Regul. Pap.* **2006**, *53*, 2788–2803. [CrossRef]
9. Demiray, T.; Andersson, G.; Busarello, L. Evaluation study for the simulation of power system transients using dynamic phasor models. In Proceedings of the 2008 IEEE/PES Transmission and Distribution Conference and Exposition: Latin America, Bogota, Colombia, 13–15 August 2008; pp. 1–6.
10. Marti, A.T.J.; Jatskevich, J. Transient Stability Analysis Using Shifted Frequency Analysis (SFA). In Proceedings of the 2018 Power Systems Computation Conference (PSCC), Dublin, Ireland, 11–15 June 2018; pp. 1–7.
11. Sanders, S.R.; Noworolski, J.M.; Liu, X.Z.; Verghese, G.C. Generalized averaging method for power conversion circuits. *IEEE Trans. Power Electron.* **1991**, *6*, 251–259. [CrossRef]
12. Holmes, D.G.; Lipo, T.A. *Pulse Width Modulation for Power Converters: Principles and Practice*; John Wiley & Sons: Hoboken, NJ, USA, 2003; Volume 18.
13. Stevic, M.; Estebsari, A.; Vogel, S.; Pons, E.; Bompard, E.; Masera, M.; Monti, A. Multi-site European framework for real-time co-simulation of power systems. *IET Gener. Transm. Distrib.* **2017**, *11*, 4126–4135. [CrossRef]
14. Monti, A.; Stevic, M.; Vogel, S.; De Doncker, R.W.; Bompard, E.; Estebsari, A.; Profumo, F.; Hovsapian, R.; Mohanpurkar, M.; Flicker, J.D.; et al. A Global Real-Time Superlab: Enabling High Penetration of Power Electronics in the Electric Grid. *IEEE Power Electron. Mag.* **2018**, *5*, 35–44. [CrossRef]

15. Mirz, M.; Vogel, S.; Reinke, G.; Monti, A. DPsim—A dynamic phasor real-time simulator for power systems. *SoftwareX* **2019**, *10*, 100253. [CrossRef]
16. Vogel, S.; Mirz, M.; Razik, L.; Monti, A. An open solution for next-generation real-time power system simulation. In Proceedings of the 2017 IEEE Conference on Energy Internet and Energy System Integration (EI2), Beijing, China, 26–28 November 2017; pp. 1–6.
17. Razik, L.; Mirz, M.; Knibbe, D.; Lankes, S.; Monti, A. Automated deserializer generation from CIM ontologies: CIM++—An easy-to-use and automated adaptable open-source library for object deserialization in C++ from documents based on user-specified UML models following the Common Information Model (CIM) standards for the energy sector. *Comput. Sci. Res. Dev.* **2018**, *33*, 93–103.
18. Maas, S.A. *Nonlinear Microwave and RF Circuits*; Artech House: London, UK, 2003.
19. Proakis, J.G.; Salehi, M.; Zhou, N.; Li, X. *Communication Systems Engineering*; Prentice Hall New Jersey: Upper Saddle River, NJ, USA, 1994; Volume 2.
20. Henschel, S. Analysis of Electromagnetic and Electromechanical Power System Transients with Dynamic Phasors. Ph.D. Thesis, University of British Columbia, Vancouver, BC, Canada, 1999.
21. Zhang, P.; Marti, J.R.; Dommel, H.W. Shifted-frequency analysis for EMTP simulation of power-system dynamics. *IEEE Trans. Circuits Syst. I Regul. Pap.* **2010**, *57*, 2564–2574. [CrossRef]
22. Demiray, T. Simulation of Power System Dynamics Using Dynamic Phasor Models. Ph.D. Thesis, ETH Zurich, Zurich, Switzerland, 2008.
23. Dommel, H.W. Digital Computer Solution of Electromagnetic Transients in Single-and Multiphase Networks. *IEEE Trans. Power Appar. Syst.* **1969**, *PAS-88*, 388–399. [CrossRef]
24. Rocabert, J.; Luna, A.; Blaabjerg, F.; Rodríguez, P. Control of Power Converters in AC Microgrids. *IEEE Trans. Power Electron.* **2012**, *27*, 4734–4749. [CrossRef]
25. Watson, N.; Arrillaga, J.; Arrillaga, J. *Power Systems Electromagnetic Transients Simulation*; IET: Stevenage, UK, 2003; Volume 39.
26. Ruehli, A.; Brennan, P. The modified nodal approach to network analysis. *IEEE Trans. Circuits Syst.* **1975**, *22*, 504–509.
27. Benigni, A.; Monti, A. A parallel approach to real-time simulation of power electronics systems. *IEEE Trans. Power Electron.* **2014**, *30*, 5192–5206. [CrossRef]
28. de Siqueira, J.C.G.; Bonatto, B.D.; Martí, J.R.; Hollman, J.A.; Dommel, H.W. Optimum time step size and maximum simulation time in EMTP-based programs. In Proceedings of the 2014 Power Systems Computation Conference, Wroclaw, Poland, 18–22 August 2014; pp. 1–7.

*Article*

# Hardware-in-the-Loop Simulation Using Real-Time Hybrid-Simulator for Dynamic Performance Test of Power Electronics Equipment in Large Power System

Jiyoung Song [1,2], Kyeon Hur [3], Jeehoon Lee [3], Hyunjae Lee [3], Jaegul Lee [1,2], Solyoung Jung [2], Jeonghoon Shin [2] and Heejin Kim [3,*]

[1] School of Electrical Engineering, Korea University, Seoul 02841, Korea; jy.song@kepco.co.kr (J.S.); jaegul.lee@kepco.co.kr (J.L.)
[2] Korea Electric Power Corporation (KEPCO), Daejeon 34056, Korea; solyoung.jung@kepco.co.kr (S.J.); jeonghoon.shin@kepco.co.kr (J.S.)
[3] School of Electrical and Electronic Engineering, Yonsei University, Seoul 34056, Korea; khur@yonsei.ac.kr (K.H.); jeehun87@yonsei.ac.kr (J.L.); newbee123@yonsei.ac.kr (H.L.)
* Correspondence: jimmykim07@yonsei.ac.kr; Tel.: +82-2-2123-8396

Received: 22 June 2020; Accepted: 31 July 2020; Published: 1 August 2020

**Abstract:** This paper presents the hardware-in-the-loop simulation for dynamic performance test (HILS-DPT) of power electronic equipment replicas using a real-time hybrid simulator (RTHS). The authors developed the procedure of HILS-DPT, and as an actual case example, the results of HILS-DPT of Static VAR Compensator (SVC) replica using RTHS is presented. RTHS is a co-simulation tool that synthesizes real-time simulator (RTS) with transient stability program to perform real-time dynamic simulation of a large power system. As power electronics applications have been increasing, the electric utilities have performed HILS-DPT of the power electronics equipment to validate the performance and investigate interactions. Because inspection tests are limited in their ability to validate its impact on the power system during various contingencies, all power electronics equipment newly installed in the Korean power system should take HILS-DPT using large-scale RTS with replicas since 2018. Although large-scaled RTS offers an accuracy improvement, it requires lots of hardware resources, time, and effort to model and simulate the equipment and power systems. Therefore, the authors performed SVC HILS-DPT using RTHS, and the result of the first practical application of RTHS present feasibility comparing the result of HILS-DPT using large-scale RTS. The authors will discuss the test results and share lessons learned from the industrial experience of HILS-DPT using RTHS.

**Keywords:** real-time hybrid-simulator (RTHS); hybrid simulation; co-simulation; hardware-in-the-loop simulation (HILS); dynamic performance test (DPT); real-time simulator (RTS); testing of replicas

---

## 1. Introduction

As the power system has become larger and more complicated, the power electronic devices, such as high-voltage direct current (HVDC) and flexible AC transmission systems (FACTS), have been increased to improve stability and controllability and flexibility of the power systems. In the Korean power system, three HVDCs and eleven FACTSs are in operation, four HVDCs are under construction, and two HVDCs and seven FACTSs are under planning [1]. Power electronics applications are located close to each other in the Korean power system, and concerns about interaction with adjacent network dynamics among the equipment have increased [1–8]. The hardware-in-the-loop tests of the power electronics equipment controller are required to demonstrate the equipment performance and validate the reliability of the equipment. Therefore, power electronics equipment planned to be newly installed in the Korean power system has had to take a hardware-in-the-loop simulation dynamic performance test (HILS-DPT) of the replica controller using a large-scale real-time simulation (RTS) since 2018 [2]. Herein, we define DPT as a test that validates controller performance using HILS, considering the power system dynamics of other equipment. Figure 1 shows the configuration of DPT. DPT should include the system-wide impact study for multiple power system scenarios with various contingencies. Therefore, DPT requires a large-scale RTS with extensive hardware resources, time, and effort for modeling and simulation. KEPCO has enough RTS hardware resources to accommodate the entire Korean network over a 154 kV transmission system and several HVDC and FACTS without equivalent network and improved the process of simulation case development for efficient modeling and simulation. However, the power system becomes larger and larger, and the system configuration also becomes more and more complicated by applying various power electronics devices, such as HVDC, FACTS, and renewables. The scale of RTS needs to be expanded to analyze the complicated and expanded future power system, and expanding the RTS requires high techniques for modeling and simulation. To overcome the issue, the authors investigated and explored the feasibility of applying to HILS-DPT.

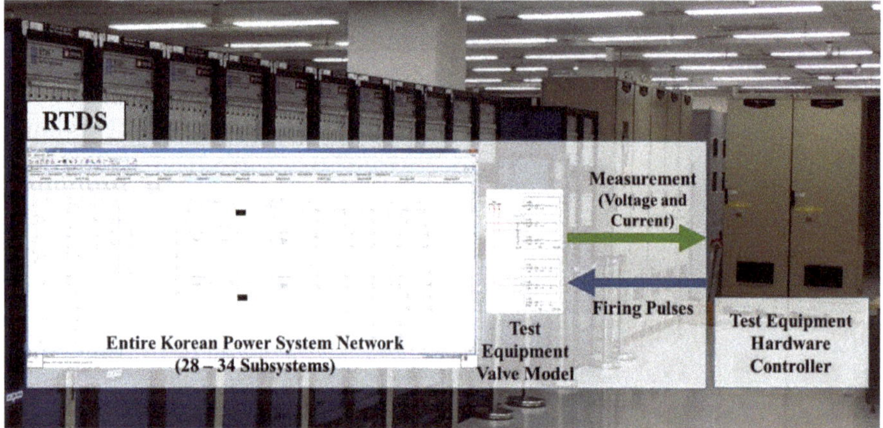


**Figure 1.** Configuration of dynamic performance test (DPT) for equipment.

Real-time Hybrid Simulator (RTHS) is a co-simulation tool that synthesizes RTS with transient stability analysis (TSA) tool to perform detailed real-time simulation of a large power system [2–5]. RTHS exchanges network power flow data of RTS and TSA program via the fiber-optic communication interface. A power system in RTHS is divided into electromagnetic transient subsystem in RTS and transient stability subsystem in the TSA program. The TSA program has the strength to build a wide area network, even though it lacks accuracy for power electronic equipment control and protection or interaction study. RTS has better accuracy of dynamic performance simulation for power electronics equipment. RTHS satisfies the accuracy of power electronics equipment dynamic performance in a large power system. Besides, RTHS with actual hardware controller and replicas can be used to study the interaction between power systems and power electronics equipment. Therefore, we will investigate the feasibility of HILS-DPT using RTHS with a practical case of Shin-Jecheon SVC compared to RTHS and large-scale RTS.

The rest of the paper is organized as follows: We will present the definition, configuration, requirement, responsibility, and industrial experience of DPT in Section 2. The procedure, benefits, and boundary determination principle of HILS-DPT using RTHS are described in Section 3. As a representative case, HILS-DPT of the Shin-Jecheon SVC controller (+675/−225 MVar) using RTHS is presented in Section 4. Conclusions are presented in Section 5.

## 2. HILS for Dynamic Performance Test in Large-Scale Power System

### 2.1. Definition and Features of HILS-DPT

DPT was one of the processes in the factory acceptance test (FAT) conducted at the manufacturer's factory, which is carried out with equivalent or small size of the power system (e.g., 1 or 2 RTDS racks). DPT in FAT focuses on the dynamic response of the equipment itself. DPT in FAT cannot demonstrate the interaction between the equipment and the large AC power system. However, those studies' needs have been increased, and the number of electric utilities performing HILS-DPT with a large power system increases. We define that HILS-DPT is a test considering the dynamics of network and impact on adjacent power electronics equipment, which cannot be achieved in FAT and Site Acceptance Test (SAT). FAT has a limitation of test network scale, and SAT has an obstacle of available test items while in energization or operation of equipment on site. Therefore, HILS-DPT is a significant alternative to overcome those limits or constraints of FAT and SAT.

HILS-DPT evaluates equipment from a power system perspective. The requirements of HILS-DPT to accurately evaluate the role of equipment in a large power system are as follows:

- A sufficiently wide area of power system should be required to reflect the system dynamics and control characteristics of an adjacent power electronics equipment.
- A power system reliability study should be carried out whether it satisfies technical specifications though contingency analysis. Actual operation strategies and schemes (e.g., special protection scheme (SPS), emergency control) need to be applied.
- Even though the parameters were approved in FAT, the controller parameter can be changed upon request of electric utility based on the results of HILS-DPT.

Thus, HILS-DPT should be tested by the electric utility, due to an appropriate study network, the contingencies of the power system, and intellectual property issues (other manufacturers model and the replica controllers).

Figure 2 shows the procedure of HILS-DPT using large-scale RTS. The large power system is modeled in RTS using initial network data of the design stage in Step 1. Modeled large-scale power system interfaces with adjacent power electronics equipment models and other external hardware controllers (e.g., replica controller, relay, external system, etc.), and operation strategy is applied to test the power system model in Step 2. Functions, interaction, and contingency tests are conducted in Step 3. In Step 4, parameters are optimized at the network data for the operation stage. HILS-DPT evaluates and demonstrates its main role when the equipment works in the power system.

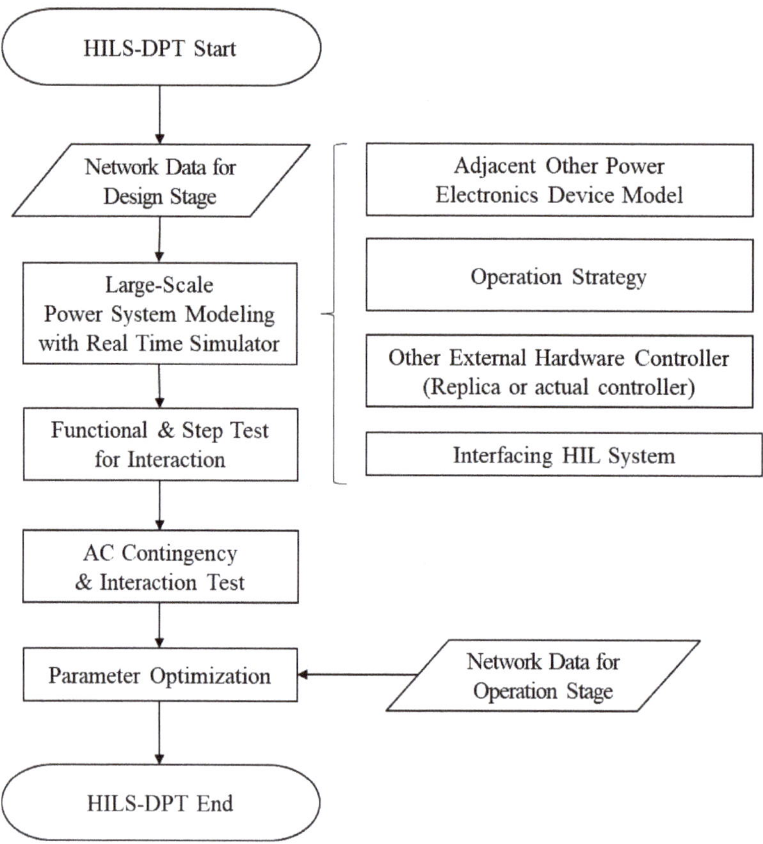


**Figure 2.** The entire procedure of hardware-in-the-loop simulation for dynamic performance test (HILS-DPT) using a large-scale real-time simulator (RTS).

### 2.2. Responsibility, Experience, and Lessons of HILS-DPT

HILS-DPT is necessary from an electric utility perspective, but it is quite a burdensome task for the manufacturer. This is because many other factors are affecting the condition of test equipment operation. For example, HILS-DPT with a large power system may be challenging to clarify the responsibility for the test results because network dynamics and adjacent power electronics devices can unintentionally distort the control characteristics of the test equipment. There may be an argument between an electric utility and manufacturer for the pass and fail, since it is difficult to evaluate quantitative criteria. Therefore, HILS-DPT is conducted as a witness test rather than an inspection test because the adjacent equipment and network are best known and responsible by the electric utility. HILS-DPT helps avoid the malfunctions that may occur during commissioning and commercial operation through the dynamic control performance test in the present system condition shortening the commissioning period by parameter tuning to the proper controller setting.

Figure 3 shows a practical case of parameter tuning for actual STATic synchronous COMpensator (STATCOM, 300 MVar, Godeok) controller. It is worth noting that HVDC (1.5 GW, Bukdangjin-Godeok, Line Commutated Converter) and STATCOM are installed at the same converter station. There are potential risks for interaction between HVDC and STATCOM. Before commissioning, HILS-DPT with a large-scale power system was conducted. The STATCOM step response considering HVDC operation at the same converter station is one of the test protocols for the interaction study. As shown in Figure 3, voltage, reactive power of STATCOM, and HVDC DC power had an oscillation and damped out slowly with initial parameter settings (blue line). STATCOM current controller gain was tuned, and the dynamic response of the STATCOM and HVDC was improved (as the green line in Figure 3). Figure 3 shows that the STATCOM controller was not well-tuned to present system conditions even though the controller setting was approved in FAT. Therefore, HILS-DPT is very useful during HVDC and FACTS projects, and its importance is to keep increasing.

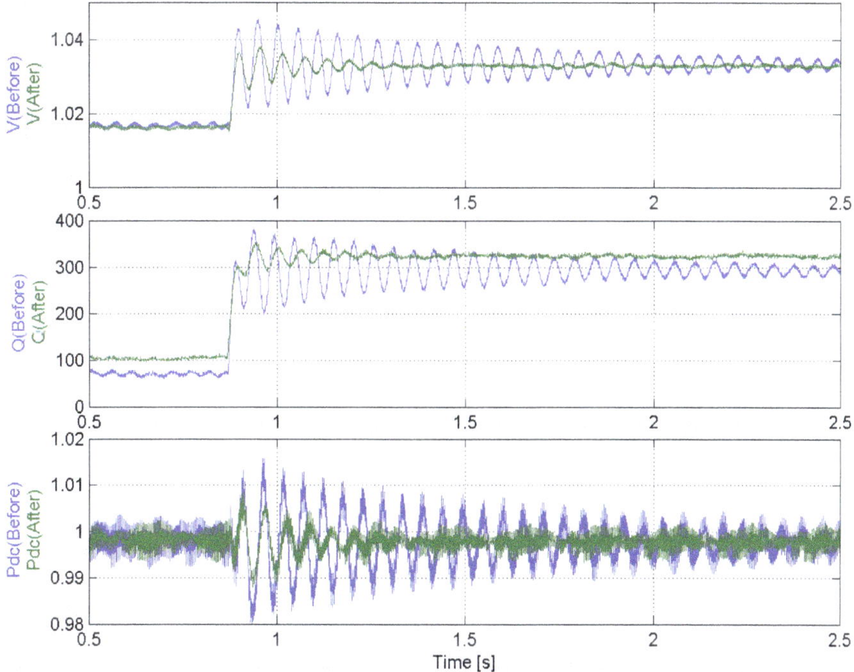

**Figure 3.** An example of parameter tuning experience during HILS-DPT; step-response test; voltage (first row), reactive power of STATCOM (second row), and active power of high-voltage direct current (HVDC; third row); before tuning (blue line) and after tuning (green line).

## 3. HILS-DPT Using RTHS

### 3.1. Features of HILS-DPT Using RTHS

A hybrid simulation is a powerful tool because power electronics applications in the power system require detailed studies with Electro-Magnetic Transient (EMT) simulation investigating the performance of the equipment, and electric utility also has needs of dynamic investigation in the large power system [3–16]. Although many hybrid simulation methods and tools exist, improving the accuracy and efficiency of a large power system simulation, those hybrid simulation tools are based on non-real-time. KEPCO, Powertech Lab, and Yonsei Univ. jointly developed RTHS combining the RTS and TSA program because there are needs of testing replicas in the KEPCO advanced real-time simulation laboratory to investigate the future power system issues [2–6]. Figure 4 shows the configuration of RTHS. RTHS has following features for HILS-DPT:

1.  Wide network dynamics

    The simulation case network using RTHS is divided into the network simulated with the RTS as the internal subsystem and the remaining network simulated with the TSA as the external subsystem. Wide area AC network dynamics, including hundreds of generators and transmission lines, can be modeled in an external subsystem. This allows us to study the system-wide impact efficiently and effectively.

2. Flexibility

The flexibility of the simulation study is enhanced because the external subsystem network topology can easily be modified with the TSA program. Various load conditions (such as peak and off-peak cases) and models (such as dynamic load model, composite load model, and constant load model), contingencies, operating points, and strategies can be applied in the external subsystem unless internal subsystem has a significant change of network topology. The dynamic load model is especially difficult to apply to large-scale RTS, due to the numerical instability and the highly required computational hardware resources. Whereas, RTHS can apply for a dynamic load model in the entire power system with the TSA program, and the power system characteristics using a dynamic load model becomes more accurate than those using a constant load model. Thus, RTHS with a dynamic load model enables the achievement of a more accurate response of power system and power electronics equipment.

3. Economical advantage

Large-scale RTS requires more than 28 racks to simulation the entire Korean power system, and the future power system will be larger and more complicated. Expanding RTS more and more is not an economical solution for future power system studies. From the KEPCO experience since 2016, RTHS requires five or fewer racks to simulate the entire Korean power system, and RTHS is free for the power system expanding issue.

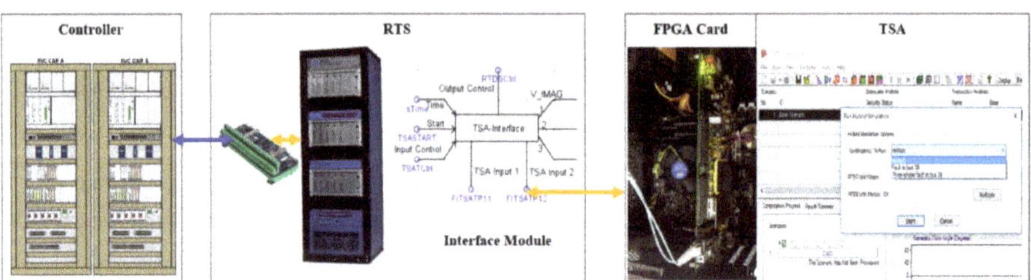

**Figure 4.** Configuration of real-time hybrid simulator (RTHS).

*3.2. The Procedure of RTHS for HILS-DPT*

Figure 5 shows the procedure of RTHS. The biggest difference between simulation using a large-scaled RTS and simulation using RTHS is the boundary area. RTHS cannot be relied upon unless the boundary buses are properly set. Setting up the boundary busses is the first step of the HILS-DPT using RTHS, and the boundary bus determination method for this practical SVC HILS-DPT will be presented in Section 3.3. The second step is the extraction of an internal and external area with TSA network data. Then, we can get two TSA network data; internal subsystem for RTS and external subsystem for TSA. In the next step, internal subsystem network data is converted to the RTS model. This RTS model includes only transmission lines, transformers, generators, and loads. Then, power electronics equipment and communication interface models are added to the RTS model manually. We call this step Power System Modeling (RTS) in Figure 5.

After configuring the base case of RTS model of an internal subsystem, various RTHS case with scenarios needs to be developed. In this step, an external hardware controller like a replica is implemented for HILS-DPT. When test equipment replica or actual controller is connected to RTS, HILS-DPT using RTHS is ready. All power electronics equipment in the RTS area must run at an operating point based on the study network data before releasing the locks of the generator governor.

Although the power flow on the boundary buses must be the same as TSA power flow, power flow mismatch between internal subsystem simulation and TSA power flow can occur due to the difference of the basic computation algorithm of the two different tools. In our practical case of SVC HILS-DPT, the mismatch was less than 100 MW. If the mismatch is greater than 1 GW, initialization failed. Once the simulation setup is completed, HILS-DPT using RTHS can be performed. If an abnormal operation of a controller or protection functions that were not found in FAT or interaction between power electronics equipment and power system is found, the root cause must be analyzed and corrected.

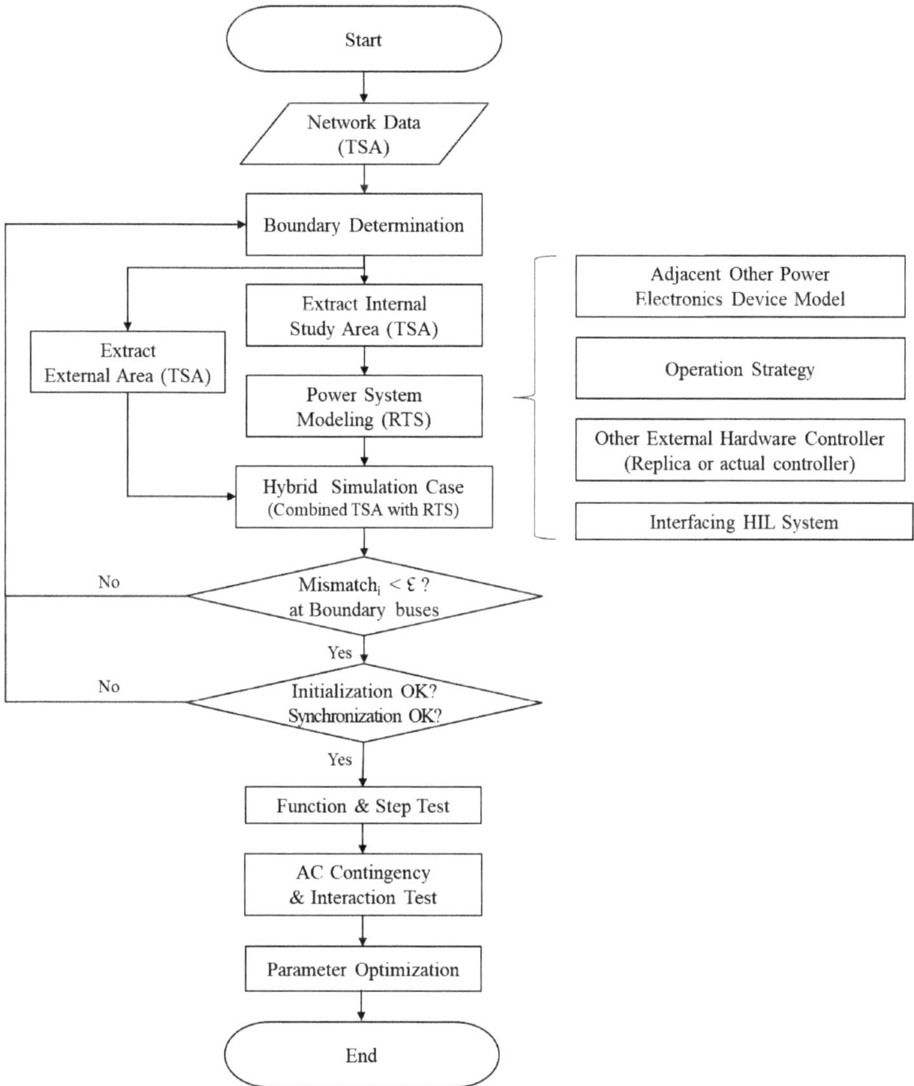


**Figure 5.** The overall procedure of RTHS for HILS-DPT.

### 3.3. Determination of Boundary Area of RTHS for HILS-DPT

The boundary buses can be set depending on the condition of the power system, fault location, fault type, and interest study area. During the RTHS development from 2016 through 2018, we concluded that RTHS requires at least three RTDS racks with six PB processor cards per rack for the internal area of the entire Korean power system to achieve a similar response of a large-scale RTS. In our experience, 4–6 racks of RTDS are required to perform HILS-DPT using RTHS under a large-scale power system. It is worth noting that it is better to set a wide internal area in the RTHS. The internal area determination algorithm based on the electrical distance between boundary buses and fault location is introduced in [9,10]. Although the wide internal area of the RTHS offers similar simulation results to the large-scale RTS, it reduces the benefits of the hybrid simulation. On the contrary, the narrow internal area of the hybrid simulation decreases accuracy. Thus, the following boundary determination principles of RTHS are applied in Shin-Jecheon SVC HILS-DPT.

1. Select boundary buses with an electrical distance of more than a certain threshold (0.4) from the bus with the most serious fault (765 kV transmission line fault).
2. Select boundary buses with an electrical distance of more than a certain threshold (0.4) from the power electronics equipment, if the power electronics are nearby the boundary.
3. Select the boundary buses without island areas.
4. Include the generators in the RTS area, if there are the generators around the boundaries.
5. Determines the number of hardware racks of the RTS system, satisfying the above conditions.
6. Expand the RTS area as wide as possible.

The flow chart of the boundary determination method for SVC HILS-DPT using RTHS is shown in Figure 6.

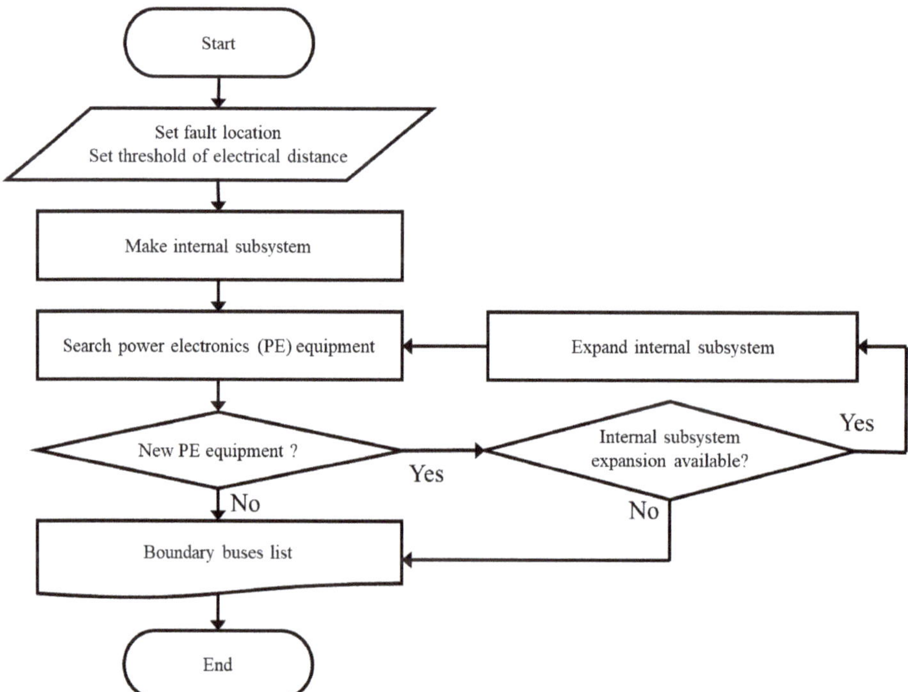


**Figure 6.** The procedure of determination of boundary buses.

The electrical distance ($D_{ij}$) between $i$ bus and $j$ bus is defined as following (1) [10]

$$D_{ij} = -\log_{10} \frac{\left(\frac{\delta V}{\delta Q}\right)_{ij}}{\left(\frac{\delta V}{\delta Q}\right)_{jj}} = -\log_{10} \frac{\Delta V_i}{\Delta V_j} \tag{1}$$

$\Delta V_i$ is voltage sensitivity at $i$ bus, and $\Delta V_j$ is voltage sensitivity at $j$ bus where the reactive power ($Q$) changes at $j$ bus. The electrical distance is calculated based on the voltage sensitivity derived from QV part of the Jacobian matrix of the power system.

## 4. A Practical Application of HILS-DPT Using RTHS; Shin-Jecheon SVC HILS-DPT

### 4.1. Background and RTHS Case Set Up

The eastern part of the Korean power system is designed to transfer electric power from a huge nuclear power plant to the load center. Total power generation from nuclear power plants is 8 GW, and 345 kV and 765 kV transmission lines transmit those power to a metropolitan area near Seoul. Because 5.4 GW of power is transmitted through the 765 kV transmission lines during peak case, the contingency of 765 kV transmission lines might threaten the stability of the entire power system. To avoid system instability, KEPCO planned to install multiple power electronics equipment around the 765 kV and 345 kV transmission lines. Shin-Jecheon SVC is among those facilities, and it is installed to improve voltage stability of 345 kV transmission line in 2018. Note that HILS-DPT of Shin-Jecheon SVC using RTHS is the first case of practical use of RTHS.

Table 1 presents one of the contingency scenarios for HILS-DPT of SVC, which tests 765 kV transmission line fault and following reclose failure. Table 2 shows the power electronics equipment location, control mode, operating point, and ratings in the test case and its modeling area. Because the HVDC is far from the fault locations and the test equipment, it is modeled in the external subsystem (TSA program). Because the rest of the power electronics equipment is located nearby the SVC, they are modeled in the internal subsystem (RTS). Figure 7 illustrates the case study map, and the internal subsystem details and simulation settings are listed in Table 3. The actual special protection scheme for 765 kV transmission line fault is applied to HILS-DPT. When 765 kV fault occurs, four Thyristor Controlled Series Compensator (TCSC)installed at the 345 kV transmission line boost their compensation level from 50% to 70%. As power flows through the 345 kV transmission lines increases, SVC and STATCOMs respond to compensate for the voltage drop.

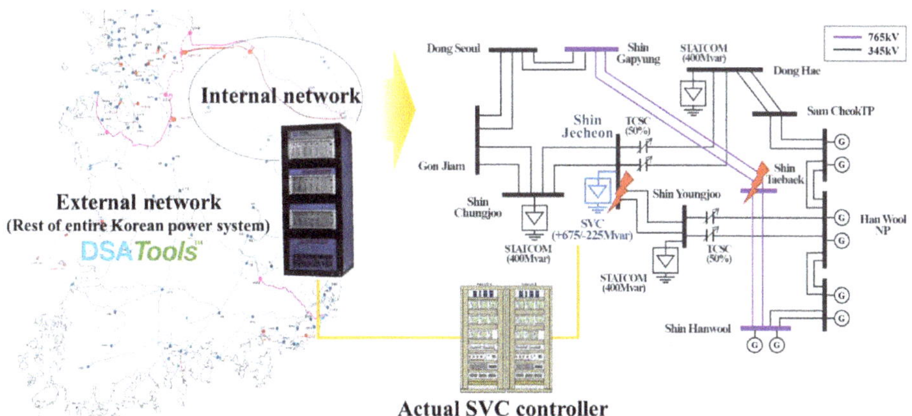


**Figure 7.** Case study map of Shin-Jecheon SVC HILS-DPT using RTHS.

**Table 1.** Contingency scenario for 765 kV transmission line fault.

| Time (Cycle) | Action |
| --- | --- |
| 0 | Fault occurs |
| 5 | 765 kV line circuit breaker open<br>TCSC compensation boost (50% → 70%) |
| 9 | Generator Trip (1.5 GW) |
| 71 | 765 kV line reclose attempted |
| 76 | Reclose failure |

**Table 2.** Power electronics equipment in the test case. SVC, Static VAR Compensator.

| Type (Modeling Region) | Location | Control Mode (Setpoint) | Ratings (Number of Equipment) |
| --- | --- | --- | --- |
| HVDC (External) | Bukdangjin Godeok | Power control (1.0 p.u.) | 1.5 GW |
| TCSC (Internal) | Sinyoungju | Impedance control (50%) | 555 MVar (2) |
|  | Shin-Jecheon | Impedance control (50%) | 595 MVar (2) |
| STATCOM (Internal) | Donghae | Q reserve control (1.0 p.u.) | ±400 MVar |
|  | Sinyoungju | Q reserve control (1.0 p.u.) | ±400 MVar |
|  | Sinchungju | Q reserve control (1.0 p.u.) | ±400 MVar |
| SVC (Internal) | Shin-Jecheon | Q reserve control (1.0 p.u.) | +675/−225 MVar |

**Table 3.** Simulation environment details of RTHS and the large-scaled RTS for Shin-Jecheon SVC HILS-DPT.

| Types | RTHS | Large-Scaled RTS |
| --- | --- | --- |
| Number of RTDS rack | 4 | 29 |
| Number of bus | 103 | 1140 |
| Equipment | 8<br>(1 SVC, 3 STATCOMs, 4 TCSCs) | 10<br>(1 HVDC, 1 SVC, 4 STATCOMs, 4 TCSCs) |
| Generators | 12 (11.12 GW) | 228 (87.5 GW) |
| Time step | 75 μsec (RTS), 4 msec (TSA Program) | 75 μsec |

The test is conducted by a large-scale RTS and RTHS to investigate the feasibility of RTHS applying for HILS-DPT. The large-scale RTS needs 29 subsystems, with six PB5 processor cards for each subsystem. RTHS uses four subsystems for the internal subsystem, and TSA program covers the rest of the power system. The internal area of RTHS includes detailed modeling of large power generators, transmission lines, STATCOMs, TCSCs, and SVC. During the development of the RTHS, we compared the Korean power systems with the RHTS and Full-RTDS and found that the results of the hybrid simulation are reliable when the electrical distance between boundary buses and the fault location is 0.4 or less in the Korean power system. Thus, the threshold for determining the boundary buses was set to 0.4 empirically, but this depends on the power system condition, types of contingencies, and HVDC and FACTS location. The energy data for each boundary bus is exchanged through fiber optic communication cable [4–6,12,13]. This test case has 17 boundary buses. SVC actual controller is shown in Figure 8. The list of DPT and contingency list of HILS-DPT are shown in Appendix A.

**Figure 8.** SVC controller interfaced with RTHS.

*4.2. Results of HILS-DPT Using RTHS and Large-Scale RTS*

This section presents the simulation results of HILS-DPT using RTHS and large-scale RTS to investigate the feasibility of RTHS applying HILS-DPT.

- Case 1: A 345 kV line fault (3-phase to ground)

The 3-phase to ground fault is applied to Shin-Jecheon bus during six cycles at 0.5 s. During the fault, the bus voltage drops to zero and SVC could not compensate it. After the fault is cleared, the bus voltage recovers immediately, and the reactive power output of SVC increased according to network recovery characteristics. Overall, the SVC dynamic performance of RTHS is almost the same as HILS-DPT using large-scale RTS. Figure 9 shows the bus voltage at Shin-Jecheon and reactive power of SVC of RTHS simulation and RTS simulation when 3-phase fault is applied. A small difference of less than 10 MVar of SVC reactive power dynamics between RTHS and RTS simulation can be seen in Figure 9. Even though the bus voltage has a small difference of less than 0.01 p.u., SVC responses to the voltage and reactive power output are slightly different from each other. The authors concluded that this amount of difference is acceptable, and HILS-DPT using RTHS is feasible.

- Case 2: A 345 kV unbalanced fault (single line to ground)

The A-phase to ground fault was applied to Shin-Jecheon bus during six cycles at 0.5 s. During the fault, the bus voltage drops to around 0.9 p.u. Figure 10 shows the bus voltage at Shin-Jecheon and reactive power of SVC of RTHS simulation and RTS simulation when a single-phase fault is applied. Even though unbalanced fault is applied at Shin-Jecheon bus, the bus voltage at Shin-Jechoen and reactive power output of SVC is almost the same. The authors concluded that this amount of difference is acceptable, and HILS-DPT using RTHS is feasible.

- Case 3: A 765 kV line fault (3-phase to ground), generator trip SPS, and emergency control

The 765 kV transmission double line fault has complicated sequences, such as TCSC boost up and nearby generators trip, as shown in Figure 11. Moreover, its impact on the network is larger than any other contingencies. The faulted location is just 3-level away from Shin-Jecheon bus, where the SVC is installed. When the fault occurred at 0.5 s, the Shin-Jecheon bus voltage dropped to around 0.7 p.u., and SVC fully outputs reactive power to compensate for the low voltage. After the fault cleared, it recovered to a lower limit of voltage deadband and swing according to the network dynamics. Figure 13 shows the bus voltage at Shin-Jecheon and reactive power of SVC of RTHS simulation and RTS simulation when 3-phase fault is applied at 765 kV transmission lines. A small difference of less than 10 MVar of SVC reactive power dynamics between RTHS and RTS simulation can be seen in Figure 13. Even in this case, HILS-DPT using RTHS and RTS have almost the same dynamic. Therefore, the authors concluded that this amount of difference is acceptable, and HILS-DPT using RTHS is feasible.

- Case 4: A 765 kV line fault (3-phase to ground), generator trip SPS, and emergency control; dynamic load model applied

As mentioned in the previous section, RTHS has the strength of flexibility for modeling of the external subsystem. For RTS simulation, it is not easy to apply various load models to the entire network, due to its computational burden. However, it depends on the study purpose; consideration of the various load models (e.g., induction motors and ZIP) in the external subsystem could be required. Figure 12 shows the result for SVC responses using RTHS when different the load model is applied, and 765 kV 3-phase fault occurs. The simulation results of the constant impedance load model (blue line) and induction motor and ZIP load model (green line) are shown in Figure 12. Because the induction motor and ZIP load model absorb more reactive power than the constant impedance load model, the voltage at Shin-Jechoen bus in Figure 12 shows that voltage drop with induction motor and ZIP load model is higher than voltage drop with constant impedance load model. This analysis might be important in future power system studies. As can be seen in Figure 12, the flexibility of the external subsystem will be great advantages of RTHS for HILS-DPT.

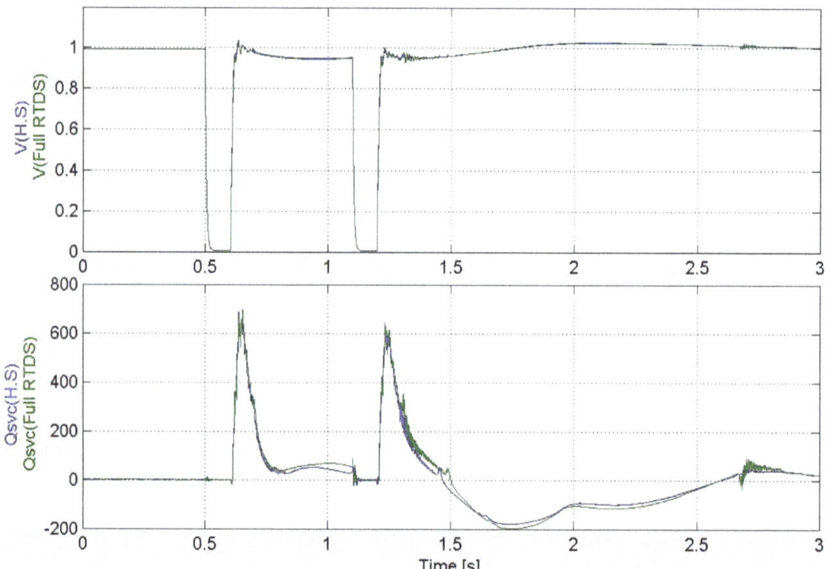

**Figure 9.** A 345 kV line fault is applied at Shin-Jecheon bus; Shin-Jecheon bus voltage (first row) and reactive power of SVC (second row); RTHS (blue line) and large-scale RTS (green line).

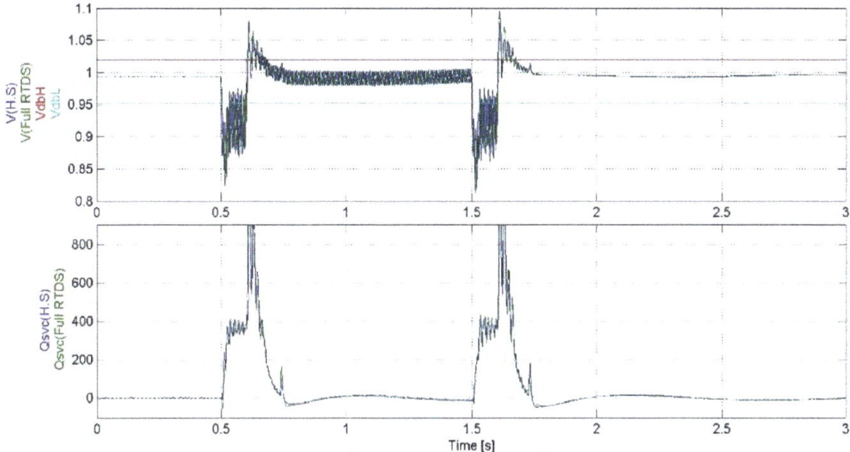

**Figure 10.** A 345 kV line unbalanced fault is applied at Shin-Jecheon bus; Shin-Jecheon bus voltage (first row) and reactive power of SVC (second row); RTHS (blue line) and large-scale RTS (green line)

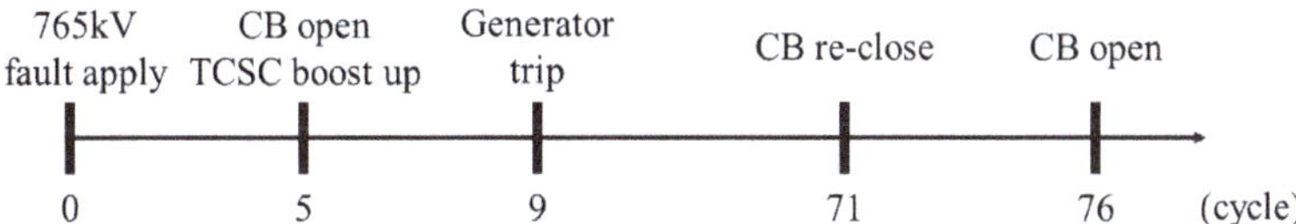

**Figure 11.** 765 kV line fault sequence considering emergency control and SPS.

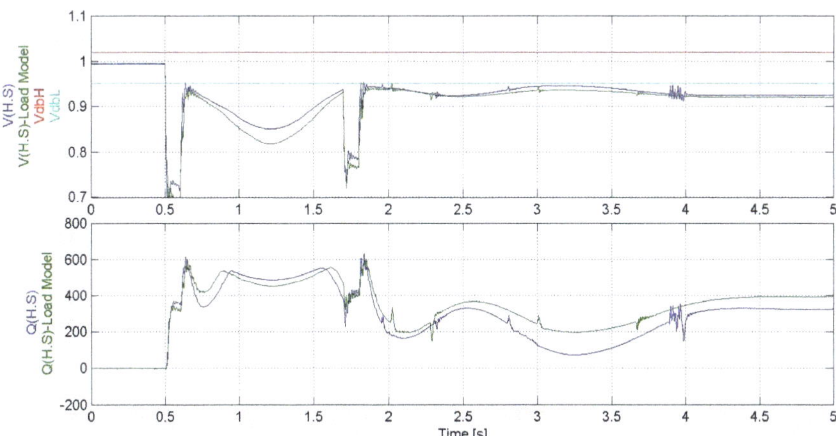

**Figure 12.** A 765 kV line fault nearby Shin-Jecheon bus considering induction motor and ZIP load model; Shin-Jecheon bus voltage (first row) and reactive power of SVC (second row); RTHS (blue line) and large-scale RTS (green line).

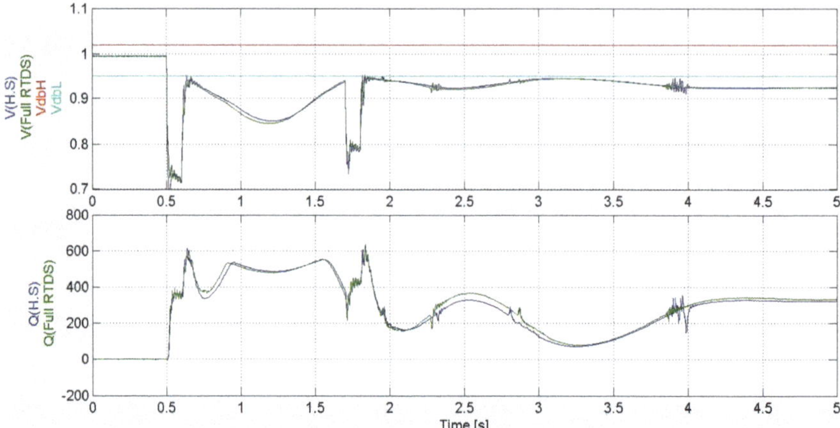

**Figure 13.** A 765 kV line fault is applied nearby Shin-Jecheon; Shin-Jecheon bus voltage (first row) and reactive power of SVC (second row); RTHS (blue line) and large-scale RTS (green line).

## 5. Conclusions

As the complexity of the modern power system grows, power system simulation and studies need to be flexible to investigate various operating points and conditions. Although a large-scale power system modeling and simulation are useful, it is less flexible than hybrid simulation as power systems become more and more complicated, and as renewable energy resources increase. That is the reason that hybrid simulation and co-simulation has been developed and expand their applications [4–18]. Hybrid simulation improves accuracy because equivalents, which represent areas of non-interest in the power system, do not completely show the responses of the actual power system.

Real-time Hybrid Simulator supports system-wide impact study efficiently and effectively because hundreds of generators and transmission lines can be modeled in the TSA program. This improves the flexibility of the simulation study because the network topologies and load conditions in TSA subsystem (which is called the external subsystem in this paper) can easily be modified. RTHS is free for the power system expanding issue for the future power system. RTHS can be used in replica test applications with real-time attributes. HILS-DPT, which test replica or actual controller of high-voltage power electronics equipment, is one of the best applications of RTHS.

HILS-DPT using RTHS validates the controls and operation of newly installed power equipment and shows its impact on the power system before its installation. It is worth noting that HILS-DPT shortens the duration of SAT and ensures SAT progress smoothly, helping utilities take over the equipment. HILS-DPT shows and validates the response of the equipment during the contingencies that cannot be performed on FAT and SAT. This improves the reliability of the equipment and allows the utility to have confidence in the equipment. In particular, HILS-DPT using RTHS can apply the saved hardware resources to the dynamic load model, while yielding similar results to the actual power system. The saved hardware resources can be used for the renewable energy source model, which is an important factor in the future power system. This will allow the interaction and influence between power electronics equipment planned to be newly installed and renewable energy sources to be studied. The authors investigated the feasibility of HILS-DPT using RTHS, and concluded RTHS is applicable for HILS-DPT. KEPCO will continue to apply more HVDC and FACTS replica controllers, and further research and development are being conducted related to its operation, and analysis technology, and RTHS will help the engineers perform tests and studies efficiently and effectively.

**Author Contributions:** Methodology, all authors; Software, all authors; writing—original draft preparation, J.S. (Jiyoung Song), K.H., J.L. (Jeehoon Lee), H.L., and H.K.; writing—review and editing, J.S. (Jiyoung Song), K.H., and H.K.; supervision, K.H. and H.K.; project administration, J.S. (Jeonghoon Shin). All authors have read and agreed the published version of the manuscript.

**Funding:** This work was supported by "Human Resources Program in Energy Technology" of the Korea Institute of Energy Technology Evaluation and Planning (KETEP) granted financial resource from the Ministry of Trade, Industry & Energy, Republic of Korea. (No.20194030202420) This research was also supported by Korea Electric Power Corporation (Grant number: R20XO02-4).

**Conflicts of Interest:** The authors declare no conflict of interest.

## Appendix A

The list of DPT is as follows:

- Start and stop sequences test

- Control modes test

- Emergency stop test

- Continuous operation test

- Protection test (overvoltage, under-voltage, under frequency, and internal fault)

- Mode change test

- Dynamic response test during an AC fault

- Verification of black-box model

The contingency list of HILS-DPT is as follows:

- 765 kV transmission line trip with TCSC boost up and generator trip for special protection scheme (2 cases)

- 345 kV transmission line trip (1 case)

- Single-phase fault and line fault at 345 kV transmission line (1 case for each)

- TCSC bypass (1 case)

- STATCOM trip (3 cases)

- Generator trip (1 case)

## References

1. Kim, H.; Kim, J.; Song, J.; Lee, J.; Han, K.; Shin, J.; Kim, T.; Hur, K. Smart and Green Substation Shaping the Electric Power Grid of Korea. *IEEE Power Energy Mag.* **2019**, *17*, 16–24. [CrossRef]
2. Kim, S.; Kim, H.; Lee, H.; Lee, J.; Lee, B.J.; Jang, G.; Lan, X.; Kim, T.; Jeon, D.; Kim, Y.; et al. Expanding power systems in the Republic of Korea feasibility studies and future challenges. *IEEE Power Energy Mag.* **2019**, *17*, 61–72. [CrossRef]
3. Hur, K.; Lee, J.; Song, J.; Ko, B.; Shin, J. Advanced Real Time Power System Simulator for Korea Electric Power Systems: Challenges and Opportunities. Presented at IEEE PES, Portland, OR, USA, August 2018; Available online: http://site.ieee.org/pes-itst/files/2018/08/2018-Panel-1.pdf (accessed on 25 July 2020).
4. Lin, X.; Zadehkhost, P.; Lee, J.; Song, J.; Ko, B.; Hur, K. TSAT-RTDS Interface-The Development of a Hybrid Simulation Tool. Presented at IEEE PES, Chicago, IL, USA, July 2017; IEEE PES. 2017. Available online: http://site.ieee.org/pes-itst/files/2018/08/2018-Panel-1.pdf (accessed on 25 July 2020).
5. RTDS Technologies Inc. IEEE PES. 2018. Available online: https://knowledge.rtds.com/hc/en-us/articles/360034295014-TSAT-RTDS-Interface/ (accessed on 25 July 2019).
6. Powertech Labs Inc. IEEE PES. 2018. Available online: https://www.dsatools.com/tsat-rtds-interface-tri/ (accessed on 25 July 2020).

7. Park, I.K.; Lee, J.; Song, J.; Kim, Y.; Kim, T. Large-scale ACDC EMT level system simulations by a real time digital simulator (RTDS) in KEPRI-KEPCO. *KEPCO J. Electr. Power Energy* **2017**, *3*, 17–21. [CrossRef]

8. Song, J.; Oh, S.; Lee, J.; Shin, J.; Jang, G. Application of the First Replica Controller in Korean Power Systems. *Energies* **2020**, *13*, 3343. [CrossRef]

9. Guo, Y.; Zhang, J.; Hu, Y.; Han, W.; Ou, K.; Jackson, G.; Zhang, Y. Analysis on the effects of STATCOM on CSG based on RTDS. In Proceedings of the IEEE PES ISGT Europe 2013, Lyngby, Denmark, 6–9 October 2013.

10. Lee, H.; Lee, J.; Ko, B.; Hur, K. Internal System Selection for TSAT-RTDS Hybrid Simulation by Electrical Distance. In Proceedings of the 24th International Conference on Electrical Engineering (ICEE 2018), Seoul, Korea, 24–28 June 2018.

11. Zhang, Y.; Gole, A.M.; Wu, W.; Zhang, B.; Sun, H. Development and Analysis of Applicability of a Hybrid Transient Simulation Platform Combining TSA and EMT Elements. *IEEE Trans. Power Syst.* **2013**, *28*, 357–366. [CrossRef]

12. Lin, X.; Gole, A.M.; Yu, M. A Wide-Band Multi-Port System Equivalent for Real-Time Digital Power System Simulators. *IEEE Trans. Power Syst.* **2019**, *24*, 237–249. [CrossRef]

13. Lin, X. System Equivalent for Real Time Digital Simulator. Ph.D. Thesis, University of Manitoba, Winnipeg, MB, Canada, 2011.

14. Hu, Y.; Wu, W.; Zhang, B.; Guo, Q. Development of an RTDS-TSA hybrid transient simulation platform with frequency dependent network equivalents. In Proceedings of the IEEE PES ISGT Europe 2013, Lyngby, Denmark, 6–9 October 2013.

15. Shu, D.; Xie, X.; Dinavahi, V.; Zhang, C.; Ye, X.; Jiang, Q. Dynamic Phasor Based Interface Model for EMT and Transient Stability Hybrid Simulations. *IEEE Trans. Power Syst.* **2018**, *33*, 3930–3939. [CrossRef]

16. Panigrahy, N.; Gopalakrishnan, K.S.; Ilamparithi, T.; Kashinath, M.V. Real-time phasor-EMT hybrid simulation for modern power distribution grids. In Proceedings of the 2016 IEEE International Conference on Power Electronics, Drives and Energy Systems (PEDES), Trivandrum, India, 14–17 December 2016.

17. Palensky, P.; Van Der Meer, A.A.; Lopez, C.D.; Joseph, A.; Pan, K. Co-simulation of intelligent power systems. *IEEE Ind. Electron. Mag.* **2017**, *11*, 34–50. [CrossRef]

18. Palensky, P.; Van Der Meer, A.A.; Lopez, C.D.; Joseph, A.; Pan, K. Applied Co-simulation of Intelligent Power Systems: Implementing Hybrid Simulators for Complex Power Systems. *IEEE Ind. Electron. Mag.* **2017**, *11*, 6–21. [CrossRef]

*Article*

# Development of a Multiphysics Real-Time Simulator for Model-Based Design of a DC Shipboard Microgrid

Fabio D'Agostino [1], Daniele Kaza [1], Michele Martelli [1], Giacomo-Piero Schiapparelli [1], Federico Silvestro [1,*] and Carlo Soldano [2]

[1] Department of Electrical, Electronic, Telecommunication Engineering and Naval Architecture (DITEN), University of Genoa, Via All'Opera Pia 11a, 16145 Genoa, Italy; fabio.dagostino@unige.it (F.D.); danielekaza@gmail.com (D.K.); michele.martelli@unige.it (M.M.); giacomo-piero.schiapparelli@edu.unige.it (G.-P.S.)

[2] ABB Marine and Ports S.p.a., Via Molo Giano, 16128 Genoa, Italy; carlo.soldano@it.abb.com

* Correspondence: federico.silvestro@unige.it

Received: 20 May 2020; Accepted: 5 July 2020; Published: 11 July 2020

**Abstract:** Recent and strict regulations in the maritime sector regarding exhaust gas emissions has led to an evolution of shipboard systems with a progressive increase of complexity, from the early utilization of electric propulsion to the realization of an integrated shipboard power system organized as a microgrid. Therefore, novel approaches, such as the model-based design, start to be experimented by industries to obtain multiphysics models able to study the impact of different designing solutions. In this context, this paper illustrates in detail the development of a multiphysics simulation framework, able to mimic the behaviour of a DC electric ship equipped with electric propulsion, rotating generators and battery energy storage systems. The simulation platform has been realized within the retrofitting project of a Ro-Ro Pax vessel, to size components and to validate control strategies before the system commissioning. It has been implemented on the Opal-RT simulator, as the core component of the future research infrastructure of the University of Genoa, which will include power converters, storage systems, and a ship bridge simulator. The proposed model includes the propulsion plant, characterized by propellers and ship dynamics, and the entire shipboard power system. Each component has been detailed together with its own regulators, such as the automatic voltage regulator of synchronous generators, the torque control of permanent magnet synchronous motors and the current control loop of power converters. The paper illustrates also details concerning the practical deployment of the proposed models within the real-time simulator, in order to share the computational effort among the available processor cores.

**Keywords:** model-based design; HIL; multi physics simulation; marine propulsion; ship dynamic; DC microgrid; shipboard power systems

---

## 1. Introduction

In recent years, complexity and interoperability requirements of industrial products and systems, have seen a significant increase. This has led to new challenges in both design and development of innovative and competitive solutions. High-tech, automotive, and aerospace industries, have been among the firsts to apply a model-based design (MBD) approach, exploiting information technologies and CAD models [1] to assist designers in every stage of the development process [2]. The main idea behind the MBD is the development of a complete system in a virtual environment, to reproduce the expected behaviour of the real system, and to predict its performances before the building. Some examples of MBD application are experienced for propulsion plant control purposes [3,4]. Figure 1a summarizes the main aspects of the MBD philosophy. More specifically, the workflow can be described in four points: (I) requirements collection; (II) physical system and control definition; (III) verification and validation; (IV) test, performed with hardware-in-the-loop (HIL) setups. As depicted in Figure 1b, these main points are not necessarily sequential, since the continuous verification after each step is the key aspect of the MBD design procedure. Several variations of this general structure are available in literature, e.g., the V-model approach is often used. It consists of an iterative process, typically used for software development and now also for microgrid design, which associates at each development step a specific test phase for validation [5,6].

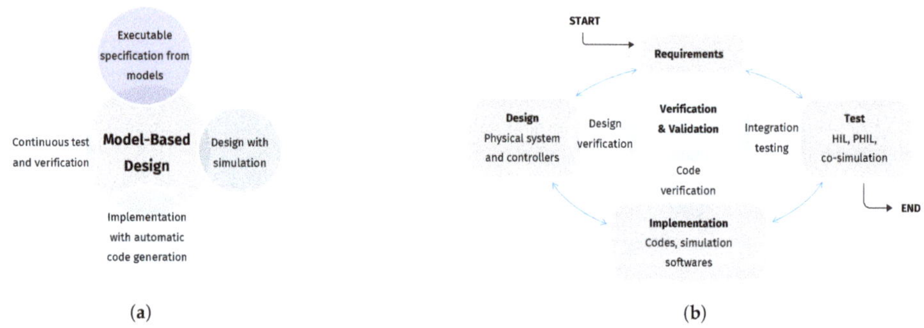


**Figure 1.** (**a**) Model-based design philosophy [1]. (**b**) Model-based design workflow scheme [2].

In general, the application of the MBD for power system and microgrid studies regards control and protection design and testing [7–9], generation sizing [5] and on-board power systems [10]. In this context the adoption of MBD can support the microgrid design phases [11], which typically are: (1) definition of the system layout and voltage levels, (2) load analysis, (3) generators and storage sizing, (4) distribution system and cables sizing, (5) fault current analysis and protection schemes selection. The availability of a model allows to validate each step in a simulated environment, especially when a microgrid is characterized by heterogeneous generation resources and technologies that increase considerably the system complexity. Finally, HIL configurations introduce additional improvements to the MBD procedure, by allowing the possibility to test real components, such as controllers or protection devices [12,13].

The installation of innovative energy sources [14–17], to reduce harmful emissions as imposed by recent regulations [18–20], has considerably increased on-board system complexity among all the assets of the marine industry. Shipboard power systems have become modern microgrids where complex technologies need to be integrated with high performance requirements. Such a complexity increase has led shipboard power systems designers to consider the use of model-based design techniques to achieve the goals of an efficient and more cost-effective product development in a simple manner [21].

In this paper, a DC microgrid multi-physics real-time model is presented. This work has been developed in the context of an industry project consisting in the retrofit of a roll-on roll-of (RORO) ferry; starting from a conventional diesel propulsion plant a new diesel–electric hybrid system has been designed and successfully tested. The aim of work is the development of a real-time simulation setup in order to support the design of a new microgrid. The ship model, studied and developed for a specific application, considering both electrical and ship dynamics, has been implemented in Matlab/Simulink environment and deployed on Opal–RT simulator [22]. Such a HIL setup , as a future development, will be integrated in a co-simulation environment provided by the ShIL research infrastructure, of the University of Genoa. The ship-in-the-loop (ShIL) research infrastructure is a project developed by DITEN, DIME and DIBRIS departments of the University of Genoa to build up a co-simulation environment power-hardware-in-the-loop (PHIL) in which shipboard microgrid, cyber range and ship dynamic behaviour tests can be driven simultaneously [23]. The project, started in 2020 is co-founded by Regione Liguria.

As it is clearly depicted in Figure 2 the laboratory is equipped with different simulators and facilities. In particular the Opal–RT simulator is connected to a power converter sized 15 kW at can provide an active front-end and a DC bus where it is possible to connect Lithium-ion battery and other DC sources. The ShIL infrastructure is composed of: four-quadrant power converters connected to Opal–RT simulator, storage systems, 5 kW fuel-cell, ship bridge simulator which replicates ship dynamics and human interaction, 5G and telecommunication simulators.

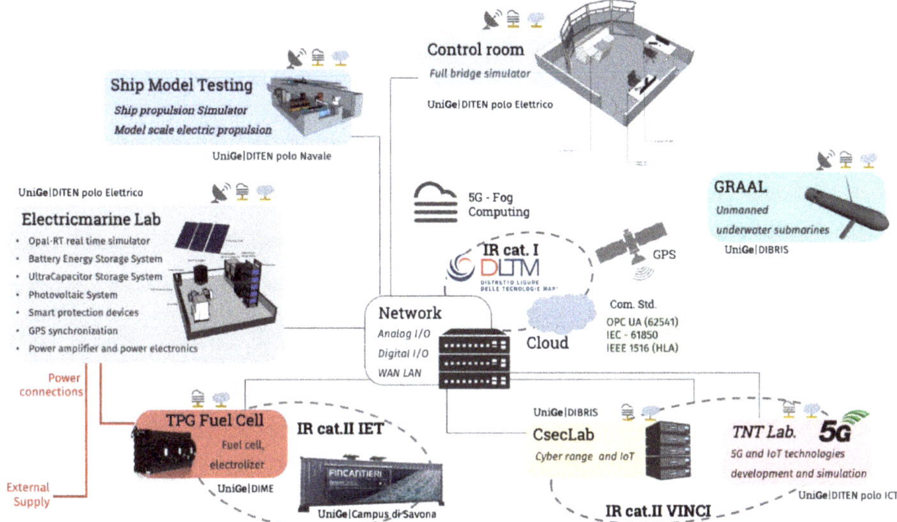


**Figure 2.** Co-simulation enviroment for HIL simulation.

The contributions of the paper are: (I) description of the model-based design methodology in the context of a revamping project of a ferry; (II) presentation of the detailed models of shipboard power system controls and components; (III) description and detailed modeling main propulsion plant machinery and ship dynamics, not usually considered coupled with power system studies [24]; (IV) critical discussion of a real-time HIL setup, developed to allow components testing; (V) integration of multi-level hierarchical controls including automation system and management of different operating modes. The latter, is essential for the integration of the ship bridge simulator in the co-simulation environment and it represents an improvement beyond the state-of-the-art.

The paper is organised as follows: Section 2 presents the main characteristics of the RORO ferry case study, highlighting design choices and requirements for the electrical and the ship components. Sections 3 and 4 describe the system models and control loops adopted. In Section 5, the model deployment in Opal-RT is presented, and results obtained by the real-time simulations are described in Section 6. Finally, conclusions and future developments are discussed in Section 7.

## 2. Revamping Project Overview

The case study used to test the proposed methodology refers to a ferry, designed for the navigation in Lago Maggiore (Italy). It is a bi-directional ship with the hull that presents a bow-stern symmetry; two propellers, located at bow and stern region, provide the thrust in both ahead and astern run. This solution is quite common for RORO ships that connects ports close each other, since it reduces the berthing time [25]. The main characteristics of the ship are summarized in Table 1.

**Table 1.** Ro-Ro ferry main dimensions.

| Flag | Italian | Draft | 2.195 m |
|---|---|---|---|
| Displacement | 377 t | Length O.A. | 45.3 m |
| Length between P.P. | 38 m | Breadth | 8.3 m |
| Total passenger capacity | 450 | Total vehicle capacity | 27 |

The older propulsion configuration is composed of two diesel engines, MTU 12V183 TE62, 310 kW, 1800 rpm, that provide torque to the two independent shaft lines. The ship electrical power supply was provided by a diesel generator, AIFO 8210M 220 V, and by 24 V uninterruptible power supply (UPS) [26,27]. In the target configuration, the diesel engines are replaced by two permanent-magnet synchronous motors (PMSMs) in a microgrid architecture. The shipboard microgrid presents a DC radial configuration organised with two main DC buses, named Fore (FOR) and After (AFT) switchboard, respectively [28]. DC systems are nowadays attractive, specially for low power range [29], since, in respect to traditional AC systems, these solutions may lead to a not negligible reduction of weight and distribution losses, improving in general the system efficiency. On the contrary, DC systems require to deal with problems related to low short-circuit currents, dc-breaker technology, and system standardization [30,31].

The electrical system is symmetric, each main bus presents a diesel generator (DG) connected to the main switchboard with an active rectifier, a battery energy storage system (BESS) interfaced with a DC-DC two quadrants converter, a propulsion system drive, a PMSM and propeller. Additionally, each DC bus bar presents a grid-forming inverter for the supply of the AC distribution system. Both the DC and AC systems are symmetric, and they could be connected alternatively, through a DC tie breaker or an AC tie breaker. Figure 3 shows the system one line diagram.

The power and energy management strategies assumed in the paper are defined in accordance with the project requirements, and tested with a specific mission profile requested by the customer. Component sizing is also defined in accordance with these guidelines, as described in Section 6.4. Nevertheless, the proposed framework can be applied to a wider set of study cases and projects [32].

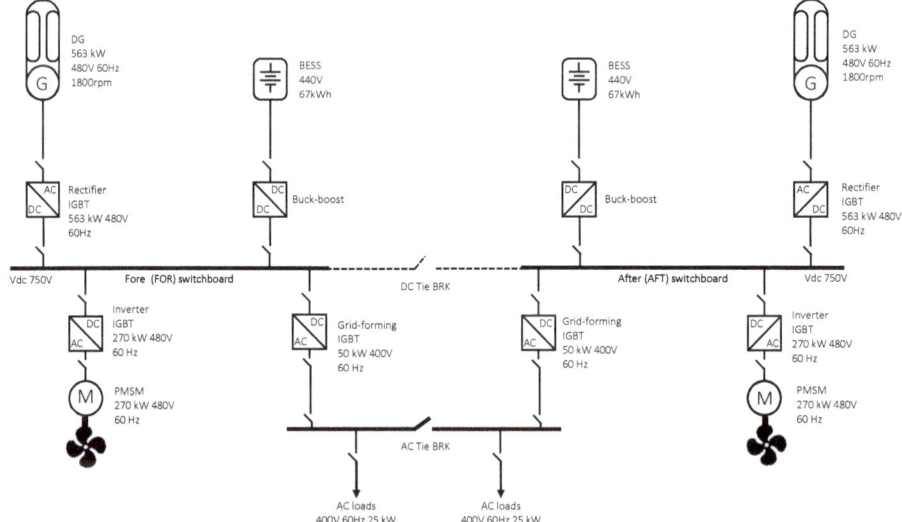


**Figure 3.** One line diagram.

## 3. Components Modeling and Control Strategies

The ferry with the new power configuration, should operates in three different operational modes: full-electric, hybrid and diesel mode. In the first case, the DGs are off and the energy to the propulsion motor is provided by BESSs. In the second case DGs and BESSs cooperates for the power supply while in the third case the BESSs are disconnected and the necessary power is provided by the DGs. In order to fulfill requirements imposed by the different navigation modes, several solutions have been developed. In this work, a DC voltage droop control scheme is considered and adopted when the operation mode requires the coordination of multiple resources (DGs and/or BESSs) [33–35]. Instead, a single DC voltage forming control scheme is used when the operative mode requires only one system providing the DC bus regulation.

Figure 4 shows a synthesis of the higher level control loops of the system. Starting from the DG, the AC voltage is controlled by an automatic voltage regulator (AVR) (IEEE-AC1A) while the speed is regulated by a governor (modelled as a standard Woodward Diesel Governor DEGOV1). The DG is connected to the DC bus through an active rectifier which implements a proportional-integral (PI) DC voltage regulator and a PI *dq*-axis current control, described in the following. Such a system is capable to operate in stand alone DC forming mode or in multiple resources droop control mode.

The BESS are connected to the DC bus via a DC-DC buck-boost converter. According to the different operating mode, the BESS can be controlled in order to regulate the injected current or the DC bus voltage, both in stand alone or droop scheme.

The propulsion system implemented consists of a PMSM supplied by a variable speed drive. The motor should provide the propeller required torque. The motor speed, and consequently the propeller rotational regime, is controlled by an upper level closed loop control cruise control, based on the error between the desired and the actual ship speed, evaluated using a one degree of freedom motion equation [36].

The AC power system is supplied by two grid-forming converters, which can be operated alone or in parallel both in grid-supporting mode [37].

Details on the adopted control strategies for the single devices are reported in the following.

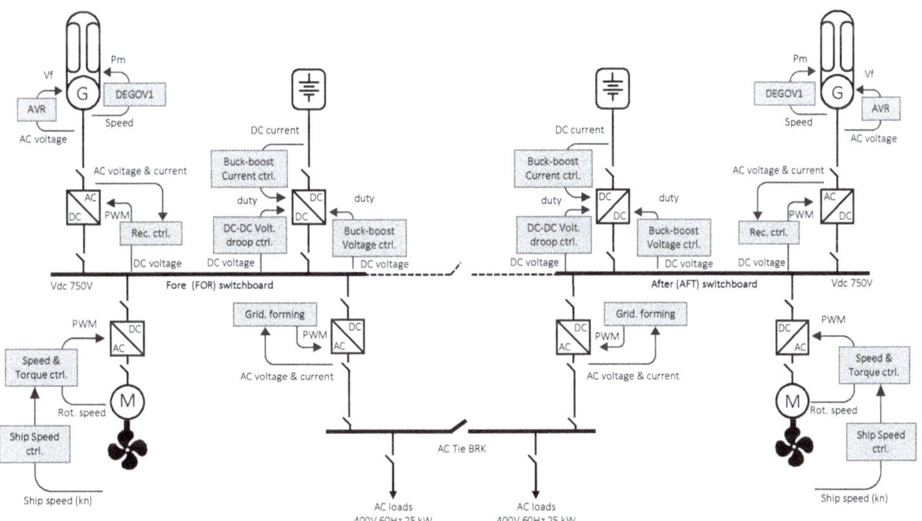


**Figure 4.** Shipboard microgrid main control loops.

### 3.1. Battery Energy Storage Systems and DC-DC Converters

The BESS model is realised using a standard Simulink SimPowerSystem (SPS) Li-ion battery model connected to the DC bus thought a DC-DC buck-boost converter, Figure 5a.

Three different control strategies can be selected. The first one, Figure 5b, consists of a PI current control with voltage droop. Specifically, the current reference signal given to the PI current control is calculated as the sum of the current set point $I_B^*$ and the voltage droop error signal, $K_{dr}(V_{dc}^* - V_{dc})$. Where $K_{dr}$ is the droop coefficient and $V_{dc}^*$, $V_{dc}$ are the voltage DC bus reference and measured signal, respectively. The duty cycle $\delta$, for the pulse width modulation (PWM) is the sum of the PI output signal and the measured battery voltage $V_B$. This control mode allows to participate in the DC bus voltage regulation and to implement state of charge (SOC) management strategies. When multiple devices are involved in the DC bus droop regulation, the power sharing, and thus the contribution of each resources, can be controlled by changing the reference current $I_B^*$ [34,38].

The second control strategy, Figure 5c, is a DC voltage forming control. The DC-DC converter is controlled alternatively either as a boost or buck converter. When the bus voltage $V_{dc}$ is lower than its reference value, $V_{dc}^*$, the converter is controlled as a boost, when the previous condition is not respected it is controlled as a buck. The duty cycles, one for each control channel, are defined by a PI regulator.

The third control strategy, Figure 5d, consists of a battery current control. This regulation strategy allows to control precisely the battery output current and it can be used for SOC management strategies.

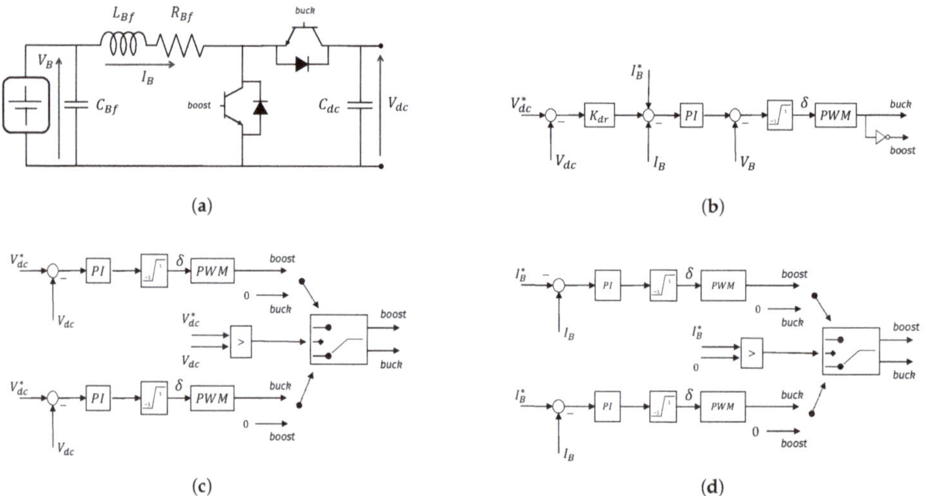

**Figure 5.** (**a**) Battery DC-DC circuit diagram. (**b**) DC voltage droop control, (**c**) DC voltage forming control, (**d**) battery current control.

### 3.2. AC-DC and DC-AC Three Phase Converters

The DG model consists of a standard SPS synchronous machine per unit model. The implemented controls are the governor, Woodward Diesel Governor DEGOV1, and the AVR IEEE-AC1A [39]. The DGs are coupled with the DC bus thought an active rectifier. The rectifier model consists of a three phase converter and a LCL filter in which the output L is realised with a transformer, Figure 6. The rectifier controls consist of a PI feed-forward current control and an outer DC voltage PI control which defines the dq-axis current references. Similarly the two DC-AC converters which provides the power supply to the AC subsystem have been implemented with the same scheme but implement a grid-forming with supporting mode control strategy [40].

The LCL filters for both cases have been designed following criteria detailed in the literature [41,42].

### 3.2.1. Converter Control Loops

The convert control loops implement a hierarchical control system organization. In this context we call device level controls the PI *dq*- current and voltage feed-forward controls, and system level controls, all the other loops which define the reference for the device level loops according to specific system requirements [43]. In this second group we consider the DC bus voltage regulation in case of the AC-DC active rectifier, and grid-forming control in case of the DC-AC converter.

The device level controls considered in this paper are *dq* PI current and voltage feed-forward controls, Figure 6. The control system design is out of the purpose of the paper. As detailed in literature, the current and voltage control loop can be modelled, considering the system of equation of the LCL filter [44–46]. Thus, control parameters can be defined according to the system eigen values and desired control performance [44,47].

The reference frame for the *dq*-transformation, can be provided by a phase locked loop (PLL) or can be generated internally by the controller according to the regulation strategy implemented by the system level controls. More specifically, power converters can be in general divided into grid following and grid forming units [43,48]. The grid following unit, can operate only with an already energised grid, since it controls active and reactive power by regulating the current output according to the voltage magnitude and phase. The active rectifier, Figure 6a, is a grid following unit. It controls active and reactive powers by properly changing the current *dq* reference in a synchronized way with the voltage phase measured by the PLL. On the other hand, the grid forming unit can be considered as voltage generators. It can control voltage magnitude ad frequency, and thus it can energize an AC grid. The DC-AC converter in Figure 6b is a grid forming unit [37].

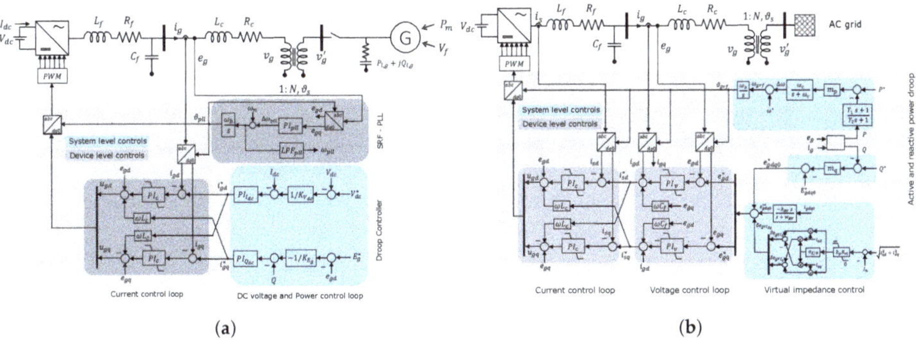

**Figure 6.** AC-DC converters single line diagrams: (**a**) rectifier control scheme (**b**) grid-forming scheme.

The diesel genset is connected to the DC bus by an active rectifier, Figure 6a. The AC section is energised by the generator and the converter has to control the active and reactive power in order to regulate the DC and AC voltage, respectively. In this specific application two droop control schemes have been adopted. The *d*-axis current reference is given by the DC current PI controller with DC voltage droop [49,50]. The *q*-axis current reference is given by a reactive power PI controller whose reference is given by AC voltage error multiplied by the AC voltage droop coefficients.

The grid forming converter, Figure 6b, is used to supply the AC distribution system. The adopted control structure is a grid forming with supporting scheme [37,40,48]. The supporting scheme is required if the converter operates in parallel with other units. More specifically, the implemented strategy corresponds to the one presented in [37]. It implements two separate control channels in order to regulate the active and reactive power sharing. The voltage phase, and the frequency, are controlled by the active power channel, which implements the so called reverse droop control applied to the active power error. While the voltage magnitude is controlled by the reactive power channel, which implements the reverse droop on the reactive power error [43]. Additionally, to improve the performance, the active power channel presents a low-pass filter which is used to avoid fast frequency variations and allows an inertial response, while the reactive power channel implements a virtual inertia control [37]. The adopted control strategy allows the systems to operate in both, AC radial or meshed grid configuration [51]. Notice that the same capability could be obtained even with virtual synchronous machines (VSM) control strategies [43,52].

*3.3. Propulsion Plant Dynamic*

The propulsion plant simulator is composed of a PMSM, a propeller, and a DC-AC drive. The motor shaft rotational speed is controlled by a PI, and the vector controller is realized by a current hysteresis control, the layout is reported in Figure 7a. The PMSM is modelled as a standard SPS which receives as input the required propeller torque computed by the propeller model. In order to have a realistic and time-dependant propeller load, the propulsion plant dynamic has been modelled using two differential equations, able to represent the transient behaviour of both shaftline and surge motion. The shaftline acceleration and deceleration are derived by the Lagrange motion equation for rotating rigid body:

$$I_P \frac{d}{dt}\Omega(t) = Q_{em}(t) - Q_D(t) - Q_F(t) \tag{1}$$

where: $I_P$ is the total polar inertia of the system, and it includes motor, shaft and propeller inertia; $\Omega$ is the rotational regime; $Q_{em}$ is the electromagnetic torque; $Q_D = Q_O/\eta_r$ is the delivered torque, while $Q_F = (1 - \eta_s)Q_D$ is the torque losses due to friction in bearings and stern tube. For the paper's purposes, both the relative rotative efficiency ($\eta_r$) and friction coefficient ($\eta_s$) have been kept constant.

The ship surge velocity ($u$) is obtain integrating the follow equation:

$$(m_{rb} + m_{ad}) \frac{d}{dt}u(t) = -R_T(t) + (1 - t_{df})T(t) - T_{env}(t). \tag{2}$$

where $m_{rb}$ is the ship displacement ($\Delta$), $m_{ad} = 0.08\Delta$ is the added mass, both expressed in kg; $R_T$ is the ship resistance in N, obtained using CFD methods. $t_{df}$ is the thrust deduction factor; $T$ is the propeller thrust and $T_{env}$ accounts for all the environmental forces due to either wind or wave, the environmental disturbances are not considered at the current stage, and they will subject to further investigations in future work.

Several methods are present in literature to evaluate the propeller performance, most of them required a high computational time not suitable to be used in a real-time simulator. For such a reason it has been decided to develop a quasi-dynamic model of the propeller, exploiting the open water diagrams. The open water diagram reports the non-dimensional thrust coefficient $K_T$, the non-dimensional torque coefficient $K_Q$, their expression are reported hereinafter.

$$K_T = \frac{T}{\rho D^4 n^2} \tag{3}$$

$$K_Q = \frac{Q_o}{\rho D^5 n^2} \tag{4}$$

where, $\rho$ is the water density (assumed equal to $1000\,\text{kg/m}^3$ since the ship operates in freshwater), $D$ is the propeller diameter expressed in m, $n$ is the propeller speed in rps (revolution per second). In addition, also the open water efficiency $\eta_o$, can be assessed as [53]:

$$\eta_o = \frac{J K_T}{2\pi K_Q} \tag{5}$$

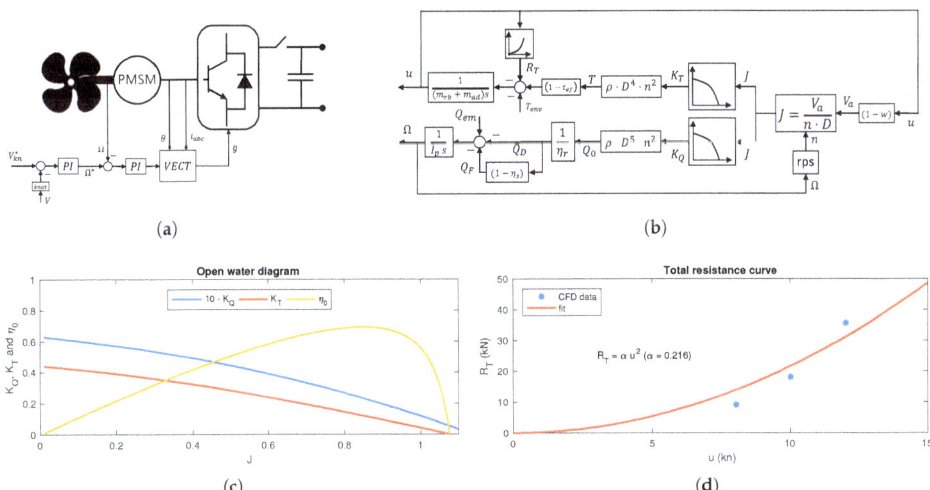

(a)                      (b)

(c)                      (d)

**Figure 7.** (**a**) Propulsion system layout; (**b**) propulsion plant model; (**c**) propeller open water diagram and (**d**) ship hull total resistance curve.

The open water diagram is shown in Figure 7c. The numerical values of these coefficients, stationary for definition, have been obtained from a 5-bladed Wageningen B series [54] and given in function of the advance coefficient $J$, computed as:

$$J(t) = \frac{V_a(t)}{n(t)D} \tag{6}$$

$$V_a(t) = (1 - w)u(t) \tag{7}$$

where: $V_a$ is the advance speed at propeller plane in $\mathrm{m\,s^{-1}}$, and $w$ is the wake fraction. As it can be seen, the advance coefficient is a time-dependant variable so the propeller transient behaviour is partially catch. Using the previous steps, it is possible to evaluate the last two terms not yet defined in (2) and (1) , $T(t)$ and $Q_0(t)$. For reader's clarity, the schematic representation of the whole propulsion model is reported in Figure 7b.

## 4. Ship Automation

The ship automation consists of the highest level control of the microgrid. At this level usually operates the algorithm which performs the energy management strategy, mission control and the human–machine interface (HMI). The development of these types of control is out of the purpose of the paper, therefore, in this study, only the minimum possible control loops have been implemented in the system. Nevertheless, the benchmark proposed will be used in the next steps of the retrofit project for testing new control algorithms and interfaces. Thus, the automation system received as command the devices on/off signals and some user-defined setpoints such as ship speed and batteries charge discharge controls. The main inputs and outputs of the implemented automation system are reported in Figure 8.

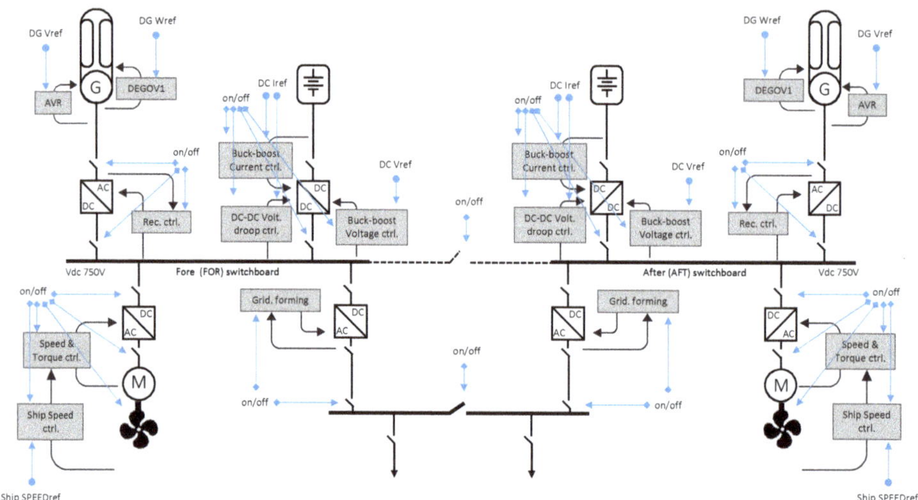

**Figure 8.** Automation controls.

## 5. Implementation on Real-Time Simulation Platform

The practical implementation of the model in Opal-RT is described in Figure 9. The simulation uses the computational power of three cores, the base model is coded in Simulink environment using the SimPowerSystems blockset and the add-on advanced real-time electro-magnetic solvers (ARTEMiS) [55]. The solver is the the state-space nodal (SSN) algorithm, a delay-free solver provided with ARTEMiS [56].

ARTEMiS precomputes and discretizes all the state-space matrix which represents the network and store them into cache memory,in order to simulate the system in real-time. The most switches are in the network the most typologies are possible and therefore the most state-space matrixes need to be stored in cache memory, causing overflows. To overcome this problem, the network is decoupled, into many sub circuits reducing the total size of the state-space matrices in memory [56]. The decoupling is realized in two different ways, the first solution exploits the fact that the transmission lines introduce a natural delay to the propagation of the waves, thus when we deal with long transmission lines, the delay naturally introduced is usually larger than the simulation step size, while in the case of short lines, which provides a delay shorter than the time steps, an additional external delay is introduced, since at the end we get at least one time step delay. The decoupling in the case of short lines is realized with the ARTEMiS StubLines, which are used in the system to divide the network in order to be simulated in parallel on different cores, Figure 9. When there are no transmission lines, a second decoupling method is provided by the state-space nodal (SSN) approach. Specifically, ARTEMiS provides the SSN solver, which combines the state-space approach with the Nodal approach which is able to divide the electrical network into smaller electrical systems without introducing any delay. This second solution for the decoupling is realised by the ARTEMiS SSN blocks which divides the system into sub circuits which can be characterised as inductive group (V-type port) considered as a voltage source connected to an inductive element, or as capacitive group (I-type port) considered a a current source connected to a capacitive element. The system decoupling strategy for the real-time (RT) simulations is shown in Figure 9.

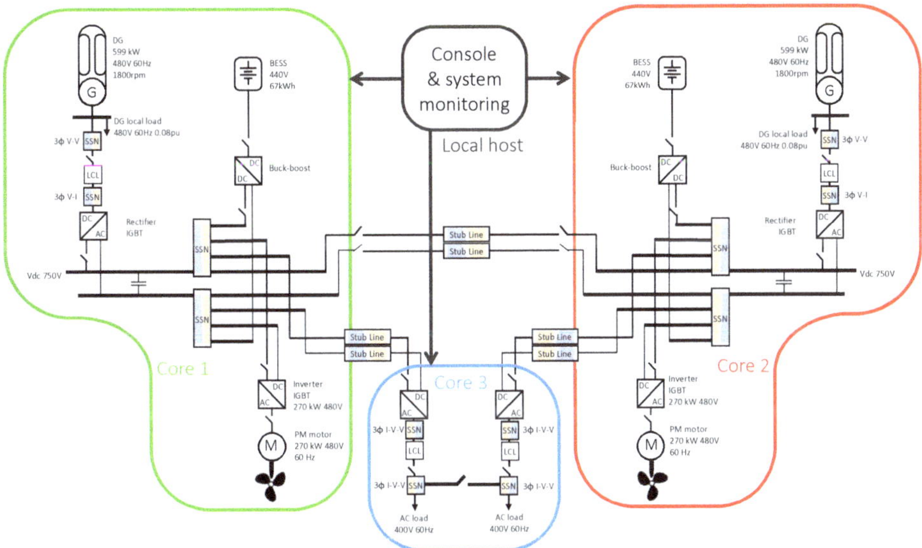

**Figure 9.** Opal-RT implementation.

The simulation step-size has been set to 30 µs and therefore with such a time step we are not able to simulate PWM with high switching frequency. In order to overcome this limitation, PWM signals are generated using the RT-EVENTS toolbox, which is capable to represent the logical state of static switches, whose value can change inside a single simulation time-step. Ten RT-EVENTS have been set for each time step, allowing the correct evaluation of PWM high frequency modulation signals. Nevertheless, in a planned upgrade of the RT simulator an FPGA board will be available for the simulation of power converters at high switching frequency.

In the described configuration the real-time model can be simulated without any overrun, meaning that at each time step the computation time of each step is less than the simulation time-step defined by the user. Numerical results are shown in the following section.

## 6. Simulation Results

The performances of the described real-time simulation setup are demonstrated with three different simulations. The first and the second ones aim to test the controls of the DC and AC system respectively. The third one, proposes a simulation of the ship in realistic design condition reproducing the the mission profile that the ship will be facing with. The main parameters of systems and controllers are reported in the Appendix A, Table A1.

### 6.1. DC Bus Regulation

All the devices in the microgrid operate to the system regulation which different strategies. Figure 10a shows the results of a simulation in which the different DC bus control strategies are exchanged. In the top graph of Figure 10a, the timeline of the different events is described. The initial configuration considers the two batteries, cooperating in droop control scheme to the regulation of the two DC switchboard connected by the DC tie breaker. Then, the two buses are disconnected, the ships accelerate to a speed of 6 kn and the two DGs are connected to the DC buses. At 180 s the two DC buses are connected again and BESS control test starts. Firstly, some set points are driven to the BESSs which operates in droop control scheme. Then, at 350 s the battery control strategy is changed to current control mode. In this configuration the BESS current is controlled directly and the voltage regulation is left to the DGs. With this control mode some set points are sent to the BESS controllers and finally at 500 s the BESS droop control strategy is restored. For the last tests, the two DGs and the FOR BESS are disconnected. In this configuration, the AFT BESS is the only resources left to feed the propulsion system in voltage forming control mode. Below of the time line plot in Figure 10a, some measurements are shown, such as, DC voltage at the two main busbars, batteries current and state of charge, generators speed and voltages and finally, generator powers.

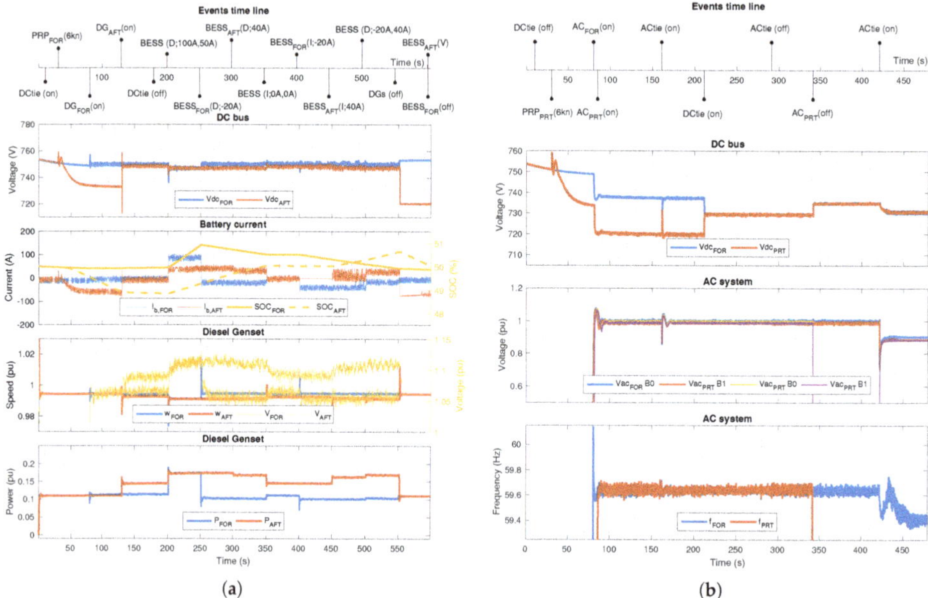

**Figure 10.** (**a**) DC and (**b**) AC control system tests.

## 6.2. AC Bus Regulation

The performance of the AC regulation system are shown in Figure 10b. The timeline in the top graph, summarize the commands sequence. Initially, the DC tie breaker is in state off, then, the two DC-AC grid forming converters are activated and the AC grid is energised. At this stage the grid is operated completely separated from starboard to port side, both AC and DC tie breakers are open. At 160 s the AC tie breaker is closed, and the two side of the grid are connected. At 160 s the DC tie breaker is also closed and the grid pass from a radial to a meshed configuration. Finally, the AC tie breaker is operated in order to disconnect the two AC grid, the port side converter is turned of and the DC tie breaker is closed again in order to reach the configuration in which all the AC supply is given by the only starboard side converter. Below of the time line plot in Figure 10b, DC voltage, AC voltage and AC frequency are shown.

## 6.3. Ship Mission

The whole system performance has been finally evaluated by simulating a realistic navigation of approximately 5 nautical miles between the port of Intra and the port of Laveno on Lago Maggiore, Italy. The mission considers a round trip of almost 1 h and it is described in the top graph of Figure 11. Specifically, mission starts in port of Intra, the ship is moored and the power system is supplied only by the two batteries with a SOC equal to 50%. After few seconds, the navigation starts; the engine set point is set to obtain 6 kn, and this speed is kept constant for almost 300 s, the time needed to reach a sufficient distance from the coast and to switch to hybrid mode. Thus, one diesel generator is turned on, the battery set-points are set 128 A (recharge), and the ship speed is increased up to 10 kn. After almost 23 min of navigation, with the DG running smoothly at half load, the ferry is close to the arrival port of Laveno, thus the battery recharge is stopped with a SOC about of 60% (suggested value by the manufactures to extend the life cycle), the diesel genset is turned off, and the ship starts the manoeuvre to enter the port in full electric mode with a speed of 6 kn. Once the port is reached, the ship is moored and after a while the way back trip starts, following the same procedure described before. For sake of completeness, in the lower sub-plot the ship speed and the propeller thrust are reported, pointing out a symmetric behavior, without any peak. Some simulation results for the shipboard power system are shown in the plots of Figure 11.

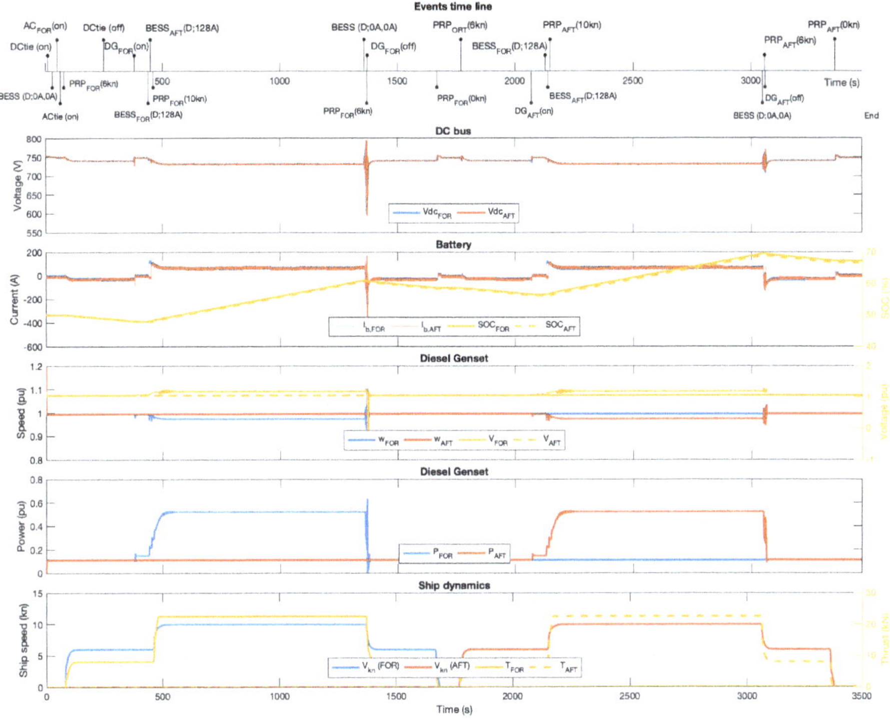

**Figure 11.** Simulation of a navigation mission between port of Intra and port of Laveno, Lago Maggiore, Italy.

Figure 12 shows the real-time simulator core monitoring outputs provided by the OpMonitor block [55]. The yellow line refers to the time step size, the time set by the user in which the simulator has to complete a cycle of measurements–computation–commands [55]. The blue line represents the number of overruns, meaning the number of times in which the time of computation exceeded the defined time step. The red line corresponds to the effective computation time, and the purple time corresponds to the time idle, meaning the difference between the step size and the computation time. As it can be seen, the simulation results to be stable, the systems is able to keep the real-time for all the mission duration. The overall loading is 49.65% of the total computation power, divided as 35.30% for core 1, 34.77% for core 2 and 78.88% for core 3.

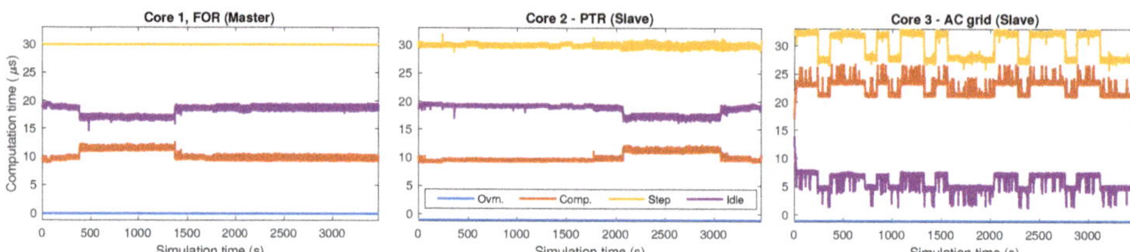

**Figure 12.** Real-time setup performance monitoring during the mission simulation.

*6.4. Observations on the Ship Management Strategies*

The mission simulation proposed in the previous section refers to the operating mode requested by the customer. Specifically, the project requirements impose (I) to be able to operate in full electric mode below 6 kn, (II) to operate the batteries between a minimum and maximum SOC in order to extend their expected life, (III) to operate in hybrid mode, with one diesel genset as main supply at cruise speed of 10 kn. Moreover, (IV) no shore connection capability is considered, thus, batteries should be kept within the limits during all the operating time and even during the time at berth. Nevertheless, other control strategies could be applied in order to improve the system efficiency [32].

## 7. Conclusions

The paper presents the development of a real-time simulation framework of a DC shipboard microgrid, realized to support the retrofitting project of a RORO Pax vessel with a model-based design approach. The shipboard microgrid model shows its capability to reproduce real operating conditions including ship dynamics within the electrical power system study. Models include propellers, prime movers, electrical machines, battery energy storage systems, power converters and regulators. The proposed architecture has been realized to allow hardware-in-the-loop configurations.

Results show that several control schemes can be implemented, to test system performances under different conditions, and to verify operational constraints during the expected ship missions. Moreover, since the simulator has been developed following a modular and parametric approach, it can be continuously tuned with experimental data to improve results fidelity, and to realize the digital twin of the ship during its entire life cycle.

Future works will regard the experimental validation of simulation results, the implementation of the shore connection, and the inclusion of protection devices for short-circuits studies .

**Author Contributions:** Conceptualization F.D., D.K., M.M., G.-P.S., F.S. and C.S.; methodology F.D., D.K., M.M., G.-P.S., F.S. and C.S., investigation G.-P.S., D.K., C.S.; supervision F.D, M.M., F.S. All authors have read and agreed to the published version of the manuscript.

**Funding:** This research received no external funding.

**Conflicts of Interest:** The authors declare no conflict of interest.

## Appendix A

The main parameters of the implemented real-time model are listed in Table A1.

Table A1. Main parameters.

| Variable Description | Value | Variable Description | Value |
|---|---|---|---|
| **Diesel Genset** | | | |
| Nominal power | 596 kW | Inertia time constant | 0.17 s |
| Nominal voltage | 480 V | AVR (gain $K_a$, time const. $T_a$) | 600 pu, 0.05 s |
| Nominal speed, freq. | 1800 rpm, 60 Hz | DEGOV1 (gain $K$, droop $K_d$) | 8 pu, 0.05 pu |
| **Propulsion motor PMSM** | | | |
| Nominal power | 270 kW | Nominal, maximum torque | 5.73 kN, 7.2 kN |
| Nominal voltage | 480 V | Nominal speed | 450 rpm |
| Nominal speed | 22.5 Hz | Nominal stator current | 347 A |
| Speed PI regulator $(K_p, K_i)$ | 10 pu, 15 pu | | |
| **AC-DC Active rectifier** | | | |
| Nominal power | 596 kW | Current PI regulator $(K_p, K_i)$ | 1 pu, 7 pu |
| Nominal AC voltage | 480 V | DC voltage PI reg. $(K_p, K_i)$ | 0.2 pu, 0.5 pu |
| Nominal DC voltage | 750 V | Reactive power PI reg. $(K_p, K_i)$ | 0.02 pu, 0.05 pu |
| Switching frequency $(f_s)$ | 1620 Hz | DC, AC volt. droop $(K_{V_{dc}}, K_{E_g})$ | 0.05 pu, 0.05 pu |
| LCL Filter $(L_1, L_2, C)$ | 1.29 pu, 0.09 pu, 0.07 pu | | |
| **Battery Energy Storage System** | | | |
| Nominal capacity | 128 A h | Switching frequency | 5000 Hz |
| Nominal voltage | 440 V | DC voltage droop $(K_{dr})$ | 1/0.05 pu |
| DC-DC stage param. $(R, L, C)$ | 0.25 Ω, 0.8 mH, 12 μF | | |
| **DC-AC grid–forming converter** | | | |
| Nominal power | 50 kVA | TVR ‡ gain $(k_{RV})$ | 0.09 |
| Transformer $(V_{t1}/V_{t2})$ | 390 V/400 V | TVR ‡ HPF * cut-off $(\omega_{RV})$ | 16.66 rad s$^{-1}$ |
| LCL Filter $(L_1, L_t, C)$ | 1.54 pu, 0.10 pu, 0.06 pu | LPF ** cut-off freq. $(\omega_c)$ | 31.41 rad s$^{-1}$ |
| P and Q droop $(m_p, m_q)$ | 0.02 pu | Virtual resistance $(R_{vi})$ | 0.6716 pu |
| VI † ratio $X_{vi}/R_{vi}$ $(\sigma_{X/R})$ | 5 | Lead-lag filter $(T_1, T_2)$ | 0.033 s, 0.011 s |
| **Ship model** | | | |
| Displacement $(\Delta)$ | 377 t | Propeller diameter (D) | 1.25 m |
| Added mass $(m_{add})$ | 30.16 t | Density $(\rho)$ | 1000 kg/m$^3$ |
| Nominal speed $(V)$ | 10 kn | Resistance at 10 kn $(R_T)$ | 18.06 kN |
| Total polar inertia $(I_{tot})$ | 383.4 kgm$^2$ | Relative rotative efficiency $(\eta_r)$ | 1.026 pu |
| Thrust deduction factor $(t_{df})$ | 0.205 pu | Shaft efficiency $(\eta_{shaft})$ | 0.98 pu |
| Wake fraction $(w)$ | 0.235 pu | Ship speed PI ctrl. $(K_p, K_i)$ | 3 pu, 1 pu |

† Virtual impedance (VI); ‡ Transient Virtual Resistor (TVR); * High-Pass Filter (HPF); ** Low-Pass Filter (LPF).

## References


1. Smith, P.F.; Prabhu, S.M.; Friedman, J. Best Practices for Establishing a Model-Based Design Culture. Technical Report, SAE Technical Paper. 2007. Available online: https://www.sae.org/publications/technical-papers/content/2007-01-0777/ (accessed on 2 March 2020).
2. Kelemenová, T.; Kelemen, M.; Miková, L.; Maxim, V.; Prada, E.; Lipták, T.; Menda, F. Model based design and HIL simulations. *Am. J. Mech. Eng.* **2013**, *1*, 276–281.
3. Martelli, M.; Figari, M. Real-Time model-based design for CODLAG propulsion control strategies. *Ocean Eng.* **2017**, *141*, 265–276. [CrossRef]
4. Vrijdag, A.; Stapersma, D.; van Terwisga, T. Systematic modelling, verification, calibration and validation of a ship propulsion simulation model. *J. Mar. Eng. Technol.* **2009**, *8*, 3–20. [CrossRef]
5. Sonnenberg, M.; Pritchard, E.; Zhu, D. Microgrid Development Using Model-Based Design. In Proceedings of the 2018 IEEE Green Technologies Conference (GreenTech), Austin, TX, USA, 4–6 April 2018; pp. 57–60.
6. Kirby, B.; Zou, L.; Cao, J.; Kamwa, I.; Heniche, A.; Dobrescu, M. Development of a predictive out of step relay using model based design. In Proceedings of the IET Conference on Reliability of Transmission and Distribution Networks (RTDN 2011), London, UK, 22–24 November 2011.
7. Panova, E.A.; Nasibullin, A.T. Development and Testing of the Adequacy of the 220/110 kV Distribution Substation Matlab Simulink Mathematical Model for Relay Protection Calculations. In Proceedings of the 2019 IEEE Russian Workshop on Power Engineering and Automation of Metallurgy Industry: Research & Practice (PEAMI), Magnitogorsk, Russia, 4–5 October 2019; pp. 134–138.
8. Zhu, W.; Shi, J.; Liu, S.; Abdelwahed, S. Development and application of a model-based collaborative analysis and design framework for microgrid power systems. *IET Gener. Transm. Distrib.* **2016**, *10*, 3201–3210. [CrossRef]
9. Zhang, Y.; Bastos, J.L.; Schulz, N.N. Model-based design of a protection scheme for shipboard power systems. In Proceedings of the 2008 IEEE Power and Energy Society General Meeting-Conversion and Delivery of Electrical Energy in the 21st Century, Pittsburgh, PA, USA, 20–24 July 2008; pp. 1–9.
10. Dufour, C.; Soghomonian, Z.; Li, W. Hardware-in-the-loop testing of modern on-board power systems using digital twins. In Proceedings of the 2018 International Symposium on Power Electronics, Electrical Drives, Automation and Motion (SPEEDAM), Amalfi, Italy, 20–22 June 2018; pp. 118–123.
11. Patel, M.R. *Shipboard Electrical Power Systems*; CRC Press: Boca Raton, FL, USA, 2011.
12. Kotsampopoulos, P.; Lagos, D.; Hatziargyriou, N.; Faruque, M.O.; Lauss, G.; Nzimako, O.; Forsyth, P.; Steurer, M.; Ponci, F.; Monti, A.; et al. A Benchmark System for Hardware-in-the-Loop Testing of Distributed Energy Resources. *IEEE Power Energy Technol. Syst. J.* **2018**, *5*, 94–103. [CrossRef]
13. Edrington, C.S.; Steurer, M.; Langston, J.; El-Mezyani, T.; Schoder, K. Role of Power Hardware in the Loop in Modeling and Simulation for Experimentation in Power and Energy Systems. *Proc. IEEE* **2015**, *103*, 2401–2409. [CrossRef]
14. Capasso, C.; Veneri, O.; Notti, E.; Sala, A.; Figari, M.; Martelli, M. Preliminary design of the hybrid propulsion architecture for the research vessel "G. Dallaporta". In Proceedings of the 2016 International Conference on Electrical Systems for Aircraft, Railway, Ship Propulsion and Road Vehicles International Transportation Electrification Conference (ESARS-ITEC), Toulouse, France, 2–4 November 2016; pp. 1–6.
15. Manouchehrinia, B.; Molloy, S.; Dong, Z.; Gulliver, A.; Gough, C. Emission and life-cycle cost analysis of hybrid and pure electric propulsion systems for fishing boats. *J. Ocean Technol.* **2018**, *13*, 66–87.
16. Martelli, M.; Figari, M. *Maritime Transportation and Harvesting of Sea Resources*; CRC Press: Boca Raton, FL, USA, 2018; Volume 1, pp. 87–93.
17. Barelli, L.; Bidini, G.; Gallorini, F.; Iantorno, F.; Pane, N.; Ottaviano, P.A.; Trombetti, L. Dynamic Modeling of a Hybrid Propulsion System for Tourist Boat. *Energies* **2018**, *11*, 2592. [CrossRef]
18. Abadie, L.M.; Goicoechea, N. Powering newly constructed vessels to comply with ECA regulations under fuel market prices uncertainty: Diesel or dual fuel engine? *Transp. Res. Part D Transp. Environ.* **2019**, *67*, 433–448. [CrossRef]
19. Halff, A.; Younes, L.; Boersma, T. The likely implications of the new IMO standards on the shipping industry. *Energy Policy* **2019**, *126*, 277–286. [CrossRef]

20. Vieira, G.; Peralta, C.; Salles, M.B.d.C.; Carmo, B.S. Reduction of $CO_2$ emissions in ships with advanced Energy Storage Systems. In Proceedings of the 2017 6th International Conference on Clean Electrical Power (ICCEP), Santa Margherita Ligure, Italy, 27–29 June 2017; pp. 564–571.

21. Hosagrahara, A.S. Measuring Productivity and Quality in Model-Based Design. U.S. Patent 7,613,589, 3 November 2009.

22. Satpathi, K.; Balijepalli, V.M.; Ukil, A. Modeling and Real-Time Scheduling of DC Platform Supply Vessel for Fuel Efficient Operation. *IEEE Trans. Transp. Electrif.* **2017**, *3*, 762–778. [CrossRef]

23. Silvestro, F.; D'Agostino, F.; Schiapparelli, G.P.; Boveri, A.; Patuelli, D.; Ragaini, E. A Collaborative Laboratory for Shipboard Microgrid: Research and Training. In Proceedings of the 2018 AEIT International Annual Conference, Bari, Italy, 3–5 October 2018; pp. 1–6.

24. Boveri, A.; D'Agostino, F.; Fidigatti, A.; Ragaini, E.; Silvestro, F. Dynamic modeling of a supply vessel power system for DP3 protection system. *IEEE Trans. Transp. Electrif.* **2016**, *2*, 570–579. [CrossRef]

25. Campora, U.; Laviola, M.; Zaccone, R. An Overall Comparison Between Natural Gas Spark Ignition and Compression Ignition Engines for a Ro-Pax Propulsion Plant. In Proceedings of the 3th International Conference on Maritime Technology and Engineering, MARTECH, Lisbon, Portugal, 4–6 July 2016; pp. 735–743.

26. Motonave Traghetto San Cristoforo. Available online: https://www.navigazionelaghi.it/doc/pdf/flotta/S.%20CRISTOFORO_Ita.pdf (accessed on 2 March 2020).

27. Italy-Milan: Conversion services of ships. Supplies—Contract notice—Restricted procedure 247237-2018, Ministero delle Infrastrutture e dei Trasporti—Gestione Governativa dei Servizi Pubblici di Navigazione sui Laghi Maggiore di Garda e di Como, 2018.

28. Ghimire, P.; Park, D.; Zadeh, M.K.; Thorstensen, J.; Pedersen, E. Shipboard Electric Power Conversion: System Architecture, Applications, Control, and Challenges [Technology Leaders]. *IEEE Electrif. Mag.* **2019**, *7*, 6–20. [CrossRef]

29. Skjong, E.; Volden, R.; Rødskar, E.; Molinas, M.; Johansen, T.A.; Cunningham, J. Past, present, and future challenges of the marine vessel's electrical power system. *IEEE Trans. Transp. Electrif.* **2016**, *2*, 522–537. [CrossRef]

30. Staudt, V.; Bartelt, R.; Heising, C. Fault Scenarios in DC Ship Grids: The advantages and disadvantages of modular multilevel converters. *IEEE Electrif. Mag.* **2015**, *3*, 40–48. [CrossRef]

31. Satpathi, K.; Ukil, A.; Nag, S.S.; Pou, J.; Zagrodnik, M.A. DC Marine Power System: Transient Behavior and Fault Management Aspects. *IEEE Trans. Ind. Inform.* **2019**, *15*, 1911–1925. [CrossRef]

32. Boveri, A.; Silvestro, F.; Molinas, M.; Skjong, E. Optimal sizing of energy storage systems for shipboard applications. *IEEE Trans. Energy Convers.* **2018**, *34*, 801–811. [CrossRef]

33. D'Agostino, F.; Grillo, S.; Schiapparelli, G.; Silvestro, F. DC Shipboard Microgrid Modeling for Fuel Cell Integration Study. In Proceedings of the IEEE PES General Meeting, Atlanta, GA, USA, 4–8 August 2019; pp. 4–8.

34. D'Agostino, F.; Massucco, S.; Schiapparelli, G.P.; Silvestro, F. Modeling and Real-Time Simulation of a DC Shipboard Microgrid. In Proceedings of the 2019 21st European Conference on Power Electronics and Applications (EPE'19 ECCE Europe), Genova, Italy, 3–5 September 2019; pp. 1–8.

35. Satpathi, K.; Ukil, A.; Thukral, N.; Zagrodnik, M.A. Modelling of DC shipboard power system. In Proceedings of the 2016 IEEE International Conference on Power Electronics, Drives and Energy Systems (PEDES), Trivandrum, India, 14–17 December 2016; pp. 1–6.

36. Alessandri, A.; Donnarumma, S.; Vignolo, S.; Figari, M.; Martelli, M.; Chiti, R.; Sebastiani, L. System control design of autopilot and speed pilot for a patrol vessel by using LMIs. Towards Green Marine Technology and Transport. In Proceedings of the 16th International Congress of the International Maritime Association of the Mediterranean, Pula, Croatia, 21–24 September 2015; pp. 577–583.

37. Qoria, T.; Cossart, Q.; Li, C.; Guillaud, X.; Colas, F.; Gruson, F.; Kestelyn, X. Deliverable 3.2: Local Control and Simulation Tools for Large Transmission Systems. MIGRATE Project. 2018; Volume 2, p. 81. Available online: https://www.h2020-migrate.eu/_Resources/Persistent/5c5beff0d5bef78799253aae9b19f50a9cb6eb9f/D3.2%20-%20Local%20control%20and%20simulation%20tools%20for%20large%20transmission%20systems.pdf (accessed on 11 July 2020).

38. Fei, G.; Ren, K.; Jun, C.; Tao, Y. Primary and secondary control in DC microgrids: A review. *J. Mod. Power Syst. Clean Energy* **2019**, *7*, 227–242.

39. Madan, M.; Meena, R. Adapting IEEE 1992 AC1A excitation system model for industrial application. In Proceedings of the 2018 International Conference on Computing, Power and Communication Technologies, GUCON 2018, Uttar Pradesh, India, 28–29 September 2019; pp. 28–31.
40. Qoria, T.; Prevost, T.; Denis, G.; Gruson, F.; Colas, F.; Guillaud, X. Power Converters Classification and Characterization in Power Transmission Systems. In Proceedings of the EPE'19 ICCE EUROPE, Genova, Italy, 3–5 September 2019.
41. Reznik, A.; Simões, M.G.; Al-Durra, A.; Muyeen, S. *LCL* filter design and performance analysis for grid-interconnected systems. *IEEE Trans. Ind. Appl.* **2013**, *50*, 1225–1232. [CrossRef]
42. Prodanovic, M.; Green, T.C. Control and filter design of three-phase inverters for high power quality grid connection. *IEEE Trans. Power Electron.* **2003**, *18*, 373–380. [CrossRef]
43. D'Agostino, F.; Massucco, S.; Schiapparelli, G.P.; Silvestro, F.; Paolone, M. Performance Comparative Assessment of Grid Connected Power Converters Control Strategies. In Proceedings of the 2020 IEEE International Conference on Industrial Electronics for Sustainable Energy Systems (IESES), Cagliari, Sardinia, 1–3 September 2020.
44. Timbus, A.; Liserre, M.; Teodorescu, R.; Rodriguez, P.; Blaabjerg, F. Evaluation of current controllers for distributed power generation systems. *IEEE Trans. Power Electron.* **2009**, *24*, 654–664. [CrossRef]
45. Dong, N.; Yang, H.; Han, J.; Zhao, R. Modeling and Parameter Design of Voltage-Controlled Inverters Based on Discrete Control. *Energies* **2018**, *11*, 2154. [CrossRef]
46. Bajracharya, C.; Molinas, M.; Suul, J.A.; Undeland, T.M. Understanding of tuning techniques of converter controllers for VSC-HVDC. In Proceedings of theNordic Workshop on Power and Industrial Electronics (NORPIE/2008), Espoo, Finland, 9–11 June 2008.
47. Aydin, O.; Akdag, A.; Stefanutti, P.; Hugo, N. Optimum controller design for a multilevel AC-DC converter system. In Proceedings of the Twentieth Annual IEEE Applied Power Electronics Conference and Exposition, APEC 2005, Austin, TX, USA, 6–10 March 2005; Volume 3, pp. 1660–1666.
48. Rocabert, J.; Luna, A.; Blaabjerg, F.; Rodriguez, P. Control of power converters in AC microgrids. *IEEE Trans. Power Electron.* **2012**, *27*, 4734–4749. [CrossRef]
49. Jin, Z.; Meng, L.; Vasquez, J.C.; Guerrero, J.M. Specialized Hierarchical Control Strategy for DC Distribution based Shipboard Microgrids: A combination of emerging DC shipboard power systems and microgrid technologies. In Proceedings of the IECON 2017—43rd Annual Conference of the IEEE Industrial Electronics Society, Beijing, China, 29 October–1 November 2017; pp. 6820–6825.
50. Guerrero, J.M.; Chandorkar, M.; Lee, T.L.; Loh, P.C. Advanced control architectures for intelligent microgrids—Part I: Decentralized and hierarchical control. *IEEE Trans. Ind. Electron.* **2012**, *60*, 1254–1262. [CrossRef]
51. Kumar, D.; Zare, F.; Ghosh, A. DC microgrid technology: System architectures, AC grid interfaces, grounding schemes, power quality, communication networks, applications, and standardizations aspects. *IEEE Access* **2017**, *5*, 12230–12256. [CrossRef]
52. D'Arco, S.; Suul, J.A.; Fosso, O.B. A Virtual Synchronous Machine implementation for distributed control of power converters in SmartGrids. *Electr. Power Syst. Res.* **2015**, *122*, 180–197. [CrossRef]
53. Carlton, J. *Marine Propellers and Propulsion*; Butterworth-Heinemann: Cambridge, MA, USA, 2018.
54. Kuiper, G. *The Wageningen Propeller Series*; Technical Report; Marin: Wageningen, The Netherlands, 1992.
55. RT-LAB Powering Real-Time Simulation. Available online: https://www.opal-rt.com (accessed on 2 March 2020).
56. Dufour, C.; Mahseredjian, J.; Bélanger, J.; Naredo, J.L. An advanced real-time electro-magnetic simulator for power systems with a simultaneous state-space nodal solver. In Proceedings of the 2010 IEEE/PES Transmission and Distribution Conference and Exposition: Latin America (T&D-LA), Sao Paulo, Brazil, 8–10 November 2010; pp. 349–358.

*Article*

# Real-Time Hardware in the Loop Simulation Methodology for Power Converters Using LabVIEW FPGA

**Leonel Estrada [1], Nimrod Vázquez [2,*], Joaquín Vaquero [3], Ángel de Castro [4] and Jaime Arau [5]**

[1]   Electronics Department, Instituto Tecnológico Superior del Sur de Guanajuato, Educación Superior 2000, Benito Juárez, 38980 Uriangato, Guanajuato, Mexico; l.estrada@itsur.edu.mx

[2]   Electronics Department, Tecnológico Nacional de México-IT de Celaya, Antonio García Cubas 600, Fovissste, 38010 Celaya, Guanajuato, Mexico

[3]   Electronics Technology Department, Universidad Rey Juan Carlos, Calle Tulipán, s/n, 28933 Móstoles, Madrid, Spain; joaquin.vaquero@urjc.es

[4]   Electronics Technology and Communications Department, Universidad Autónoma de Madrid, Ciudad Universitaria de Cantoblanco, 28049 Madrid, Spain; angel.decastro@uam.es

[5]   Electronics Engineering Department, Tecnológico Nacional de México-CENIDET, Interior Internado, Palmira, 62490 Cuernavaca, Morelos, Mexico; jarau@cenidet.edu.mx

*   Correspondence: n.vazquez@ieee.org

Received: 27 November 2019; Accepted: 7 January 2020; Published: 13 January 2020

**Abstract:** Nowadays, the use of the hardware in the loop (HIL) simulation has gained popularity among researchers all over the world. One of its main applications is the simulation of power electronics converters. However, the equipment designed for this purpose is difficult to acquire for some universities or research centers, so ad-hoc solutions for the implementation of HIL simulation in low-cost hardware for power electronics converters is a novel research topic. However, the information regarding implementation is written at a high technical level and in a specific language that is not easy for non-expert users to understand. In this paper, a systematic methodology using LabVIEW software (LabVIEW 2018) for HIL simulation is shown. A fast and easy implementation of power converter topologies is obtained by means of the differential equations that define each state of the power converter. Five simple steps are considered: designing the converter, modeling the converter, solving the model using a numerical method, programming an off-line simulation of the model using fixed-point representation, and implementing the solution of the model in a Field-Programmable Gate Array (FPGA). This methodology is intended for people with no experience in the use of languages as Very High-Speed Integrated Circuit Hardware Description Language (VHDL) for Real-Time Simulation (RTS) and HIL simulation. In order to prove the methodology's effectiveness and easiness, two converters were simulated—a buck converter and a three-phase Voltage Source Inverter (VSI)—and compared with the simulation of commercial software (PSIM® v9.0) and a real power converter.

**Keywords:** design methodology; FPGA; hardware in the loop; LabVIEW; real-time simulation; power converters

---

## 1. Introduction

A hardware in the loop (HIL) simulation is the implementation of a system model in embedded hardware, which represents part of a real system. The main requirement of HIL simulation is that it has to be in real-time [1]. HIL simulation plays a significant role in the development of technology for many applications, presenting advantages such as the short time to market for new products; low cost of prototyping; and risk reduction of damaging test equipment and, more importantly, of harming people during testing.

There are commercial HIL real-time simulators available, such as RTDS® (Winnipeg, MB, Canada), OPAL-RT® (Montreal, QC, Canada), Typhoon® (Somerville, MA, USA), and dSPACE® (Paderborn, Germany), to name a few. The software used by these simulators allows the user to implement the system through a graphical interface; these types of equipment are used for the simulation of complex models and are tools that help designers and engineers all over the world. A review of the state-of-the-art of Real-Time Simulation (RTS) technologies, both hardware and software, is presented in [2]. Many examples of the use of this technology can be found in the literature, including simulating power electronics converters. In [3], the simulation of a Voltage Source Converter (VSC), used in a High Voltage Direct Current (HVDC) system for distributed generation and power quality regulation, is presented. In [4], it is used to test a new sliding mode controller for a standalone system based on photovoltaic (PV) generation. In [5], it is used for evaluating modular multilevel converters. Additionally, in [6], the authors propose an ultralow latency platform for the RTS, which can be used for complex power electronics systems and, as a case study for this particular platform, a driver for a Permanent Magnet Synchronous Generator (PMSG) is simulated. As shown in these references, RTS is employed to test the controller used in power electronics converters and all simulations use a commercial HIL system.

A drawback of these products is their high cost, which can amount to tens of thousands of dollars. As a result, many universities or research centers cannot afford the acquisition of this kind of equipment. Due to this issue, some researchers have developed ad-hoc solutions using low-cost hardware.

For research purposes, three types of hardware are mainly used to achieve RTS: Digital Signal Processors (DSP), Graphic Processing Units (GPU), and Field-Programmable Gate Arrays (FPGA). At present, FPGA is the most used hardware for RTS [7]. The use of FPGA for HIL testing of power converters can be found in the references, such as [8–14]. This is due to its inherent characteristic of parallel processing, which allows the fast resolution of many equations simultaneously, with short times of integration, in the order of tens of nanoseconds.

Often, the preferred language for the implementation of the system model is Very High-Speed Integrated Circuit Hardware Description Language (VHDL). This language is the best option when the objective is to minimize the FPGA resources and allows the time step to be minimized. However, the use of VHDL requires an expert designer, capable of optimizing the capabilities and the implementation of a proper FPGA application. Some FPGA manufacturers are beginning to offer a more user-friendly Integrated Development Environment (IDE) tool for their FPGA. These IDEs permit the development of FPGA applications using high-level programming languages such as C. In addition, some manufacturers use third-party software, such as MATLAB or LabVIEW, for FPGA applications [15,16]. Some authors have proposed the fast development of power electronics models for HIL simulations using these programs. In [17], the authors develop a new RTS method to test nonlinear control techniques and a methodology to implement the RTS is described, but it is based in a Digital Signal Processor (DSP), which is slower than an FPGA. In [18], the authors propose a methodology for the development of a fast HIL simulation based in MATLAB, but it includes just a small part of the work and is not detailed enough. In [19], the authors suggest fast HIL development based in MATLAB; however, the methodology is only for the design of PV converters. In [20], a high-performance real-time simulator for power electronics based in a novel massively parallel computational engine implemented in low-cost FPGA hardware using VHDL is presented, but there is not enough detail in the implementation. In [21], the HIL implementation of a validation system for an energy management system used in microgrids is presented. This HIL simulation is performed on a PC and has an overall time step of 100 ms, which is slower than the one reached using an FPGA. As a major advantage, the use of a PC means a lower implementation cost.

The company National Instruments (NI) offers reconfigurable industrial controllers, known as Reconfigurable Input Output (RIO) controllers. This platform has huge potential for the development of control systems for power electronics applications and for the development of HIL systems. It is based on an FPGA combined with a microprocessor, although recently, RIO systems were implemented using System on Chip (SoC) technology, which has been demonstrated to work properly in industrial environments.

LabVIEW is the name of the generic NI software development tool. There is a specific module for developing FPGA applications: the LabVIEW FPGA module. It facilitates the generation of FPGA applications using a graphical language that helps non-expert users to create complex and efficient FPGA applications [22].

In this paper, a step-by-step methodology for HIL power converter simulation using the LabVIEW FPGA module is shown. The main difference between this approach and those reported in the literature is that, in the latter, the methodologies used to perform this type of simulation are not clearly explained or do not include enough details for implementation; the advantage of the proposal is that it employs the pipelined technique of LabVIEW and results in a small time step. This methodology is intended to be simple, in order to allow researchers with no experience in languages such as VHDL for RTS to carry out HIL simulations with a reduced cost compared to other tools, but with an accuracy high enough to obtain an appropriate real converter model. In the proposal, LabVIEW is used to allow relatively easy and fast implementation. The methodology consists of five steps: The first step is the design of the power converter; in the second one, the converter model is obtained; during the third step, the model is solved using a numerical method; the fourth step consists of programming an off-line simulation of the model using fixed-point representation; and finally, implementation of the numerical method in the FPGA is conducted. In order to verify the results, a comparison of simulation software such as PSIM, the HIL simulation proposed, and a real converter was carried out using a buck converter and a three-phase voltage source inverter (VSI) as a case study.

## 2. Proposed Methodology

Figure 1 shows the flow diagram of the suggested methodology. As mentioned before, it is composed of five steps. A buck converter and a three-phase VSI are used as a case study. The first step is the design of the converter. Several ways to design both converters can be found in the literature. The second step consists of modeling the power converter. For this purpose, the voltage and current differential equations for every converter switching state are used. In the third step, the differential equations obtained in step two are solved using numerical methods; in this case, by means of the Euler method. During the fourth step, a simulation of the converter using a software tool is carried out, in order to know the best fixed-point representation for the variables involved in the converter, as suggested in [23]. In the last step, FPGA implementation is conducted using the pipelining technique and the single-cycle LabVIEW FPGA mathematical functions. If the simulation result is not satisfactory, the fixed-point representation should be changed and step four should be returned to.

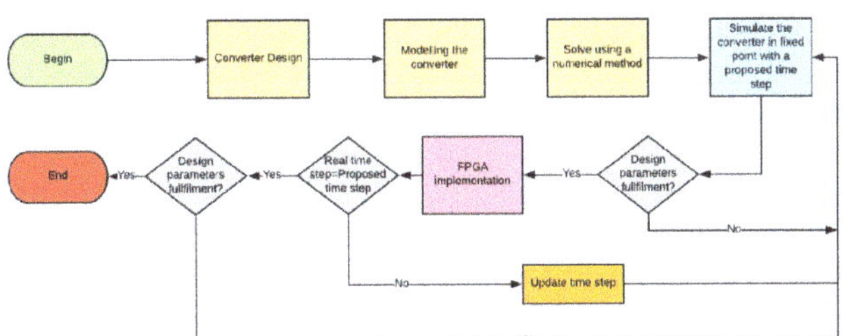


**Figure 1.** Flow diagram of the methodology used to develop the hardware in the loop (HIL) simulation.

### 2.1. Power Converter Design (Step One)

As mentioned above, the buck converter (Figure 2a) and three-phase VSI (Figure 2b) were selected to test the methodology. In order to design the buck converter, the steps proposed in [24] were considered.

The design parameters of the buck converter and the three-phase VSI are shown in Tables 1 and 2, respectively, while their calculated passive components are shown in Tables 3 and 4, respectively.

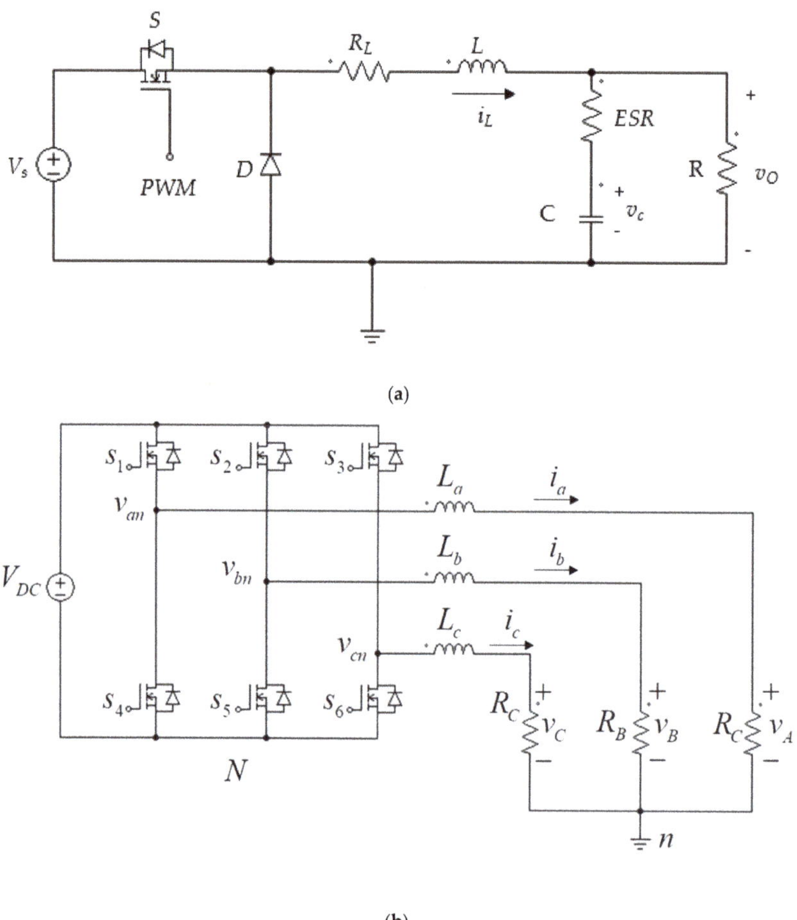


**Figure 2.** Topologies studied: (**a**) Buck converter; (**b**) Three-phase inverter.

**Table 1.** Design parameters of the buck converter.

| Parameter | Value | | | Units |
|---|---|---|---|---|
| | Min | Nom | Max | |
| Switching Frequency | | 40 | | kHz |
| Input Voltage | 20 | 24 | 28 | V |
| Output Power | 2 | | 20 | W |
| Output Voltage Ripple Percentage | | 1 | | % |
| Output Voltage | | 12 | | V |
| Efficiency | | 85 | | % |

**Table 2.** Design parameters of the three-phase Voltage Source Inverter (VSI).

| | Value | | | Units |
|---|---|---|---|---|
| | **Min** | **Nom** | **Max** | |
| Switching Frequency | | 10 | | kHz |
| DC Voltage | | 250 | | V |
| Output Power | 50 | 250 | 500 | W |
| Cutoff Frequency | 750 | 800 | 850 | Hz |

**Table 3.** Buck converter passive components.

| Parameter | Value | Units |
|---|---|---|
| $C$ | 10 | µF |
| $L$ | 500 | µH |
| $R_L$ | 0.75 | Ω |

**Table 4.** Three-phase VSI passive components.

| Parameter | Value | Units |
|---|---|---|
| $L_x$ | 7 | mH |
| $R_x$ | 35 | Ω |

*2.2. Power Converter Modeling (Step Two)*

2.2.1. Buck Converter Model

In order to model the converter, the following considerations were made: the switch ($S$) presents an on-resistance ($R_{DS(ON)}$) and the diode ($D$) is considered as a DC voltage when it is conducting ($V_D$). The buck converter presents three switching states, depending on the control signal and the inductor current. The first state occurs when the switch is turned on and the diode is not conducting, the second state starts when the switch is turned off and the diode begins conducting, and the last state is presented when the inductor current is equal to zero. However, when the power converter is working in Continue Conduction Mode (CCM), only the two first states occur. The differential equations are obtained by using Kirchhoff Law's.

Differential equations for the first state are:

$$\frac{di_L}{dt} = \frac{V_s - i_L\left(R_L + R_{DS(on)}\right) - v_o}{L}$$
$$\frac{dv_C}{dt} = \frac{1}{C}\left(i_L - \frac{v_o}{R}\right) \tag{1}$$
$$v_o = \frac{R}{R + ESR}\left(i_L ESR + v_c\right)$$

where $v_C$ is the capacitor voltage, $i_L$ is the inductor current, $V_s$ is the input voltage, $R$ is the load resistance, $L$ is the inductance, $R_L$ is the inductance series resistance, $C$ is the capacitance, $v_o$ is the output voltage, $ESR$ is the equivalent series resistance of the capacitor, and $R_{DS(ON)}$ is the on-resistance of the switch.

Differential equations for the second state are:

$$\frac{di_L}{dt} = \frac{-i_L R_L - v_o - V_D}{L}$$
$$\frac{dv_C}{dt} = \frac{1}{C}\left(i_L - \frac{v_o}{R}\right) \tag{2}$$
$$v_o = \frac{R}{R + ESR}\left(i_L ESR + v_c\right)$$

Differential equations for the third state are:

$$\frac{di_L}{dt} = 0$$
$$\frac{dv_C}{dt} = -\frac{v_o}{CR}$$
$$v_0 = \frac{v_C R}{R + ESR}$$

(3)

### 2.2.2. Three-Phase Voltage Source Inverter Model

The three-phase VSI considered in this study is shown in Figure 2b. The following considerations were made in order to simplify the model: in this case, all switches are considered as ideal switches and the inductance series resistance is neglected. Certainly, if a better representation is desired, the non-idealities should be considered. The obtained system model is:

$$\frac{di_a}{dt} = \frac{v_{an} - v_A}{L_a}$$
$$\frac{di_b}{dt} = \frac{v_{bn} - v_B}{L_b}$$
$$\frac{di_c}{dt} = \frac{v_{cn} - v_C}{L_c}$$

(4)

where $v_{xn}$ is the phase 'x' inverter voltage, $v_X$ is the phase 'X' voltage, $i_x$ is the phase 'x' inverter current, and $L_x$ is the phase 'x' inductance.

The inverter output voltage is determined by the switching state or vector applied to the three-phase VSI. Each vector is determined by the switches' position. It is considered that the control signals of complementary switches of each inverter leg are the opposite, and then just the upper switches' control signals ($s_1$, $s_2$, and $s_3$) are used in the analysis; a logic 0 means that the device is turned "off" and a logic 1 means that it is turned "on".

The effective voltage applied by the inverter is obtained considering the common-mode voltage, as follows:

$$v_{an} = v_{aN} - v_{Nn}$$
$$v_{bn} = v_{bN} - v_{Nn}$$
$$v_{cn} = v_{cN} - v_{Nn}$$

(5)

where $v_{xN}$ is the inverter voltage referred to node N and $v_{Nn}$ is the common-mode voltage.

The voltage of the inverter referred to the node N is determined by

$$v_{aN} = s_1 V_{DC}$$
$$v_{bN} = s_2 V_{DC}$$
$$v_{cN} = s_3 V_{DC}$$

(6)

where $s_x$ is the control signal of the switch and $V_{DC}$ is the inverter input DC voltage.

The common-mode voltage is defined as:

$$V_{Nn} = \frac{v_{aN} + v_{bN} + v_{cN}}{3} = \frac{(s_1 + s_2 + s_3) V_{DC}}{3}$$

(7)

Table 5 summarizes the values of the three-phase VSI voltages, depending on the vector applied.

Considering the Equations (4) and (7) and Table 5, the dynamic behavior of the system under a switching state is fully determined. For example, if vector $V_1$ is selected, the following equations are obtained:

$$\frac{di_a}{dt} = \frac{(2V_{DC}/3) - v_A}{L_a}$$
$$\frac{di_b}{dt} = \frac{(-V_{DC}/3) - v_B}{L_b}$$
$$\frac{di_c}{dt} = \frac{(-V_{DC}/3) - v_C}{L_c}$$

(8)

A similar procedure may be followed to obtain the equations for any other switching state represented by its corresponding $V_x$ vector.

**Table 5.** Vectors and voltages of the three-phase VSI.

| Vector | Switches | | | Voltages | | | |
|---|---|---|---|---|---|---|---|
| | $s_1$ | $s_2$ | $s_3$ | $v_{aN}$ | $v_{bN}$ | $v_{cN}$ | $v_{Nn}$ |
| $V_0$ | 0 | 0 | 0 | 0 | 0 | 0 | 0 |
| $V_1$ | 1 | 0 | 0 | $V_{DC}$ | 0 | 0 | $V_{DC}/3$ |
| $V_2$ | 1 | 1 | 0 | $V_{DC}$ | $V_{DC}$ | 0 | $2V_{DC}/3$ |
| $V_3$ | 0 | 1 | 0 | 0 | $V_{DC}$ | 0 | $V_{DC}/3$ |
| $V_4$ | 0 | 1 | 1 | 0 | $V_{DC}$ | $V_{DC}$ | $2V_{DC}/3$ |
| $V_5$ | 0 | 0 | 1 | 0 | 0 | $V_{DC}$ | $V_{DC}/3$ |
| $V_6$ | 1 | 0 | 1 | $V_{DC}$ | 0 | $V_{DC}$ | $2V_{DC}/3$ |
| $V_7$ | 1 | 1 | 1 | $V_{DC}$ | $V_{DC}$ | $V_{DC}$ | $V_{DC}$ |

### 2.3. Solving the System Using a Numerical Method (Step Three)

In order to solve differential equations like (1)–(3), a numerical method should be used. Many different methods exist. The forward Euler method is used due to its algorithm simplicity, which leads to less use of hardware resources, allows for a smaller time step simulation, and is easy to implement. Convergence and stability are not an issue since the obtained time step is very small, in the order of hundreds of nanoseconds [25]. Some other integration methods, as the 2nd order Runge-Kutta approach, are more complex, resource-demanding, and should imply a more difficult implementation. These factors could be an obstacle for non-expert users or developers. Commercial HIL systems implement basic integration algorithms, such as trapezoidal or backward Euler algorithms, to find a solution [2].

Generally, the numerical solution for a differential equation can be found using the following algorithm:

$$y_{k+1} = y_k + \theta h \tag{9}$$

where $y_{k+1}$ is the next solution of the method, $y_k$ is the current solution of the method, $\theta$ is the slope of the differential equation, and $h$ is a fixed time step.

Graphically, the method is shown in Figure 3.

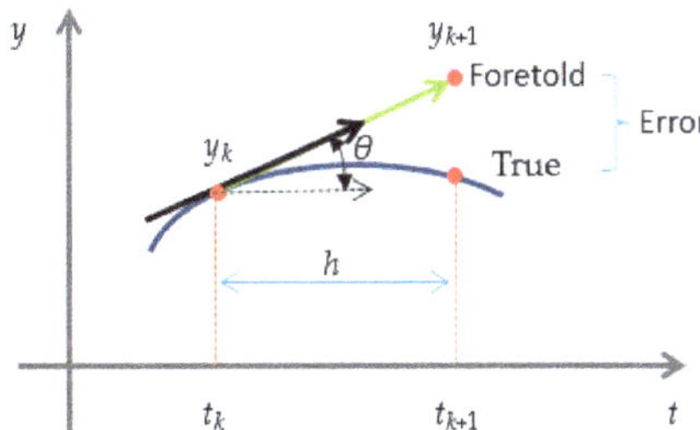


**Figure 3.** Graphic representation of the general solution using numerical methods.

The difference among the mathematical methods is the slope calculation. For the forward Euler method, the first derivative offers a direct estimation of the slope at point $t_k$, so it can be derived as:

$$\theta = \frac{dy}{dt} = f(t_k, y_k) \tag{10}$$

The algorithm for the forward Euler method is:

$$y_{k+1} = y_k + f(t_k, y_k)h \tag{11}$$

In order to verify the concept, in the following sections, a buck converter case study will be analyzed in detail. A similar approach can be used for the three-phase VSI.

Using (11) to solve Equation (1), the solution for $i_{L(k+1)}$ and $v_{C(k+1)}$ is:

$$
\begin{aligned}
i_{L(k+1)} &= i_{L(k)} + \tfrac{h}{L}\left(V_{s(k)} - i_{L(k)}\left(R_L + R_{DS(ON)}\right) - v_{o(k)}\right) \\
v_{C(k+1)} &= v_{C(k)} + \tfrac{h}{C}\left(i_{L(k)} - \tfrac{v_{o(k)}}{R}\right) \\
v_{o(k)} &= \tfrac{R}{R+ESR}\left(i_{L(k)}ESR + v_{C(k)}\right)
\end{aligned}
\tag{12}
$$

where $i_{L(k+1)}$ is the next inductor current, $i_{L(k)}$ is the present inductor current, $v_{o(k)}$ is the present output voltage, $v_{C(k+1)}$ is the next capacitor voltage, $v_{C(k)}$ is the present capacitor voltage, and $V_{s(k)}$ is the current input voltage.

The solution of Equation (2) is then:

$$
\begin{aligned}
i_{L(k+1)} &= i_{L(k)} - \tfrac{h}{L}\left(i_{L(k)}R_L + v_{o(k)} + V_{D(K)}\right) \\
v_{C(k+1)} &= v_{C(k)} + \tfrac{h}{C}\left(i_{L(k)} - \tfrac{v_{o(k)}}{R}\right) \\
v_{o(k)} &= \tfrac{R}{R+ESR}\left(i_{L(k)}ESR + v_{C(k)}\right)
\end{aligned}
\tag{13}
$$

The solution to Equation (3) is:

$$
\begin{aligned}
i_{L(k+1)} &= 0 \\
v_{C(k+1)} &= v_{C(k)} - \tfrac{h}{C}\left(\tfrac{v_{o(k)}}{R}\right) \\
v_{o(k)} &= \tfrac{v_{C(k)}R}{R+ESR}
\end{aligned}
\tag{14}
$$

These equations can be programmed using any commercial software and using different numerical representations. In this paper, the use of fixed-point representation is strongly recommended since the implementation of a floating-point number in the FPGA is very resource-demanding [26].

### 2.4. Simulation of the Solution Using a Numerical Method (Step Four)

A simulation of the Equations (12)–(14) using fixed-point representation was carried out in the LabVIEW environment in order to determine the best numeric representation. The following step is the implementation of the discrete model of the buck converter. Its flow diagram is shown in Figure 4.

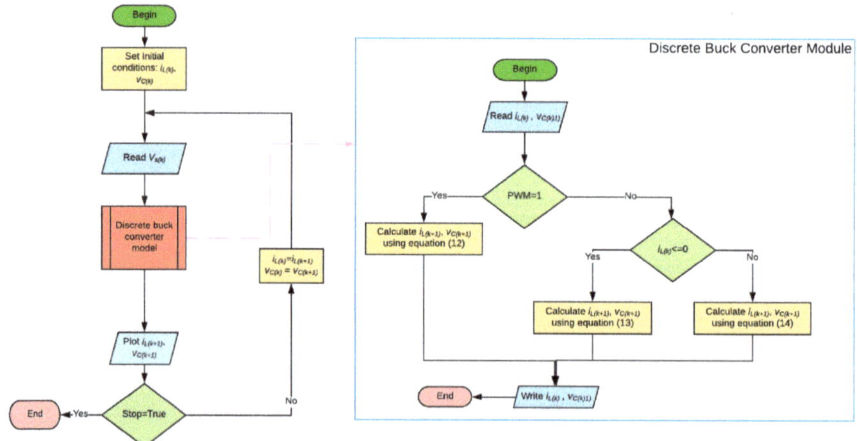


**Figure 4.** Flow diagram for the implementation of the iterative discrete buck converter model.

Figure 5 shows the discrete model programmed in LabVIEW. It can be seen that the equation terms for *h/C* and *h/L* are considered constants and their values were previously calculated for the simulation.

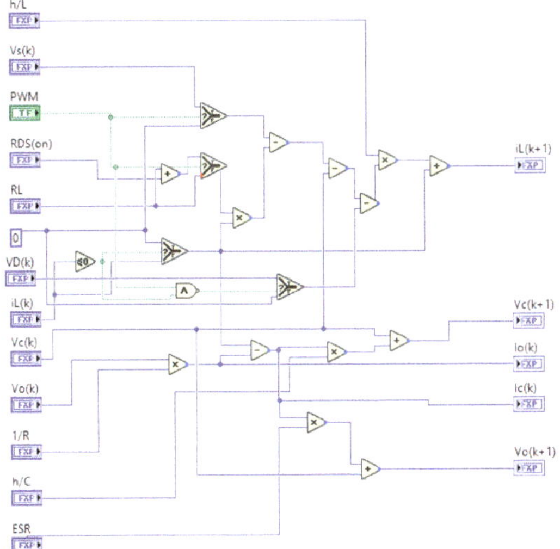

**Figure 5.** Discrete buck converter model programming in LabVIEW.

In Figure 6, the flow diagram for the main routine that solves the buck converter discrete model is shown. In Figure 7, the implementation using LabVIEW is presented. Figure 7a shows a block diagram of the LabVIEW program and in Figure 7b, its front panel is illustrated, where all parameters of the converter model can be updated online. The only unknown parameter is the time step *h*. This parameter will be determined after the FPGA implementation is conducted. For the simulation, a final time step *h* = 200 ns was selected. A value for the time step *h* must be proposed for the first simulation.

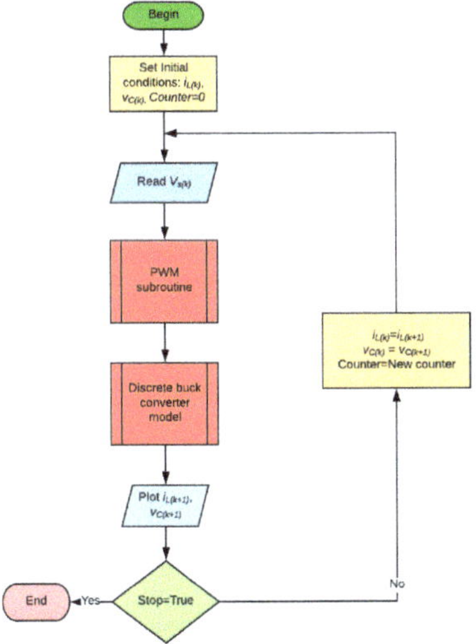

**Figure 6.** Flow diagram of the off-line simulation of the buck converter.

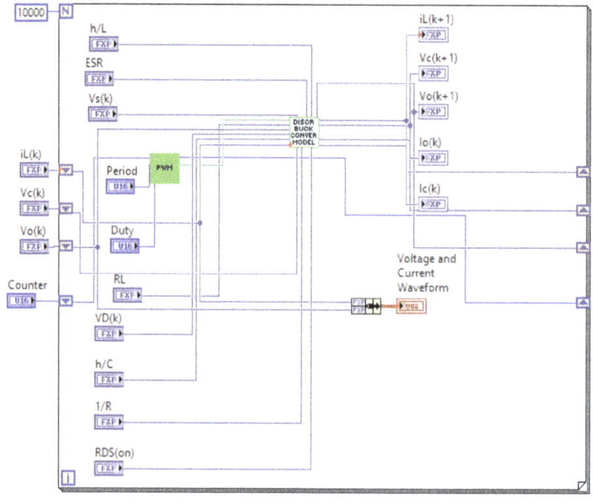

(**a**)

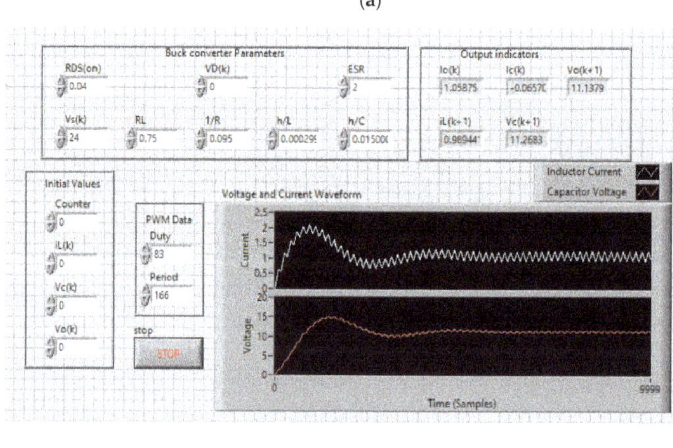

(**b**)

**Figure 7.** Block diagram of the programmed buck converter. (**a**) Discrete model; (**b**) Front panel of the simulated buck converter.

### 2.5. FPGA Implementation (Step Five)

FPGA implementation consists of four modules. In the first module, the difference equations are programmed using the fixed-point representation obtained in the previous step. Figure 8 shows the Equations (12)–(14) implemented in the FPGA using LabVIEW predefined DSP blocks, which perform an operation in a single clock cycle. In LabVIEW software, the developed applications are called Virtual Instrumentation (VI), and the potential of LabVIEW software lies in the hierarchical nature of the VI. After a VI is created, it can be used in the Block Diagram of another VI called a SubVI, which is a section of code that can be called from another program in a higher hierarchical position (similar to a subroutine in high-level programming). There is no limit to the number of layers in the hierarchy. This module is created as a SubVI, which will be called from the main VI. In order to minimize the execution time required to solve the difference equations, the pipeline technique is used. Pipelining is a technique used to increase the clock rate and throughput of an FPGA. Pipelined designs take advantage of the parallel processing capabilities of the FPGA to increase the efficiency of sequential code. To implement a pipeline, the code is divided into discrete steps and the outputs of each step are connected to the inputs of the next step through shift registers. The execution diagram of the pipelining technique is shown in Figure 9a. It can be noticed that the first valid output is obtained after four clock cycles and a valid output is then obtained every clock cycle. In Figure 9b, the execution is shown without a pipeline, in which a new output is obtained after four clock cycles, but it is always necessary to wait for four cycles in order to obtain a valid output. Therefore, the pipelining technique helps code run faster.

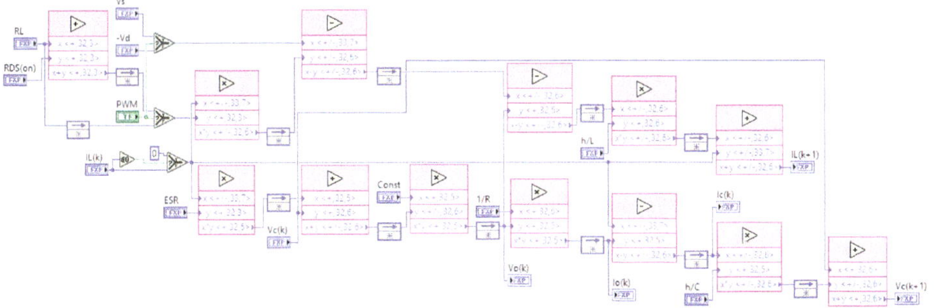

**Figure 8.** Implementation of Equations (12)–(14) in Field-Programmable Gate Arrays (FPGA) using LabVIEW predefined Digital Signal Processor (DSP) blocks.

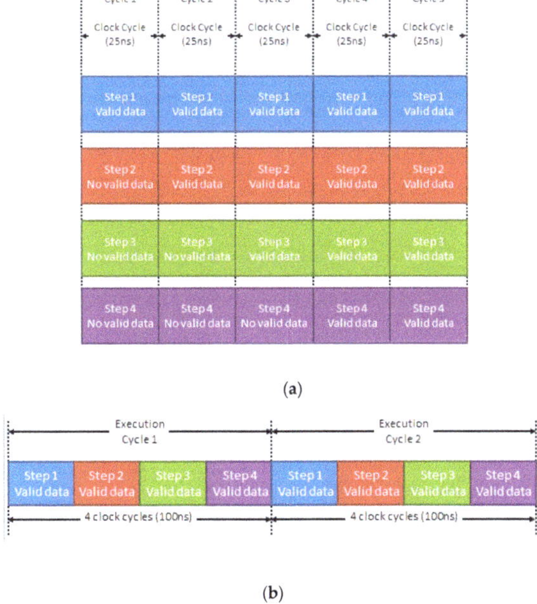

**Figure 9.** Execution diagram of the buck converter discrete model (**a**) using a pipeline and (**b**) without a pipeline.

The second module allows data obtained from the difference equations to be sent to a Digital-Analog Converter (DAC). This hardware converts the digital value of a variable to the voltage (Figure 10), and it is used to observe the response in an oscilloscope. It can be noticed that signal conditioning must be performed for the voltage capacitor value, in order to scale it with a gain of 0.5. This is conducted with the purpose of maintaining the capacitor voltage inside the full-scale voltage of the DAC converter. Therefore, every volt of the capacitor will be shown in the oscilloscope as 0.5 volts.

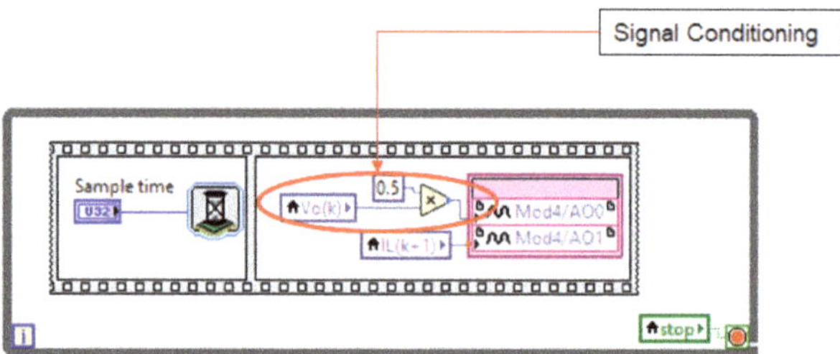

**Figure 10.** Program used to make the analog representation of $v_{o(k)}$ and $i_{L(k)}$ values at the Digital-Analog Converter (DAC) output available.

The third module is shown in Figure 11 and its function is to calculate the elapsed clock cycles per iteration. It consists of a LabVIEW function that counts the clock cycles during an iteration and subtracts them from the previous cycle's count value. In this way, the number of cycles that the program takes to calculate one solution of the model equations is obtained.

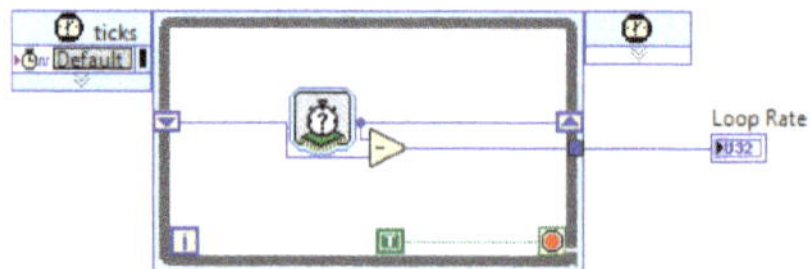

**Figure 11.** Program employed to obtain the clock cycles per iteration.

It can be observed that each part of the code is executed in parallel, reducing the simulation time. In Figure 12, the main VI is shown.

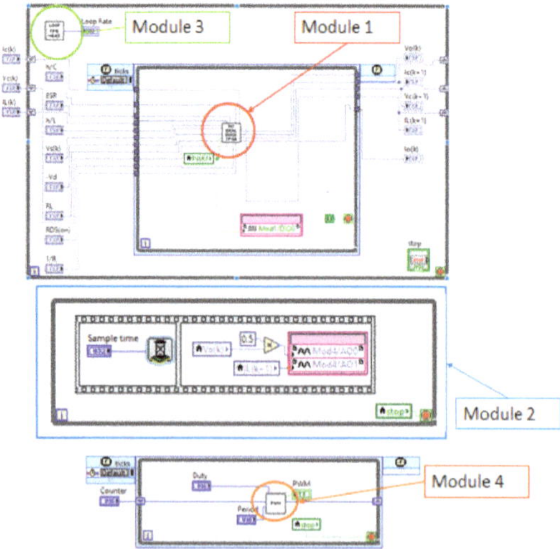


**Figure 12.** The main code of the simulation system, with the different modules.

## 3. Results

In order to verify the proposed methodology, a real-time HIL simulation was implemented and compared against an off-line simulation using PSIM as a reference simulator and also against a real converter. Table 6 indicates the measured parameters of the buck converter.

**Table 6.** Buck converter measured parameters.

| Parameter | Value | Units |
| --- | --- | --- |
| $C$ | 10 | μF |
| $L$ | 500 | μH |
| $R_L$ | 0.75 | Ω |
| $R_{DS(ON)}$ | 0.04 | Ω |
| $V_D$ | 0.1 | V |
| ESR | 2 | Ω |

The hardware used for the RTS is implemented in an NI cRIO-9067 that includes an FPGA Zynq-7020 using a 40 MHz clock. In order to convert the RTS into an HIL simulation, the addition of two modules is necessary: one is a digital I/O (NI9401) used for acquiring the control signals and the other is a DAC module (NI6292) used to generate the output signals. In Figure 13, the front panel of the main VI program for the buck converter is shown. It can be noticed that the loop rate is 6 clock cycles, so the time required to find every solution of the numerical method, with a 40 MHz clock, is 150 ns. The same parameters, which are shown in Figure 7b for the off-line simulation, were used to perform the real-time simulation.

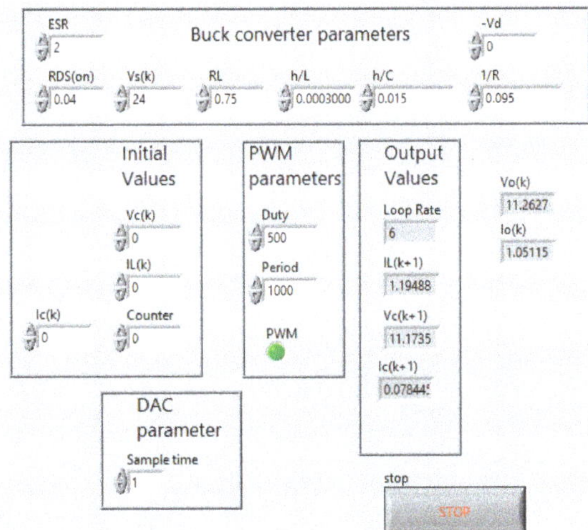


**Figure 13.** Front panel of the real-time simulation.

In Figures 14 and 15, the result of $i_L$ and $v_O$ for the real buck converter, the proposed methodology, and the PSIM simulation can be observed. It can be seen that the signal obtained from the converter and the signal obtained from the Real-Time simulation have a similar behavior, such as the shape, $\Delta i_L$ or $\Delta v_O$, and steady-state value.

In Table 7, the Mean Absolute Error (MAE) from the real converter signals against the Real-Time simulated signals and PSIM waveforms is calculated. These results show that RTS is very accurate compared with a real buck converter and is similar to the off-line simulation by using PSIM with the same parameters.

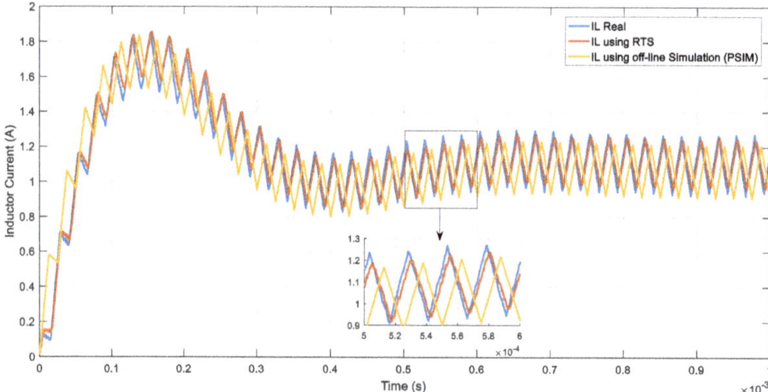

**Figure 14.** Inductor current of the real buck converter: real-time simulation and PSIM.

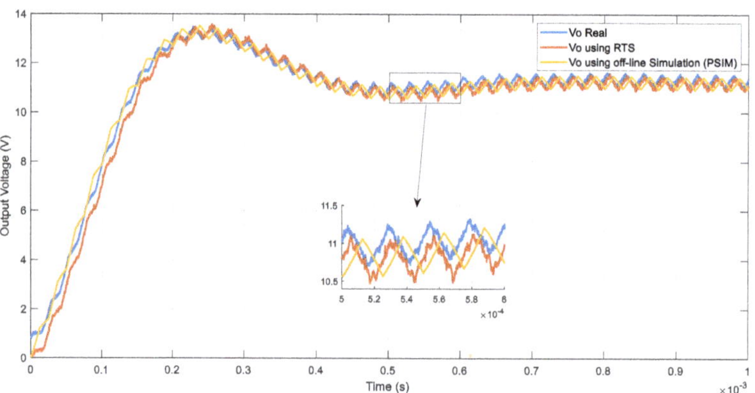

**Figure 15.** Output voltage of the real buck converter: real-time simulation and PSIM.

**Table 7.** Mean Absolute Error (MAE) between the real buck converter and simulation platform.

| Signal | MAE | | Units |
|--------|-----|-----|-------|
| | **Real Converter and RT Simulation** | **Real Converter and PSIM** | |
| $i_L$ | 0.0434 | 0.152 | A |
| $v_o$ | 0.243 | 0.251 | V |

In Figure 16, the inductor current per phase for the three-phase VSI can be seen. For this converter, the time required to solve the model is 750 ns due to the model's complexity.

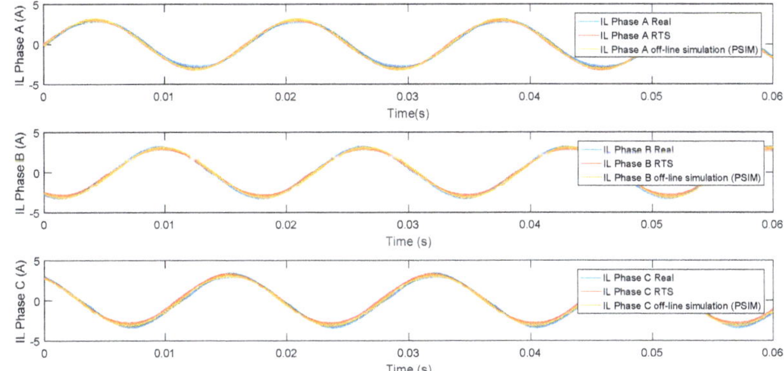

**Figure 16.** Phase current of the three-phase VSI, real-time simulation and PSIM.

As mentioned above, in order to make the RTS an HIL simulation, the calculated output signals can be externally monitored by means of the DAC module outputs. Figure 17 shows an oscilloscope image comparing the signals obtained from the real buck converter and the HIL simulation. Figure 18 is a zoom under a steady-state, in which a time delay of 4 µs between the real converter and the HIL simulation can be observed, which corresponds to the conversion time of the DAC converter used in this test.

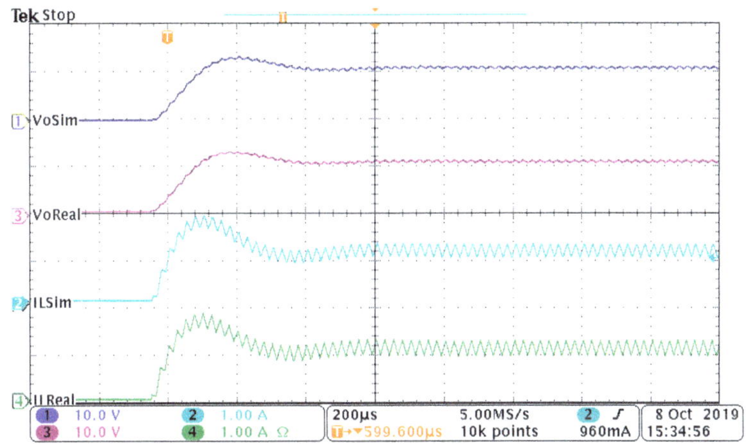

**Figure 17.** Output signals of the real converter and HIL simulation, simulated voltage (Channel 1), simulated current (Channel 2), real voltage (Channel 3), and real current (Channel 4).

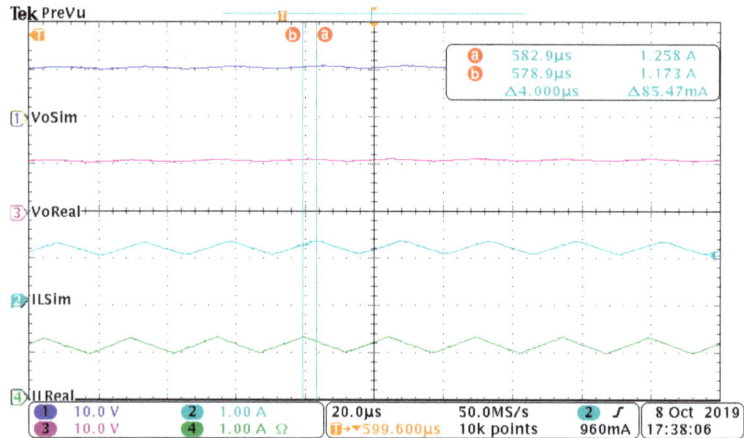

**Figure 18.** Output signals of the real converter and HIL simulation with a time delay, simulated voltage (Channel 1), simulated current (Channel 2), real voltage (Channel 3), and real current (Channel 4).

In Figure 19, the response of the real three-phase VSI and HIL simulation can be observed. The test was conducted under a modulation index change (0.5 to 0.9). In Figure 20, a time zoom is shown; as can be noticed, the behavior of the HIL simulation is close to the real currents ($i_a$ and $i_b$).

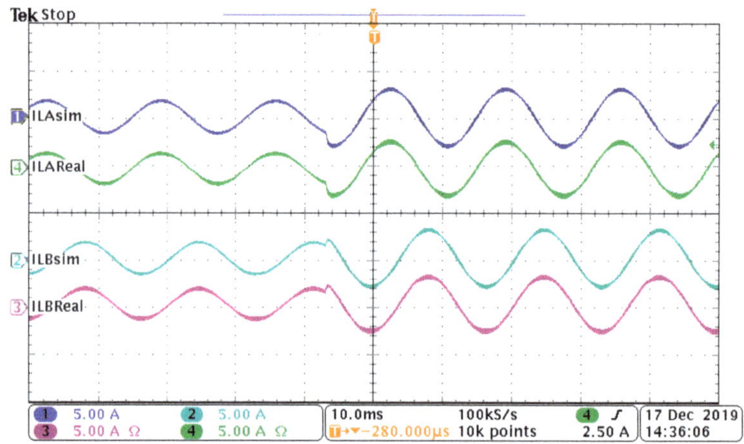

**Figure 19.** Output currents ($i_a$ and $i_b$) of the three-phase VSI and HIL simulation, simulated $i_a$ current (Channel 1), simulated $i_b$ current (Channel 2), real $i_b$ current (Channel 3), and $i_a$ real current (Channel 4).

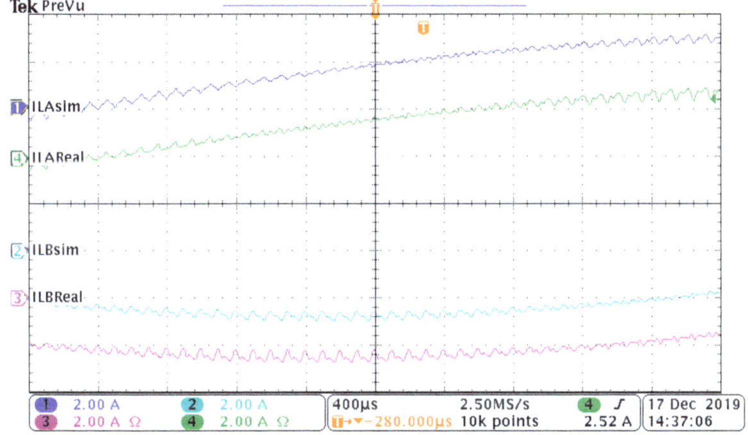

**Figure 20.** Zoomed output currents ($i_a$ and $i_b$) of the three-phase VSI and HIL simulation, simulated $i_a$ current (Channel 1), simulated $i_b$ current (Channel 2), real $i_b$ current (Channel 3), and $i_a$ real current (Channel 4).

These results show that the methodology proposed in this article can be used for a real-time HIL simulation of power converters. This may help to produce a fast and cost-effective test cycle for the development of new controllers. In addition, the proposed methodology for real-time HIL simulation using LabVIEW can be used by students and researchers with basic knowledge regarding numerical methods and LabVIEW programming. The paper is focused on providing an understandable methodology for people with no experience in HIL simulation and text-based languages. The proposal is very easy to follow due to its simplicity; however, accurate results can be obtained.

Another advantage of the proposal is the use of low-cost FPGA hardware (Table 8). The HIL simulation was tested using the CompactRIO platform, which certainly has a lower cost compared to other HIL platforms. However, a MyRIO board could also be employed to obtain similar results; the difference would be the DAC converter included, which is slower than the hardware available in the CompactRIO. In the case of the MyRIO board, switching frequencies should be limited to 20 kHz or below.

**Table 8.** HIL platform prices.

| Hardware | Description | Price (USD) |
| --- | --- | --- |
| cRIO-9067 | Digital platform | $3611 |
| NI-9262 | DAC module | $1372 |
| NI-9401 | Digital I/O | $355 |

An additional advantage is the creation of SubVI that can be shared between the community of scientists and researchers around the world and can enrich this work with other contributions. Once a SubVI is created, it can be used as many times as desired with a simple copy and paste operation.

## 4. Conclusions

In this paper, an understandable methodology for users with no experience in HIL simulation is presented, in order to simulate power converters using a real-time HIL technique. This methodology is based on NI hardware and software, allowing any non-VHDL expert to optimize power electronics converters and their controllers. Besides the simplicity of the proposal, accurate results are obtained.

Using the hardware and following the detailed methodology proposed here, a good HIL simulation for dc/dc converters and three-phase inverters was obtained, the time step of 150 ns for a buck converter and 750 ns for a three-phase VSI was achieved, and both cases used the main clock of 40 MHz. The experimental result and the measured error (Table 7) allowed us to establish that the HIL simulation developed with the use of the proposed methodology is accurate and close to the real converters, so the HIL simulator can be employed satisfactorily by using LabVIEW.

The advantages provided by the proposed method are that the simulation is more realistic since it can be used to evaluate a controller in real-time. Additionally, it is safe for components and people, and the cost of controller implementation and testing, and the development time are reduced. The proposed methodology facilitates understanding of the programming process and it is also an easy and cost-effective way to validate power electronics controllers.

As a drawback of the proposed methodology and tool, it should be mentioned that the modeling technique used would be complex for converters with many switches, which determines the number of equations. A higher number of equations implies more programming, greater hardware resources use, and a long time to calculate a solution. This may mean that the implementation does not fit in the FPGA target and the time step is not short enough to minimize the error presented by the Euler method.

As future work, the connection of more than one digital platform could be made, in order to increase the capacity to simulate more complex systems. Furthermore, a study of different integration methods could be pursued, in order to compare them and verify their performance. Additionally, a bank of power converter models in LabVIEW software may be generated, which could be shared between researchers and students.

**Author Contributions:** Conceptualization, Á.d.C.; Investigation, L.E.; Methodology, J.A.; Supervision, N.V. and J.V.; Writing—original draft, L.E.; Writing—review & editing, N.V., J.V. and Á.d.C. All authors have read and agreed to the published version of the manuscript.

**Funding:** This research was partially funded by the PROMINT-CM: S2018/EMT-4366 program from the Comunidad de Madrid, Spain, and also partially funded by CONACyT, Mexico.

**Conflicts of Interest:** The author declares no conflict of interest.

## References

1. Sarhadi, P.; Yousefpour, S. State of the art hardware in the loop modeling and simulation with its applications in design, development and implementation of system and control software. *Int. J. Dynam. Control.* **2015**, *3*, 470–479. [CrossRef]
2. Faruque, O.M.; Strasser, T.; Lauss, G.; Jalili-Marandi, V.; Forsyth, P.; Dufour, C.; Dinavahi, V.; Monti, A.; Kotsampopoulos, P.; Strunz, K. Real-Time Simulation Technologies for Power Systems Design, Testing, and Analysis. *IEEE Power Energy Technol. Syst. J.* **2015**, *2*, 63–73. [CrossRef]
3. Faruque, O.M.; Dinavahi, V. Hardware-in-the-Loop Simulation of Power Electronic Systems Using Adaptive Discretization. *IEEE Trans. Ind. Electron.* **2010**, *57*, 1146–1158. [CrossRef]
4. Rezkallah, M.; Hamadi, A.; Chandra, A.; Singh, B. Real-Time HIL Implementation of Sliding Mode Control for Standalone System Based on PV Array Without Using Dumpload. *IEEE Trans. Sustain. Energy* **2015**, *6*, 1389–1398. [CrossRef]
5. Matar, M.; Paradis, D.; Iravani, R. Real-Time simulation of modular multilevel converters for controller hardware-in-the-loop testing. *IET Power Electron.* **2016**, *9*, 42–50. [CrossRef]
6. Vekić, M.S.; Grabić, S.U.; Majstorović, D.P.; Čelanović, I.L.; Čelanović, N.L.; Katić, V.A. Ultralow Latency HIL Platform for Rapid Development of Complex Power Electronics Systems. *IEEE Trans. Power Electron.* **2012**, *27*, 4436–4444. [CrossRef]
7. Rodríguez-Andina, J.J.; Valdés-Peña, M.D.; Moure, M.J. Advanced Features and Industrial Applications of FPGAs—A Review. *IEEE Trans. Ind. Inform.* **2015**, *11*, 853–864. [CrossRef]
8. Li, W.; Belanger, J. An Equivalent Circuit Method for Modelling and Simulation of Modular Multilevel Converters in Real-Time HIL Test Bench. *IEEE Trans. Power Deliv.* **2016**, *31*, 2401–2409.
9. Karimi, S.; Poure, P.; Saadate, S. An HIL-Based Reconfigurable Platform for Design, Implementation, and Verification of Electrical System Digital Controllers. *IEEE Trans. Ind. Electron.* **2010**, *57*, 1226–1236. [CrossRef]

10. Jiménez, Ó.; Lucía, Ó.; Urriza, I.; Barragan, L.A.; Navarro, D.; Dinavahi, V. Implementation of an FPGA-Based Online Hardware-in-the-Loop Emulator Using High-Level Synthesis Tools for Resonant Power Converters Applied to Induction Heating Appliances. *IEEE Trans. Ind. Electron.* **2015**, *62*, 2206–2214. [CrossRef]
11. Matar, M.; Iravani, R. FPGA Implementation of the Power Electronic Converter Model for Real-Time Simulation of Electromagnetic Transients. *IEEE Trans. Power Deliv.* **2010**, *25*, 852–860. [CrossRef]
12. Wang, W.; Shen, Z.; Dinavahi, V. Physics-Based Device-Level Power Electronic Circuit Hardware Emulation on FPGA. *IEEE Trans. Ind. Inform.* **2014**, *10*, 2166–2179. [CrossRef]
13. Lucía, Ó.; Urriza, I.; Barragan, L.A.; Navarro, D.; Jiménez, Ó.; Burdío, J.M. Real-Time FPGA-Based Hardware-in-the-Loop Simulation Test Bench Applied to Multiple-Output Power Converters. *IEEE Trans. Ind. Appl.* **2011**, *47*, 853–860. [CrossRef]
14. Yan, Q.; Tasiu, I.; Chen, H.; Zhang, Y.; Wu, S.; Liu, Z. Design and Hardware-in-the-Loop Implementation of Fuzzy-Based Proportional-Integral Control for the Traction Line-Side Converter of a High-Speed Train. *Energies* **2019**, *12*, 4094. [CrossRef]
15. Math Works. Available online: https://la.mathworks.com/products/matlab.html (accessed on 26 October 2018).
16. National Instruments. Available online: http://www.ni.com/es-mx.html (accessed on 26 October 2018).
17. Rosa, A.; Silva, M.; Campos, M.; Santana, R.; Rodrigues, W.; Morais, L. SHIL and DHIL Simulations of Nonlinear Control Methods Applied for Power Converters Using Embedded Systems. *Electronics* **2018**, *7*, 241. [CrossRef]
18. Abbas, G.; Gu, J.; Farooq, U.; Abid, M.I.; Raza, A.; Asad, M.U.; Balas, V.E.; Balas, M.E. Optimized Digital Controllers for Switching-Mode DC-DC Step-Down Converter. *Electronics* **2018**, *7*, 412. [CrossRef]
19. Selvamuthukumaran, R.; Gupta, R. Rapid prototyping of power electronics converters for photovoltaic system application using Xilinx System Generator. *IET Power Electron.* **2014**, *7*, 2269–2278. [CrossRef]
20. Matar, M.; Karimi, H.; Etemadi, A.; Iravani, R. A High Performance Real-Time Simulator for Controllers Hardware-in-the-Loop Testing. *Energies* **2012**, *5*, 1713–1733. [CrossRef]
21. Casolino, G.M.; Russo, M.; Varilone, P.; Pescosolido, D. Hardware-in-the-Loop Validation of Energy Management Systems for Microgrids: A Short Overview and a Case Study. *Energies* **2018**, *11*, 2978. [CrossRef]
22. Ponce-Cruz, P.; Molina, A.; MacCleery, B. *Fuzzy Logic Type 1 and Type 2 Based on LabVIEW™ FPGA*, 1st ed.; Springer: New York, NY, USA, 2016.
23. Goñi, O.; Sanchez, A.; Todorovich, E. Resolution Analysis of Switching Converter Models for Hardware-in-the-Loop. *IEEE Trans. Ind. Inform.* **2014**, *10*, 1162–1170. [CrossRef]
24. Kazimierczuk, M.K. *Pulse-width Modulated DC–DC Power Converters*, 1st ed.; John Wiley and Sons: Chennai, India, 2008.
25. Butcher, J.C. *Numerical Methods for Ordinary Differential Equations*, 2nd ed.; John Wiley and Sons: Padstow, Cornwall, UK, 2008.
26. Sanchez, A.; Todorovich, E.; Castro, A. Exploring Limits of Floating-Point Resolution for Hardware-In-the-Loop Implemented with FPGAs. *Electronics* **2018**, *7*, 219. [CrossRef]

*Article*

# Proving a Concept of Flexible Under-Frequency Load Shedding with Hardware-in-the-Loop Testing

Denis Sodin [1,2,*], Rajne Ilievska [3], Andrej Čampa [1,2], Miha Smolnikar [1,2] and Urban Rudez [3]

[1] Institute Jozef Stefan, Jamova 39, 1000 Ljubljana, Slovenia; andrej.campa@ijs.si (A.Č.); miha.smolnikar@ijs.si (M.S.)

[2] ComSensus d.o.o., Brezje pri Dobu 8, 1233 Dob, Slovenia

[3] Laboratory of Electric Power Supply, Faculty of Electrical Engineering, University of Ljubljana, Trzaska 25, 1000 Ljubljana, Slovenia; rajne.ilievska@fe.uni-lj.si (R.I.); urban.rudez@fe.uni-lj.si (U.R.)

* Correspondence: denis.sodin@ijs.si

Received: 19 June 2020; Accepted: 10 July 2020; Published: 13 July 2020

**Abstract:** It is widely recognized that in the transition from conventional electrical power systems (EPSs) towards smart grids, electrical voltage frequency will be greatly affected. This is why this research is extremely valuable, especially since rate-of-change-of-frequency (*RoCoF*) is often considered as a potential means of resolving newly arisen problems, but is often challenged in practice due to the noise and its oscillating character. In this paper, the authors further developed and tested one of the new technologies related to under-frequency load shedding (UFLS) protection. Since the basic idea was to enhance the selected technology's readiness level, a hardware-in-the-loop (HIL) setup with an RTDS was assembled. The under-frequency technology was implemented in an intelligent electronic device (IED) and included in the HIL setup. The IED acted as one of several protection devices, representing a last-resort system protection scheme. All main contributions of this research deal with using *RoCoF* in an innovative UFLS scheme under test: (i) appropriate selection and parameterization of *RoCoF* filtering techniques does not worsen under-frequency load shedding during fast-occurring events, (ii) locally measured *RoCoF* can be effectively used for bringing a high level of flexibility to a system-wide scheme, and (iii) diversity of relays and *RoCoF*-measuring techniques is an advantage, not a drawback.

**Keywords:** under-frequency load shedding; intelligent electronic device; proof of concept; hardware-in-the-loop testing; real-time digital simulator; frequency stability margin; rate-of-change-of-frequency

---

## 1. Introduction

In most of the developed world, EPSs are known for their robustness, efficiency and extremely high reliability. It comes as no surprise that throughout the decades, this reputation eventually resulted in most of the population taking the electrical energy supply for granted. If it were not for the significant environmental impact of associated activities, such as electrical energy generation and its transmission, no noteworthy changes would be imposed on EPSs in the near future. However, the reality of climate-related concerns has driven the need for technological development to such a level that the conventional electrical energy supply paradigm was placed in front of a challenge none of us had ever faced before. Among others, the sector of electrical energy generation is the most notable example. In a relatively short period of time, significant portions of conventional centralized generating capacities were replaced by distributed renewable energy sources, mostly interfacing with EPSs via semi-conductive power converters [1]. This can be seen as both a blessing and a curse, since power converters are indeed able to provide a convenient fast-acting intervention, but unfortunately not inherently.

The transition towards converter-based generation caused the vast amount of energy temporarily stored within rotating masses of conventional power plants in the form of rotational kinetic energy to decrease quickly [2]. This energy serves as a buffer when sudden active power imbalances occur in the EPS, providing a limited energy source for a temporary load supply [3]. Withdrawing the rotating energy causes generators to decelerate. The deceleration rate depends on the amount of power imbalance and the overall stored rotational kinetic energy. As a consequence, conventional EPSs are not prone to significant *RoCoF*. Consequently, the more we advance toward the smart grid paradigm, the more frequency (among several others quantities), measured from electrical signals (voltage, current), becomes affected [4,5].

For decades, frequency stability was of no concern to transmission system operators (TSOs). An evident proof of this is the applied UFLS schemes in most developed networks around the world. Apart from a few exceptions in countries whose isolated EPS of smaller dimensions is asynchronously connected to larger interconnections, the great majority of them still use an approach introduced decades ago (known as the conventional or traditional UFLS concept). The research community recognized the problem of neglecting UFLS development quite some time ago, which resulted in an impressive number of scientific publications (e.g., the latest [6–11]). Yet, most of those published methods appear complex and require the use of advanced mathematical techniques, several of which are non-transparent. Many of the methods rely either on real-time communication between a vast number of protective devices all across the EPS [12,13] or the assumption of having accurate enough information about the involved EPS inertia [14–16]. As a result, apart from a few individual pilot projects, no significant progress has been made in this field so far, especially concerning practical applicability.

In this paper, we present a procedure for proving the concept of an innovative and practically feasible UFLS [17] in terms of HIL simulations. For this purpose, a real-time digital simulator (RTDS) was applied together with an IED selected for running a local UFLS algorithm. IED is a commonly used term in the electric power industry to describe microprocessor-based controllers of EPS equipment [18]. During the HIL testing of the innovative UFLS method, we have developed a few improvements related to a more straightforward deployment of the method: (i) improved filtering for better resolution and filtering of anomalies, (ii) a proof that despite the *RoCoF* filtering, introduced time delays do not diminish the speed of UFLS, and (iii) easy deployment of the method on the existing IED-based devices. For the reader's convenience, a brief summary of the new UFLS methodology from [17] is provided in Section 2, together with the IED-related implementation challenges and provided solutions for the applicability of the method in a real environment. This is followed by the specifics of the experimental HIL setup in Section 3 and the simulation results in Section 4. Finally, the conclusions are drawn in Section 5.

## 2. Methods

### 2.1. Innovative Use of Rate-of-Change-of-Frequency

The UFLS method, summarized in this section, was previously published in [17] and evaluated in an off-line EPS-dynamics simulation tool PSS Netomac, which is a part of the PSS SINCAL Platform [19]. From this point on, it will be referred to as *innovative UFLS*. Since it is of great importance for the reader to be acquainted with its main philosophy, it seems reasonable to provide its brief explanation in this paper as well. After all, this paper is specifically about proving that the concept works in an environment as close as possible to real-life conditions and providing the required improvements for final deployment.

The first cornerstone of the innovative UFLS method [17] is the on-line calculation of the frequency stability margin $M(t)$ at all frequency relay locations across the EPS. $M(t)$ is a time-dependent variable that is, in its essence, the worst-case estimated time before the EPS frequency $f(t)$ is expected to violate the predefined frequency stability limit ($f_{\text{LIM}}$). For this purpose, an assumption is made that if none of the frequency control/protection functionalities intervene, the $f(t)$ variation will follow the same trend as indicated by the respective $RoCoF(t)$ measurement at the same instant. The local nature of the EPS frequency dynamics that follows an active power imbalance $\Delta P$ incident means that in a selected point in time, the calculated $M(t)$ is different in every busbar in the system (due to the oscillating nature of the EPS [20]). In Figure 1a, the blue and yellow curves depict the EPS frequency response to two unequal $\Delta P$ incidents ($\Delta P_1$ and $\Delta P_2$ respectively) assumed to have taken place in the same busbar. At moment $t = t_1$, respective $RoCoF(t_1)$ indicates different $M(t_1)$ values, depending on the underlying $\Delta P$ incident. It is clear from Figure 1a that $\Delta P_1$ caused a less serious frequency excursion compared to $\Delta P_2$. This manifests in a larger $M(t_1)$ value, an indicator of having more time before $f_{LIM}$ is reached.

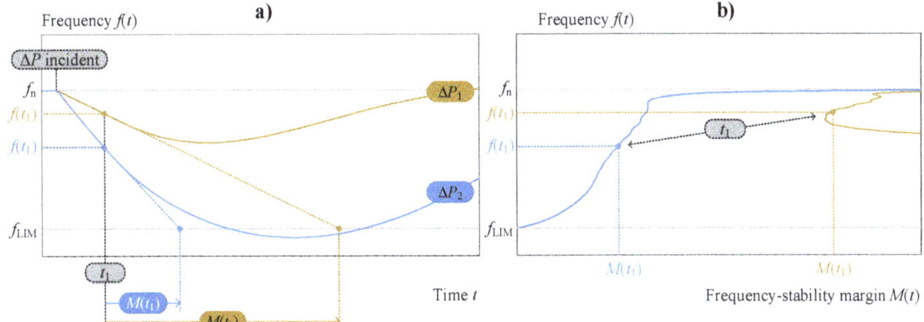


**Figure 1.** EPS frequency $f(t)$ response to different $\Delta P$ incidents (**a**) and the corresponding frequency versus frequency stability margin diagram (**b**).

The second cornerstone of [17] is an innovative representation of conditions in the $f(t)$ versus $M(t)$ diagram (Figure 1b). After the $\Delta P$ incident, the operating point in the $f(t)$-$M(t)$ diagram begins to drift along the corresponding trajectory that points towards one of two possible directions: (i) coordinates ($M = 0, f = f_{LIM}$) in the lower left part of the diagram (blue curve) or (ii) far away from the ordinate axis in the right-hand side of the diagram (yellow curve). This conclusion makes one able to define a new, double-criteria UFLS tripping function for each UFLS stage. The first criterion remains identical to conventional UFLS, being the violation of a predefined frequency threshold $f_{thr}$. The second criterion, on the other hand, is a violation of a predefined stability threshold $M_{thr}$ and is monitored independently of the first. Only when both tripping criteria are simultaneously fulfilled, the trip signal is generated. A simplified logical diagram of the process is presented in Figure 2 since the detailed version is a subject to an international patent application submitted by the University of Ljubljana.

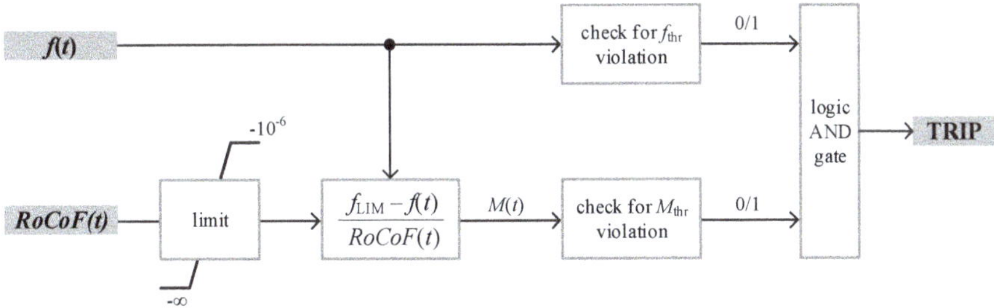


**Figure 2.** A simplified logical diagram of the innovative UFLS process.

For the reader's convenience, it appears reasonable at this point to describe the simplified logical diagram of the innovative UFLS process in Figure 2. Since the main task of UFLS is to handle frequency-decaying situations, the input $RoCoF(t)$ value is limited to only negative values (indicating a frequency drop). Next, the calculation of a frequency stability margin $M(t)$ requires the division with a $RoCoF(t)$ value, which is a consequence of applying a simple linear relationship graphically described in Figure 1a. Since in stable cases the $RoCoF(t)$ value is constantly somewhere around zero, it is reasonable to provide an upper limit of $RoCoF(t)$ to a value slightly below zero. Our analysis showed that this simple precaution successfully stabilizes the division process. In the continuation, the frequency and frequency stability thresholds are checked independently for a potential violation and the trip signal is generated only when both criteria are met simultaneously (logic AND gate).

The most important feature of the described UFLS concept is that any $RoCoF$-related variety (be it the diverse local frequency dynamics in the network or the diverse relay equipment performing $RoCoF$ calculation procedures and filtering techniques) is inherently converted into an important advantage. The resulting variety of calculated $M(t)$ values across the network causes the $M_{thr}$ criterion not being violated simultaneously in different locations. This can be seen as if one is dealing with a much larger number of UFLS stages, which can eventually manifest as a fine-tuning of UFLS (further explained in the continuation of this paper).

### 2.2. Intelligent Electronic Device

For the purpose of RTDS testing of the innovative UFLS methodology presented in Section 2.1, we used an IED based on our in-house developed and fully functional phasor measurement unit (PMU). Therefore, to avoid any misunderstandings, from this point onward this device will be referred to as PMU-IED. In the first step, the PMU-IED was tested for whether its primary function was suitable for use in the innovative UFLS. For evaluation purposes, we used one of the commercially available PMUs, referred to as PMU-COMM. Both PMUs are compliant with the latest revision of IEEE Std C37.118.1-2011 [21]. The advantages of deriving an IED from a PMU are mainly the compelling features that the PMU-IED offers, such as the 200 Hz reporting rate ($F_S$), high accuracy and short response times, input/output (I/O) ports and real-time processing. By modifying the external quantities according to the needs of the innovative UFLS method, we obtained a fully functional IED device that is ready for deployment.

In the first set of tests, we have extensively tested the response of both the PMU-IED and the PMU-COMM along with an internal software-based PMU (since the RSCAD software we are referring to is a part of the RTDS simulator, the internal PMU-8 model is denoted as PMU-RTDS). For the PMU-IED and PMU-COMM, we used maximal reporting rates of 200 Hz and 50 Hz respectively. For benchmarking purposes, the PMU-RTDS was set to have a 200 Hz reporting rate. The first thing that was noticed was that the PMU-RTDS represents an ideal case where any kind of environmental noise does not influence the measured quantities, which is not the case for the PMU-IED and PMU-COMM devices. To support this claim, a simulation of a significant amount of active power deficit, which causes the $RoCoF$ to suddenly change from the value of zero to approximately $-0.4$ Hz/s, is presented in Figure 3a. As expected, the PMU-RTDS generates the $RoCoF$ result almost completely without noise (red curve in Figure 3a), since the only limiting feature is the finite precision of representing floating-point numbers and the implemented frequency estimation method in the hardware. On the other hand, both physical devices under test, the PMU-IED (cyan curve) and the PMU-COMM (black curve), show a high degree of fluctuation/noise around expected values. This is because, in its essence, $RoCoF$ is the derivate of frequency. As a result, any small and fast variations of frequency are most likely to be amplified in $RoCoF$ [22]. As expected, these fluctuations have a high impact on the innovative UFLS (see Figure 3b). Therefore, $RoCoF$ filtering needed to be addressed first.

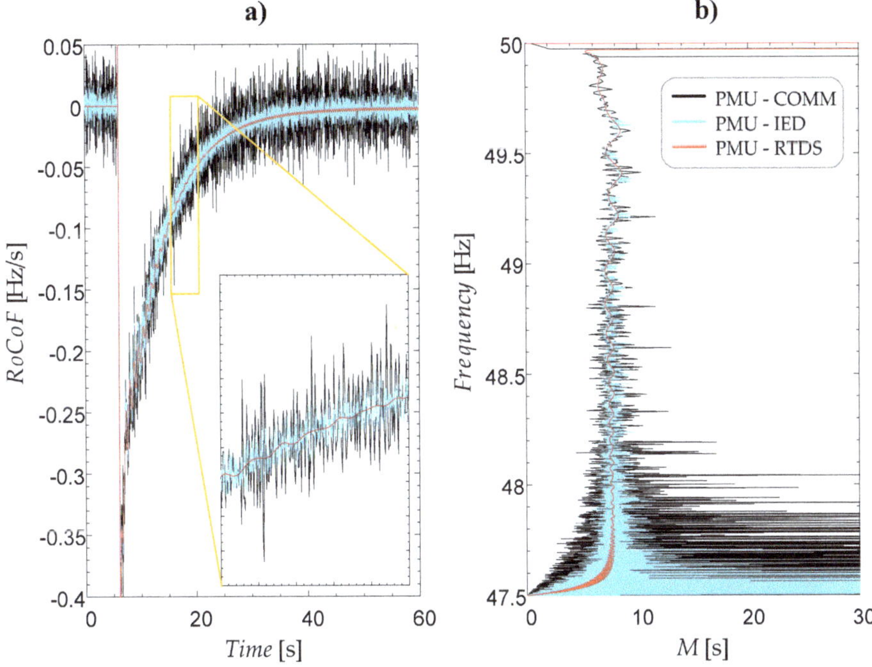

**Figure 3.** The *RoCoF(t)* report of three applied sources (PMU-IED, PMU-COMM and PMU-RTDS) after an RTDS simulation of an active power deficit (**a**) and a corresponding frequency versus frequency stability margin diagram (**b**).

*2.3. Rate-of-Change-of-Frequency Filtering*

By definition, *RoCoF* is a time derivate of the frequency $f$:

$$RoCoF = \frac{df}{dt} \approx \frac{\Delta f}{\Delta t} = \Delta f \cdot F_S \tag{1}$$

Thus, high fluctuations are expected in *RoCoF* in a noisy environment when one is dealing with a high reporting rate $F_S$. To decrease the noise, we could lower $F_S$, but this would lead to a slower response of the UFLS system depending on *RoCoF*. This is why it is of utmost importance to find the optimal response of the UFLS system while still getting meaningful values for *RoCoF*. In Figure 4, the 200-ms-long windowing function was added to the *RoCoF* calculation (to both the PMU-IED and the PMU-COMM) that averages all samples (moving average) in the window. Averaging reduces the system response capabilities since, with a longer window, it passes only the DC component in its limit in the frequency domain. As seen in Figure 4, the moving average filter greatly improves the *RoCoF* reported by both devices. On the other hand, this is achieved on account of having a higher output delay (around half of the window length, i.e., ≈100 ms, which can be noticed as a shift of the PMU-IED and PMU-COMM curves with respect to the PMU-RTDS), which is also reflected in the *f(t)-M(t)* diagram (Figure 4b).

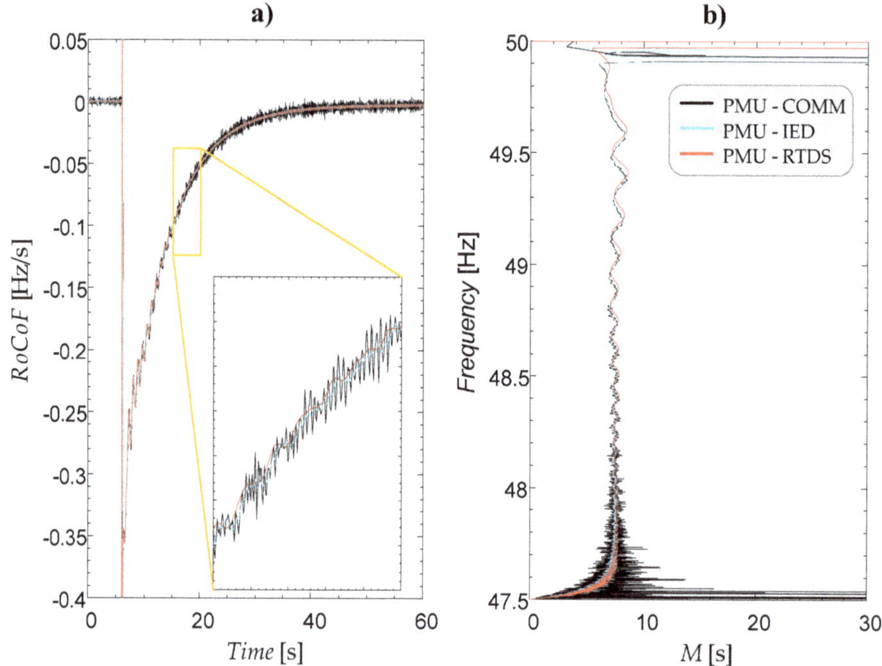



**Figure 4.** The improvement in reported *RoCoF(t)* after applying a moving average filter to the PMU-IED and the PMU-COMM (**a**), and a corresponding frequency versus frequency stability margin diagram (**b**).

As expected, the PMU-COMM shows lesser improvement compared to the PMU-IED since it has a lower $F_S$ (50 Hz) and the noise reduction is related to the square root of points used in the window [23].

In the next phase, we put our focus on the *RoCoF* report dynamics after a discrete active power balancing event occurs in the EPS (either initial event causing frequency to deviate from the nominal value or a UFLS intervention). In this case, the simulated power imbalance incident was twice as large, so the *RoCoF* reaches values up to −0.80 Hz/s. Such events taking place result in momentarily distorted measured electrical quantities (voltages, currents) and, consequently, both the frequency and *RoCoF* as well. For the sake of clarity, a set of raw measurements (without the moving average filtering described above) are presented in Figure 5 for the case where the conventional UFLS scheme operates at thresholds of 49.0, 48.8, and 48.6 Hz after the EPS was subject to a power imbalance incident.

After each event related to power balance, numerical oscillations are present in both the frequency and *RoCoF* reports (Figure 5a), which is manifested in the *f(t)-M(t)* diagram (Figure 5b) as a significant deviation from the expected trajectory. These momentary oscillations needed to be filtered out in frequency and especially in *RoCoF*. The performed analysis showed that after every event, the *RoCoF* momentarily increased in amplitude for several orders. Thus, one of the non-linear filtering techniques is required; otherwise (by using a moving average or similar linear-based filter), the unwanted oscillation will disperse along the time axis, resulting in the need for a long linear-based filter that would generate undesirable delays. Eventually, the use of such *RoCoF* in innovative UFLS might cause unexpected load tripping (shedding) and fail the power re-balancing.

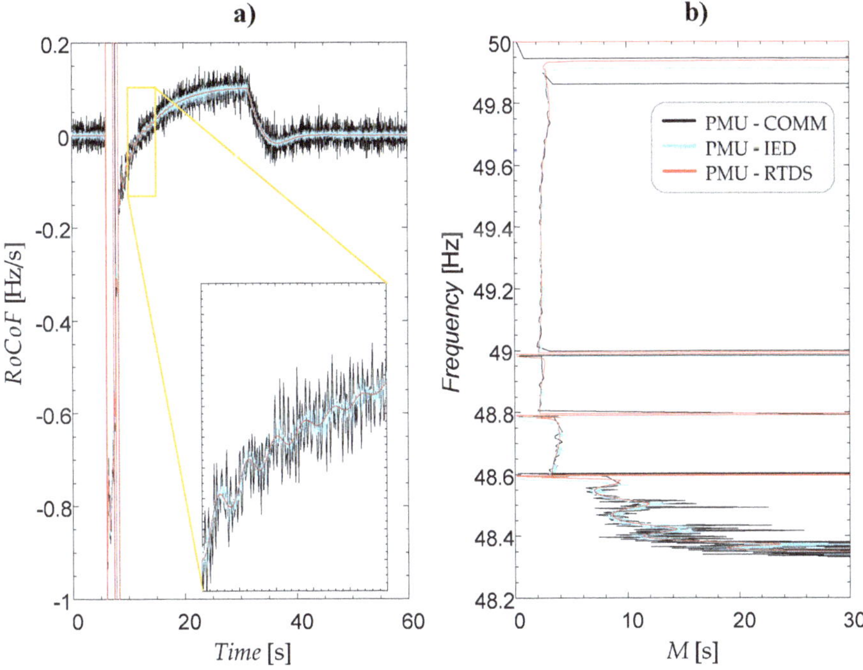



**Figure 5.** The *RoCoF(t)* report of three applied sources (PMU-IED, PMU-COMM and PMU-RTDS) after the RTDS simulation of an active power deficit and three UFLS interventions (**a**) and a corresponding frequency versus frequency stability margin diagram (**b**).

After careful consideration, we implemented the median filter whose impact on *RoCoF* reports along with the moving average filtering is depicted in Figure 6. Figure 6b shows that the described approach is successful in filtering out all disruptive anomalies and noise without scarifying the response time. Finally, the PMU-IED was modified accordingly by changing the default output filter (the one that complies with the IEEE C37.118 standard [24,25]) with a combination of filters as shown in Figure 7. A thorough investigation revealed that the depicted sequence of filtering represents an optimal combination. First, the moving average filter (window length of 40 ms) improves the *RoCoF* resolution and decreases the noise. Next, the median filter (window length of 110 ms), which proved itself to be much more efficient with an existing 40 ms pre-filtering by the moving average filter, handles the numerical anomalies. Finally, the second part of the moving average filtering (window length of 100 ms) can further improve the resolution of the measurement for the sake of time delay.

As far as the frequency is concerned, a median filter (window length of 110 ms) was adequate to solve the issue without causing any unwanted UFLS behavior.

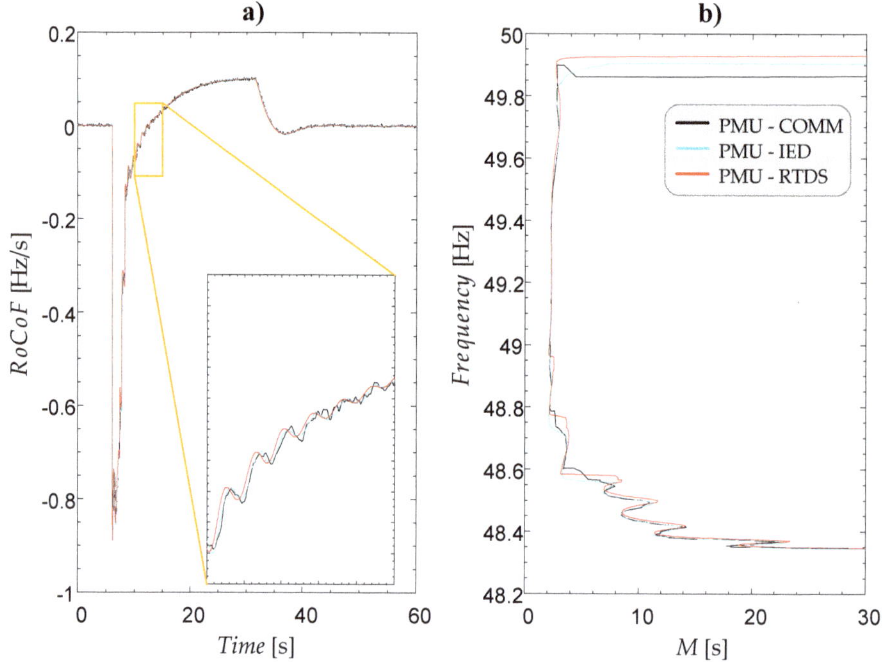

Figure 6. The improvement in reported *RoCoF* dynamics after applying a median filter to the PMU-IED and the PMU-COMM (**a**), a corresponding frequency versus frequency stability margin diagram (**b**).

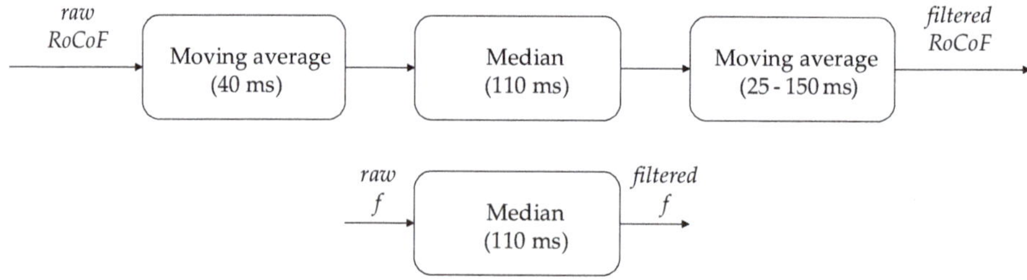

Figure 7. The configuration of filtering applied to the reported frequency *f* and *RoCoF*.

## 3. Experimental Setup

### 3.1. HIL Setup—Overview

The assembled HIL setup consists of five hardware elements: (i) an RTDS simulator, (ii) an Omicron CMS-156 amplifier, (iii) a PMU-IED, (iv) a global-positioning system (GPS) antenna, and (v) a personal computer (PC) which runs the *RSCAD* software and represents a human-machine interface for accessing the simulated part of the setup. An overview of a HIL arrangement is shown in Figure 8.

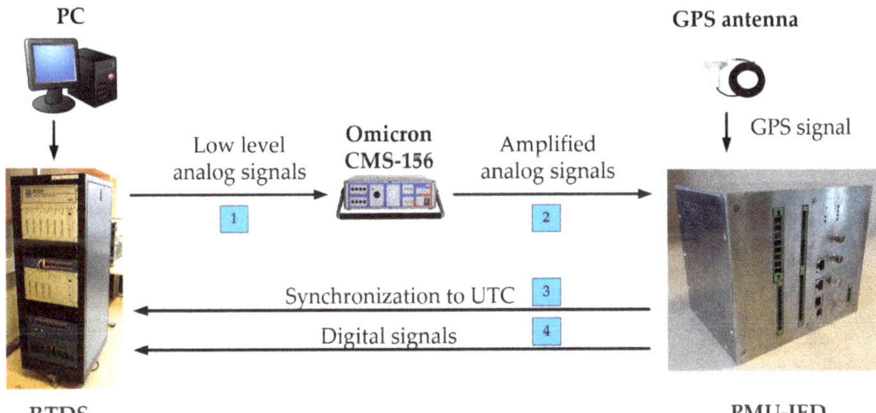


**Figure 8.** A graphical representation of a real-time digital simulation setup.

### 3.1.1. Real-Time Digital Simulator

The RTDS simulates the operation of EPSs using parallel processing architecture that is specifically designed to solve the transient electromagnetic simulation (EMT) algorithm in real-time. The RSCAD interface software allows the user to perform all the necessary steps to prepare and run simulations as well as visualize the results. Apart from processor cards, which present the main driving core of the simulator itself, the RTDS contains analogue and digital input/output channels that enable communication with external physical devices. This is achieved by means of a specialized analogue output card that is used to produce analogue waveforms (see blue-shaded annotation 1 in Figure 8) expected by the PMU-IED terminals. However, it is worth noting that the GTAO card produces a low-level analogue signal with a maximum of ±10 V peak. Even though some IEDs can accept such low-level signals directly (including the PMU-IED), this is not generally the case. To create a more generalized testing platform, low-level signals were scaled up with an amplifier (see blue-shaded annotation 2 in Figure 8) before being fed to the PMU-IED. In contrast, for the purposes of the UFLS scheme, the shedding signal generated by the PMU-IED is in a digital format (see blue-shaded annotation 4 in Figure 8). A specialized digital input card in the RTDS simulator was therefore used to bring the shedding/trip command back into the simulation and close the testing loop.

Another very important aspect of HIL testing is to keep all participating devices in synchronism. A so-called GTSYNC card ensures that the simulator time-step remains locked to the GPS signal, using either the IEEE 1588 precision time protocol (PTP), 1 pulse per second (PPS) or an inter-range instrumentation group time code format B (IRIG-B) signal. The proposed test setup used the PPS signal provided by the PMU-IED as the synchronization source (see blue-shaded annotation 3 in Figure 8).

### 3.1.2. Omicron CMS-156 Amplifier

The omicron CMS-156 amplifier features a three-phase voltage with neutral outputs and a three-phase current with neutral outputs that are galvanically isolated from the inputs. The amplifier can produce three current signals of up to 25 A each and three voltage signals of up to 250 V each. At a nominal frequency of 50/60 Hz, the amplifier introduces 1.88°/2.26° of phase lag for current outputs and 1.95°/2.34° of phase lag for voltage outputs respectively. For the purposes of HIL testing, phase lags need to be properly accounted for. In the proposed HIL setup, the calibration offset angles of the PMU-IED were set accordingly.

### 3.2. HIL Setup—Intelligent Electronic Device

A physical PMU-IED, described in Section 2.2, requires a three-phase voltage input that is used for calculating the frequency $f(t)$ and $RoCoF(t)$. The voltages that have to be appropriately amplified (see Section 3.1) are supplied to the PMU-IED by the RTDS simulator. On the output side, the PMU-IED device enables six digital outputs (logical type 0/1) used as trip signals for each of the six UFLS stages (see Table 1). These signals control the circuit-breaker models within the RTDS, influencing the connection status of an individual EPS load.

**Table 1.** Conventional UFLS setting (frequency thresholds $f_{thr}$) along with supplemented frequency stability margin ($M_{thr}$) thresholds.

| UFLS Stage Number | $f_{thr}$ [Hz] | $M_{thr}$ [s] | EPS Load Decrease [%] |
|:---:|:---:|:---:|:---:|
| 1. | 49.0 | 6.0 | 10 |
| 2. | 48.8 | 5.0 | 10 |
| 3. | 48.6 | 4.0 | 10 |
| 4. | 48.4 | 3.0 | 10 |
| 5. | 48.2 | 2.0 | 10 |
| 6. | 48.1 | 1.0 | 5 |

However, our laboratory capacities enabled us to include only one physical PMU-IED device in the HIL experiments. In order for the overall UFLS scheme to function correctly, we had to build a RTDS computer model of all PMU-IED devices that were not physically present in the laboratory. For this purpose, a computer model of innovative UFLS was created in *C-builder*, running on $f(t)$ and $RoCoF(t)$ measurements provided by the PMU-RTDS internally within the RTDS.

The RTDS simulator used in this paper can be handled through an RSCAD software environment [26]. Apart from modelling elementary EPS elements (main high-voltage components, protection and control functions), RSCAD enables the user to create custom-built components in a module C-builder. Once created, any user-defined component is used in the same manner as any other component from the RSCAD library.

To create a representative copy of the PMU-IED in the RTDS simulator, the same amount of UFLS stages had to be allowed in the model as well. To make it more general, we decided to allow the user to specify an arbitrary number of UFLS stages (up to six). A corresponding number of outputs is also enabled. Apart from that, a $f_{LIM}$ and all threshold values ($f_{thr}$ and $M_{thr}$) can be provided separately, together with window widths for filtering described in Figure 7.

To verify the computer model of the PMU-IED running innovative UFLS logic, we performed a large number of tests. Eventually, we were able to prove a satisfying compliance of the physical and the modelled PMU-IED.

### 3.3. HIL Setup—Electric Power System

We adopted an IEEE 9-bus benchmark EPS as a basis, whose parameters and other specifics can be found in [3]. This benchmark model consists of three synchronous generators (denoted by G1, G2 and G3 in Figure 9) together with corresponding step-up transformers, six transmission lines and three equivalent loads (denoted by L5, L6 and L8). All machines are controlled by their respective governor and excitation controllers. The total installed generating power is 567.5 MVA and 10% of each machine's rated power is selected to attribute to frequency control purposes, whereas loads are modelled as a constant impedance.

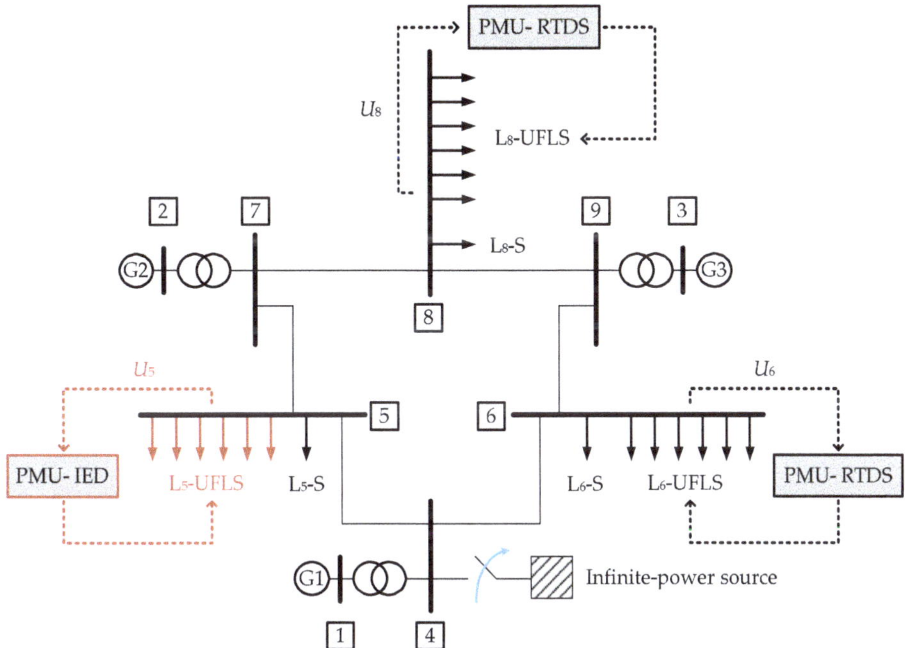


**Figure 9.** A single-line diagram of a IEEE 9-bus test system [3], used for the RTDS HIL testing.

To use this benchmark model for UFLS testing, several modifications had to be applied (Figure 9). First, each of the three equivalent loads (L5, L6 and L8) was split into seven partitions. Six of them correspond to individual UFLS stages (see Table 1) denoted by $L_x$-UFLS where subscript "x" represents the number of the corresponding busbar. The seventh partition, however, represents the remaining aggregation of all the consumers not included in the UFLS scheme ($L_x$-S). Second, an infinite power source was additionally introduced to the model, busbar 4, to be more specific. Its disconnection was considered the main event causing active power deficit conditions in the newly formed island. With this, many different power conditions can be simulated by changing the steady-state production of all three generating units prior to the main event. An infinite power source in all cases supplies/consumes enough power to meet the power balance in a steady-state. Once disconnected, an EPS island is formed with a certain imbalance between the production and the consumption of active and reactive powers. In this paper, circumstances with a lack of active power were simulated, which cause an EPS frequency to decay and consequently trigger UFLS.

Lastly, devices performing UFLS were included in the model. A single physical PMU-IED was fed by voltage $U_5$ on busbar 5 and was able to disconnect loads $L_5$-UFLS according to Table 1. The other two load busbars (6 and 8) were monitored and controlled with a PMU-RTDS component within a simulated environment, fed by voltages $U_6$ and $U_8$, respectively.

## 4. Results

Both the conventional six-stage UFLS scheme currently used in Slovenia and the innovative UFLS scheme were applied in simulation scenarios. Specifics of both UFLS settings are provided in Table 1. Frequency thresholds $f_{thr}$ are taken from the Slovenian Grid code [27], and additional frequency stability margin thresholds $M_{thr}$ are selected with a one-second span between them since the analysis showed there are no special differences when selected otherwise.

In the base case, steady-state conditions before the main event were set so that the total loading of the benchmark IEEE 9-bus system model (Figure 9) was 315 MW and the overall generating power was 319.6 MW. Apart from the base case, this research included 65 additional simulated cases (66 altogether), in which the total generation capacity before the main event was gradually decreased by 1% per case. In this way, the amount of power provided by the power source (and consequently power mismatch between the generation and the consumption) in the moment of a simulated switching event increased with each consecutive case. This was repeated to the point where all available frequency control and protection mechanisms (primary frequency control and UFLS) were insufficient to prevent frequency decay bellow the $f_{LIM}$.

In Figure 10, an overview of the results extracted from the entire set of 66 simulated cases is provided. Cases are listed on the horizontal axis. Conventional UFLS results are depicted with cyan markers, whereas black markers correspond to innovative UFLS. Figure 10a shows the percentage of the EPS de-loading due to UFLS. Conventional UFLS is activated in 79% of all cases (52 out of 66). Among these, innovative UFLS keeps more load supplied in 81% of cases (42 out of 52) while still being successful in achieving frequency stabilization. This is marked by a green-shaded area between the two curves in Figure 10a.

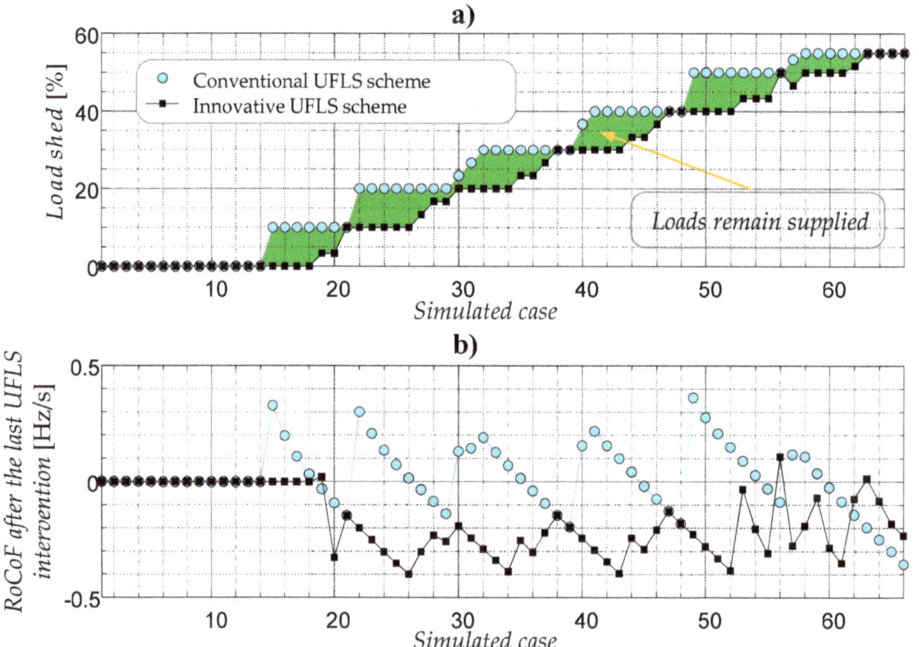


**Figure 10.** Overview of HIL results extracted from the entire set of 66 simulated cases; the total amount of disconnected consumption in percentages (**a**) and recorded RoCoF after the last UFLS intervention (**b**).

Another thing worth mentioning is that with a conventional UFLS scheme, shedding in all three busses will nearly always take place simultaneously (92% of cases/48 out of 52 cases) and, consequently, the entire UFLS stage will be tripped. However, partial UFLS activation is much more frequent for the innovative UFLS scheme (31% of cases/16 out of 52), as can be observed in Figure 10a. The reason for this is that even though the $f_{thr}$ and $M_{thr}$ values are set to same values for all three devices participating in a UFLS scheme, the variety of calculated $M(t)$ values is much higher across the network than the variety of calculated $f(t)$ values. With three load busses, we therefore introduced three "substages" for each of the UFLS stages, which can be seen as fine-tuning power imbalance with UFLS. Furthermore, in systems with more load busses, even finer tuning could be achieved due to those inherently introduced substages.

However, observing the amount of disconnected load is only one of several ways to highlight the superiority of innovative UFLS. When evaluating the efficiency of UFLS, different authors took different approaches in the existing literature (established in [28]), ranging from ranking the schemes according to the time required before the frequency is returned back to the nominal value, to comparing post-event steady-state frequency offset or by observing frequency overshoots that might appear after UFLS is activated. It was concluded in [28] that there is no unified scoring metric for this. One thing is for certain though: the main objective of UFLS, by its definition, is to prevent a further frequency drop [29] or, in other words, restore the balance [30] between generated and consumed active power in the EPS in due time. Plainly speaking, this means that UFLS is to stabilize the frequency and bring *RoCoF* to a value as close as possible to zero.

For this reason, Figure 10b shows the *RoCoF* value after the last UFLS intervention took place for each case (this might be any of the stages, depending on the imbalance conditions). This observation is independent of any other influential mechanisms, especially specifics of the governor and its control, that often have a significant impact on frequency response after UFLS is activated. The value of zero was dedicated to *RoCoF* results when UFLS was not activated at all. When conventional UFLS was activated, the re-balancing was unsuccessful in 52% of the cases (29 out of 52), since the last shedding resulted in having a surplus of power generation. One might treat this as if dealing with a flip of a coin, i.e., pure guessing. On the other hand, innovative UFLS paused its intervention in the time since it successfully recognized that the frequency is about to be stabilized without triggering the last stage. The *RoCoF* values after the last UFLS intervention are kept below zero (average value around −0.25 Hz/s) in 98% of the cases (46 out of 47) and frequency control is left to continuously fine-tune the power balance. At this point, it should be stressed that the paused stage triggering does not mean that the stage is permanently blocked. Quite the opposite; shedding hibernates until conditions for its triggering are met. This means that if the available frequency control is exhausted, the stage will be triggered later on, once the $M(t)$ value becomes low enough. An example of such conditions is simulated case no. 56. However, as can be seen from Figure 10, such situations are extremely rare and yet, they nevertheless successfully solve under-frequency conditions.

For simulated case no. 49, a time-domain response of the EPS frequency and *RoCoF* is depicted in Figure 11a and Figure 11b respectively. From observing the diagrams, it becomes evident that at the moment of reaching the frequency threshold of the fifth UFLS stage, innovative UFLS detects that there is still enough time available before frequency instability is endangered ($M \cong 8$ s, Figure 12). As a result, the fifth UFLS stage is paused, which allows the frequency control to finish the frequency stabilization process by itself. If that was not the case, the fifth UFLS stage would have been triggered later on when the frequency would have approached the selected frequency stability limit $f_{\text{LIM}}$ and, consequently, the $M$ criterion of the fifth stage would have been triggered.

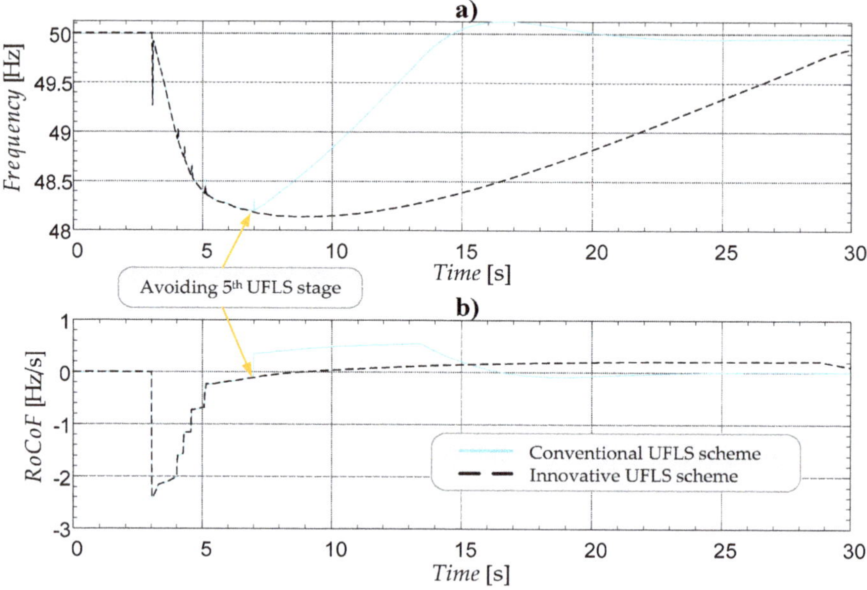



**Figure 11.** A time-domain response of the EPS frequency *f(t)* (**a**) and *RoCoF*(t) (**b**), corresponding to simulation case 49.

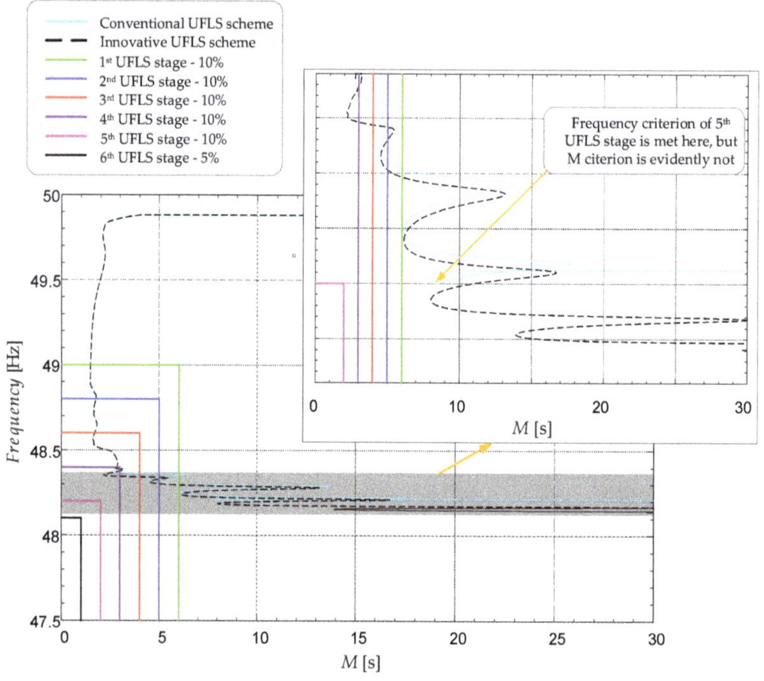



**Figure 12.** Frequency *f(t)* versus frequency stability margin *M(t)* diagram of simulation case 49.

There is another essential thing that has to be discussed. When the frequency is decreasing with a significant *RoCoF(t)*, a prompt triggering of UFLS is required since any additional time delay might be unacceptable. This is inherently achieved in conventional UFLS (triggering solely according to the frequency criterion $f_{thr}$) and it is vital to be aware that this is also the case with innovative UFLS. The initial concern with the innovative UFLS scheme was related to extreme scenarios, in which the curve on the *f(t)-M(t)* plane is expected to make a rather abrupt jump from the upper right corner (steady-state conditions) to the upper left corner of the diagram once power imbalance appears. The delayed recognition of such critical conditions due to two additional moving average filters (Figure 7) could, therefore, translate into $f_{thr}$ of a certain stage being violated before the corresponding $M_{thr}$. This would in turn be observed as shedding at lower frequency compared to a conventional UFLS scheme. However, our testing proved that even when dealing with an extreme initial *RoCoF* ($\cong$−10 Hz/s), the $M_{thr}$ criterion is still violated sufficiently prior to the $f_{thr}$ criterion. In Table 2, the actual frequency at which the $M_{thr}$ was met is given for each UFLS stage. Evidently, *M* criteria of all UFLS stages were met before the frequency even reached the $f_{thr}$ value of the first UFLS stage (0.483 Hz margin in a worst-case scenario—see the last column in Table 2). This proves that the innovative UFLS scheme would respond to such an extreme event identically as the conventional scheme, with no additional time delay.

**Table 2.** Frequency values at which individual $M_{thr}$ thresholds are violated (*RoCoF* = −10 Hz/s).

| UFLS Stage Number | $f_{thr}$ [Hz] | Frequency at Which $M_{thr}$ Criterion Is Met [Hz] | Margin [Hz] |
|---|---|---|---|
| 1. | 49.0 | 49.4830 | 0.4830 |
| 2. | 48.8 | 49.4830 | 0.6830 |
| 3. | 48.6 | 49.4623 | 0.8623 |
| 4. | 48.4 | 49.4623 | 1.0623 |
| 5. | 48.2 | 49.4364 | 1.2364 |
| 6. | 48.1 | 49.3415 | 1.2415 |

## 5. Conclusions

In the most common viewpoint of smart grids, the decentralization of electric power generation is usually accompanied by the centralization of protection functions. In this paper, we proved the concept of a *RoCoF*-based innovative UFLS which does not require centralization, yet still provides a high level of efficiency and flexibility. The main conclusion of this paper is that the innovative UFLS is deemed feasible and robust in practical applications. A proposed transition from conventional towards innovative UFLS is extremely simple; (i) an exceptionally simple logic based on *RoCoF* has to be incorporated in each under-frequency relay and (ii) a second criterion has to be added (parallel to the existing frequency threshold) for real-time monitoring. This new criterion is based on a patent-pending frequency stability margin calculated in real-time from the *RoCoF* and frequency values.

HIL simulations proved that innovative UFLS is feasible for real-world scenarios and that applying *RoCoF* does not affect its robustness in any of the tested conditions if the correct filtering techniques are selected. The main advantages of the innovative UFLS can be observed in moderate *RoCoF* conditions, being such from the very moment of a power deficit occurrence or decreased later on by UFLS intervention.

To summarize the main contributions of this research which all relate to the practical aspects of the implementation of an innovative UFLS scheme under test: (i) a multi-stage *RoCoF* filtering procedure that enables using *RoCoF* for local frequency stability margin calculation and monitoring in real-time for UFLS purposes. Filtering improves the resolution (linear filter stages) and filters out the anomalies (non-linear filter stage) in *RoCoF* measurements, (ii) a proof that despite *RoCoF* filtering, introduced time delays do not diminish the speed of UFLS operation during fast-occurring events (tested for −10 Hz/s) as long as *RoCoF* is used in an appropriate manner as suggested in [17], (iii) a proof that

IEDs with specifications similar to the PMU can be used for UFLS, which decreases cost (i.e., using the same device for EPS observability and UFLS protection) and opens up numerous possibilities, including building business and market models for UFLS as a part of the end consumer response and other market opportunities related to Smart Grids, and (iv) a new criterion for comparison between several UFLS methods, being the remaining *RoCoF* after UFLS stops intervening.

The analysis revealed that the diversity of IEDs (e.g., relays) and *RoCoF*-measuring techniques present in an EPS is beneficial to the UFLS method under test. Further optimization of *RoCoF* filtering is possible to reduce time delays, but one has to keep in mind that optimization, according to the noise requirements, has to be appropriately considered. Future work will be directed towards the development of the UFLS concept in which bulk load shedding is left to conventional UFLS, whereas handling the remaining power imbalance is left to smart devices (e.g., smart meters) located at end consumers and, therefore, widespread within the entire EPS.

## 6. Patents

This work is a subject to a pending International Patent Application No. PCT/EP2018/059048 filed on 9 April 2018.

**Author Contributions:** Conceptualization, U.R. and D.S.; methodology, U.R., A.Č. and D.S.; software, R.I. and D.S.; validation, U.R. and D.S.; formal analysis, A.Č. and D.S.; investigation, R.I. and D.S.; resources, U.R. and M.S.; data curation, D.S.; writing—original draft preparation, U.R., A.Č. and D.S.; writing—review and editing, U.R., A.C. and D.S.; visualization, U.R. and D.S.; supervision, U.R. and M.S.; project administration, U.R. and M.S.; funding acquisition, U.R. and M.S. All authors have read and agreed to the published version of the manuscript.

**Funding:** This research was partly funded by the Slovenian Research Agency, grant number P2-0356 and J2-9232, the European Commission through the H2020 project RESOLVD (Grant no. 773715), and the European Regional Development Fund through the OIS-AIR project with the financial support of the ADRION programme.

**Conflicts of Interest:** The authors declare no conflict of interest.

## References

1. H2020 Migrate. Available online: https://www.h2020-migrate.eu/ (accessed on 14 May 2020).
2. Winter, W.; Elkington, K.; Bareux, G.; Kostevc, J. Pushing the Limits: Europe's New Grid: Innovative Tools to Combat Transmission Bottlenecks and Reduced Inertia. *IEEE Power Energy. Mag.* **2015**, *13*, 60–74. [CrossRef]
3. Anderson, P.M. *Power System Control and Stability*, 2nd ed.; Wiley-IEEE Press: Piscataway, NJ, USA, 2008; ISBN 978-0-471-23862-1.
4. Pulgar-Painemal, H.; Wang, Y.; Silva-Saravia, H. On Inertia Distribution, Inter-Area Oscillations and Location of Electronically-Interfaced Resources. *IEEE Trans. Power Syst.* **2018**, *33*, 995–1003. [CrossRef]
5. Bidram, A.; Davoudi, A. Hierarchical Structure of Microgrids Control System. *IEEE Trans. Smart Grid* **2012**, *3*, 1963–1976. [CrossRef]
6. Li, S.; Tang, F.; Shao, Y.; Liao, Q. Adaptive Under-Frequency Load Shedding Scheme in System Integrated with High Wind Power Penetration: Impacts and Improvements. *Energies* **2017**, *10*, 1331. [CrossRef]
7. Małkowski, R.; Nieznański, J. Underfrequency Load Shedding: An Innovative Algorithm Based on Fuzzy Logic. *Energies* **2020**, *13*, 1456. [CrossRef]
8. Sigrist, L. A UFLS Scheme for Small Isolated Power Systems Using Rate-of-Change of Frequency. *IEEE Trans. Power Syst.* **2015**, *30*, 2192–2193. [CrossRef]
9. Hong, Y.-Y.; Nguyen, M.-T. Multiobjective Multiscenario Under-Frequency Load Shedding in a Standalone Power System. *IEEE Syst. J.* **2020**, *14*, 2759–2769. [CrossRef]
10. Banijamali, S.S.; Amraee, T. Semi-Adaptive Setting of Under Frequency Load Shedding Relays Considering Credible Generation Outage Scenarios. *IEEE Trans. Power Deliv.* **2019**, *34*, 1098–1108. [CrossRef]
11. Potel, B.; Cadoux, F.; De Alvaro Garcia, L.; Debusschere, V. Clustering-based method for the feeder selection to improve the characteristics of load shedding. *IET Smart Grid* **2019**, *2*, 659–668. [CrossRef]
12. Shekari, T.; Gholami, A.; Aminifar, F.; Sanaye-Pasand, M. An Adaptive Wide-Area Load Shedding Scheme Incorporating Power System Real-Time Limitations. *IEEE Syst. J.* **2018**, *12*, 759–767. [CrossRef]

13. Zuo, Y.; Frigo, G.; Derviškadić, A.; Paolone, M. Impact of Synchrophasor Estimation Algorithms in ROCOF-Based Under-Frequency Load-Shedding. *IEEE Trans. Power Syst.* **2020**, *35*, 1305–1316. [CrossRef]

14. He, P.; Wen, B.; Wang, H. Decentralized Adaptive Under Frequency Load Shedding Scheme Based on Load Information. *IEEE Access* **2019**, *7*, 52007–52014. [CrossRef]

15. Tofis, Y.; Timotheou, S.; Kyriakides, E. Minimal Load Shedding Using the Swing Equation. *IEEE Trans. Power Syst.* **2017**, *32*, 2466–2467. [CrossRef]

16. Jallad, J.; Mekhilef, S.; Mokhlis, H.; Laghari, J.A. Improved UFLS with consideration of power deficit during shedding process and flexible load selection. *IET Renew. Power Gener.* **2018**, *12*, 565–575. [CrossRef]

17. Rudez, U.; Mihalic, R. RoCoF-based Improvement of Conventional Under-Frequency Load Shedding. In Proceedings of the 2019 IEEE Milan PowerTech, Milan, Italy, 23–27 June 2019; pp. 1–5.

18. Maglia, M.F.; Olama, M.M.; Ollis, B. Hardware-in-the-loop Testing of Signal Temporal Logic Frequency Control in a Microgrid. In Proceedings of the 2020 IEEE Power Energy Society Innovative Smart Grid Technologies Conference (ISGT), Washington, DC, USA, 17–20 February 2020; pp. 1–5.

19. SimTec—PSS SINCAL Platform. Available online: http://www.simtec-gmbh.at/sites_en/platform.asp (accessed on 15 June 2020).

20. Rudez, U.; Mihalic, R. Monitoring the First Frequency Derivative to Improve Adaptive Underfrequency Load-Shedding Schemes. *IEEE Trans. Power Syst.* **2011**, *26*, 839–846. [CrossRef]

21. *IEEE/IEC Approved Draft International Standard—Measuring Relays and Protection Equipment—Part 118-1: Synchrophasor for Power System—Measurements*; IEEE: Piscataway, NJ, USA, 1 January 2018; pp. 1–75, ISBN 978-1-5044-4682-2.

22. Eriksson, R.; Modig, N.; Elkington, K. Synthetic inertia versus fast frequency response: A definition. *IET Renew. Power Gener.* **2018**, *12*, 507–514. [CrossRef]

23. Smith, S.W. *The Scientist & Engineer's Guide to Digital Signal Processing*, 1st ed.; California Technical Pub: San Diego, CA, USA, 1997; ISBN 978-0-9660176-3-2.

24. *IEEE Standard for Synchrophasor Measurements for Power Systems. IEEE Std. C371181-2011*; IEEE: Piscataway, NJ, USA, 28 December 2011; pp. 1–61, ISBN 978-0-7381-6811-1. [CrossRef]

25. *IEEE Standard for Synchrophasor Data Transfer for Power Systems. IEEE Std C371182-2011*; IEEE: Piscataway, NJ, USA, 28 December 2011; pp. 1–53, ISBN 978-0-7381-6813-5. [CrossRef]

26. RSCAD: Real-Time Simulation Software Package. Available online: http://knowledge.rtds.com/hc/en-us/articles/360046352893 (accessed on 16 June 2020).

27. ELES d.o.o. *Sistemska Obratovalna Navodila Za Prenosni Sistem Električne Energije Republike Slovenije*; ELES d.o.o.: Ljubljana, Slovenia, 2016.

28. Santos, A.Q.; Monaro, R.M.; Coury, D.V.; Oleskovicz, M. New scoring metric for load shedding in multi-control area systems. *Transm. Distrib. IET Gener.* **2017**, *11*, 1179–1186. [CrossRef]

29. Matthewman, S.; Byrd, H. Blackouts: A Sociology of Electrical Power Failure. *Soc. Space* **2013**, *2*, 31–55.

30. Sigrist, L.; Rouco, L.; Echavarren, F.M. A review of the state of the art of UFLS schemes for isolated power systems. *Int. J. Electr. Power Energy Syst.* **2018**, *99*, 525–539. [CrossRef]

*Article*

# Application of the First Replica Controller in Korean Power Systems

**Jiyoung Song [1,2], Seungchan Oh [2], Jaegul Lee [1,2], Jeonghoon Shin [2] and Gilsoo Jang [1,***

[1] School of Electrical Engineering, Korea University, Anam-ro, Sungbuk-gu, Seoul 02841, Korea; jy.song@kepco.co.kr (J.S.); jaegul.lee@kepco.co.kr (J.L.)

[2] Korea Electric Power Corporation (KEPCO), Munji-ro, Daejeon 34056, Korea; seungchan.oh@kepco.co.kr (S.O.); jeonghoon.shin@kepco.co.kr (J.S.)

* Correspondence: gjang@korea.ac.kr; Tel.: +82-2-3290-3246

Received: 7 May 2020; Accepted: 28 June 2020; Published: 30 June 2020

**Abstract:** The purpose of this paper is to introduce, examine, and evaluate the industrial experiences and effectiveness of a Thyristor Controlled Series Compensator (TCSC) replica controller installed in Korea in 2019 through a review of its configuration, test platform, and practical application, and further to propose operational guidelines for replica controllers. Four representative practical cases were conducted: a Dynamic Performance Test (DPT) under a sufficiently large-scale power system prior to the Site Acceptance Test (SAT), pre-verification for on-site controller modification during operation stage, parameter tuning to mitigate the control interaction, and time domain simulation for Sub-Synchronous Torsional Interaction (SSTI). None of these four cases can be performed in a Factory Acceptance Test (FAT) or on-site. Therefore, TCSC control performance was accurately verified under the entire Korean power system based on a large-scale real-time simulator, which demonstrated its effectiveness as a powerful tool for operations including multiple power electronics devices. Our review herein of these four practical cases is expected to show the usefulness of replica controllers, to demonstrate their strength to deal with practical field events, and to contribute to the further expansion of the application area from a perspective of electric utility.

**Keywords:** replica controller; TCSC; HIL; DPT; real-time simulation; testing; control and protection; large-scale power system

---

## 1. Introduction

With the increasing complexity of power systems and the growth in renewable energy resources, rapidly increasing numbers of power electronic equipment, such as large capacity high-voltage direct current (HVDC) and flexible AC transmission system (FACTS), are being installed to enhance the power system stability and acceptance limit [1–3]. Power systems with multiple, large-capacity HVDC and FACTS are capable of fast, active, and flexible operation, unlike previous passive power systems based on conventional synchronous generators. However, this trend requires a higher level of analysis and operation technology due to the possible severe impacts on power systems and potential unexpected problems such as control interaction [4–6].

The Electromagnetic Transient (EMT) tool is commonly used to analyze power electronics devices, and real-time simulators are used to connect with the actual controller based on EMT [7–9]. Recently, as the area needed to be studied and the number of power electronics devices have increased, the scale of real-time simulators has correspondingly expanded. Typically, Korea, China, Taiwan, and Canada have established environments of large-scale real-time simulators capable of conducting over 1000 buses, and have organized expert departments for analyzing HVDC and FACTS. China Southern Power Grid (CSG) established 33 racks of Real-Time Digital Simulator (RTDS) in 2012, which can contain an entire CSG network above 220 kV. One of the applications of large-scale RTDS is to playback for several contingencies in real world, such as single-phase to ground fault, main protection faults, and HVDC block. In Operador Nacional do Sistema Elétrico (ONS) in Brazil acquired their 10 racks of RTDS in 2009 to test Rio Madeira HVDC link hardware controller, which is the longest HVDC link in the world. Manitoba Hydro in Canada operates 20 racks of RTDS to study the impact of HVDC links and hydro generation. Southern California Edison (SCE) test control and protection systems, equipment interoperability, and performance under numerous contingency scenarios. Recently, Saudi Electricity Company (SEC) expanded its simulator and now operates more than 40 racks of RTDS. Its purpose is to determine the optimal locations for load shedding for voltage collapse protection, and to study the behavior of the actual controller and the interaction of FACTS with their future HVDC links. Such large-scale real-time expansion is a world-wide trend for various applications, especially for equipment testing under wide area power systems. Additionally, Korea installed 34 racks of RTDS in 2017 to test control and protection systems and to study the interaction of HVDC with FACTS and the network of the entire Korean power system [10,11].

Meanwhile, Hardware In the Loop Simulation (HILS) is to test a performance of an external device and to validate an implementation of a new algorithm or control scheme embedded in actual hardware that is interfaced with a real-time simulator [12]. Moreover, it is mainly applied to one-time tests to approve the pre-performance of an external single device such as protective relay [13,14]. Typically, the device is certified and installed in the field if the performance test is approved in a testing institute or laboratory. However, complicated control and protection systems like HVDC and FACTS should be investigated regularly for continuous analysis of site issues, even after the Site Acceptance Test (SAT). China, France, Canada, and England have established and operated HVDC replica controllers, which have fully replicated the actual on-site controller since early 2010 [15–18]. China has installed all of CSG's HVDC replica controller for study, validation, and maintenance. England established National HVDC Centre to support Caithness–Moray multiterminal HVDC project, which involves multivendors. Therefore, the replica controller was also installed to mainly study interoperability for each converter. Réseau de Transport d'Électricité (RTE) in France owns and installed IFA2000 link replica controller for the purpose of maintenance in 2017 during a refurbishment project. Additionally, in Korea, Korea Electric Power Corporation (KEPCO) will establish replica controller for of all of the HVDC project and FACTS nearby the HVDC converter station.

The replica controller consists of the same hardware board and software as the actual controller. Even though the measurement level can vary depending on the configuration, the internal communication is also the same as that of an actual controller, except for the cooling and redundancy system [16,19]. Therefore, with HILS connecting the replica controller to the real-time simulator, the simulator can reproduce and playback the issues arising in an actual controller. On the contrary, the issues arising in the replica controller may occur in the actual controller. This strength has raised the importance of the replica controller due to its advantage to perform various pre-tests and postanalyses, and to investigate alternatives that cannot be conducted with an actual field controller. In addition, its other applications such as software update of on-site controller, maintenance, and operator training are known to be very wide.

However, the use cases and experiences of the replica controller are not well shared or published. In particular, the procedure of wide area network modeling for the large-scale real-time simulator that is the basis to interface with the replica controller has not been revealed in detail. In this paper, we introduce a replica controller planned by KEPCO, describe its purpose of adaptation, present some practical cases, demonstrate its effectiveness from an electric utility perspective, and suggest further operational directions to operate replica controllers in the future.

## 2. Replica Controller in the Korean Power System

### 2.1. Planning for HVDC and FACTS in the Korean Power System

Recently, Korea has installed several HVDC and FACTS on its mainland to enhance the power system flexibility and to improve the stability for a 765 kV transmission line fault [20,21]. Since 765 kV line fault significantly impacts on the network, a special control scheme is embedded in HVDC and FACTS to support power system stability during such line fault. For example, Thyristor Controlled Series Compensators (TCSCs) are operated for impedance compensation to reduce 765 kV line loading rate in steady state. In addition, a special control scheme is embedded to boost the impedance compensation level from 50% to 70% for 5 s to improve the transit stability during 765 kV line fault [22].

To maintain a reactive power margin, three static synchronous compensators (STATCOMs) and static var compensator (SVC) are operated in reactive power reserve control mode at both ends of TCSC. In 2025 and 2026, two Line Commutated Converter (LCC) 4GW bi-pole HVDCs will be built nearby the 765 kV line. Figure 1 shows HVDC and FACTS planned for installation in the Korean power system by 2022.

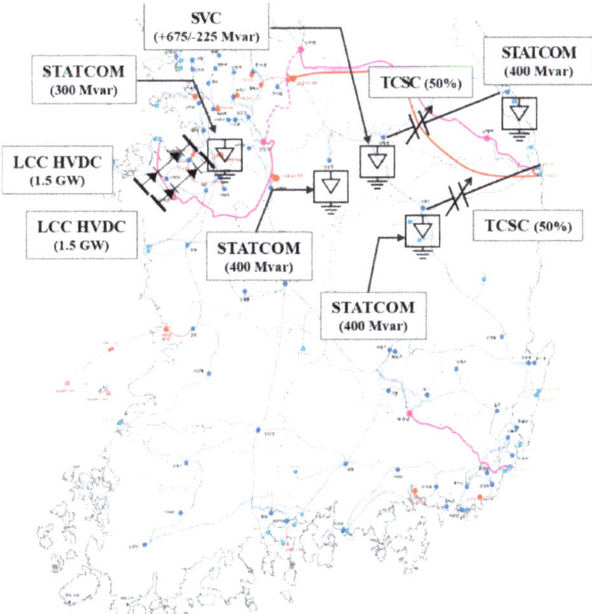


**Figure 1.** HVDC, FACTS in the Korean power system.

A total of 13 HVDC and FACTS will be built in the Korean power system within the next decade, which has increased the need for replica controllers.

### 2.2. Large-Scale Real-Time Simulator and Wide Area Network Modeling to Interface with the Replica Controller

KEPCO set up the world's largest RTDS in 2017 to analyze power systems accurately, including multiple HVDC and FACTS at the EMT level [23]. These include 34 racks of RTDS that are capable of accommodating any kind of future HVDC, FACTS, and transmission systems over 154 kV without equivalent network. This large-scale simulator operation requires significant know-how and operational technology. Therefore, KEPCO has developed and operated a preprocessing in-house tool for stable large-scale power system simulations that now can model, stabilize, and verify a large-scale RTDS case of over 1000 buses in about 2 days [24]. Wide area network modeling with real-time simulator (RTS) is very important, since it is the basis for analysis interfaced with replica controller and for study interaction of adjacent other power electronic devices with network dynamics. This paper suggests the use of data conversion functions in RTS in case of large-scale power system modeling of RTS owing to human error of dealing with huge network data. Most of RTS support data conversion function Transient Stability Analysis (TSA) tool to RTS. Procedure of wide area network modeling with RTS interfaced with replica controller is shown in Figure 2.

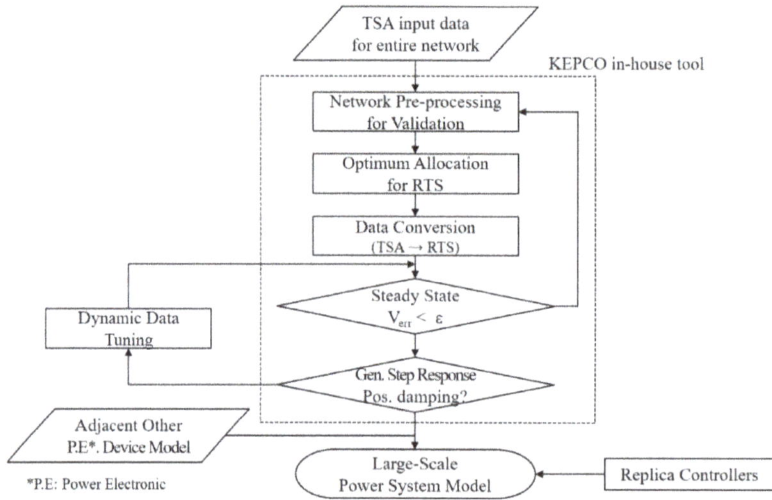


**Figure 2.** Flow chart of wide area network modeling procedure for real-time simulator (RTS) interfaced with replica controller.

It facilitates detailed analysis of the controller to reveal the dynamic characteristic of a wide area network and the interaction between the facilities and power systems at the EMT level. Figure 3 shows the real-time simulation laboratory established in KEPCO.

**Figure 3.** Real-time Digital Simulator in the Korea Electric Power Corporation (KEPCO) Research Institute.

### 2.3. Purpose of the Replica Controller

The main purposes of the replica controller are performing the tests that are unavailable in the field, overcoming the limitations of Software In the Loop Simulation (SILS), and performing more precise and accurate analysis [25–28]. These purposes may vary internationally depending on the configuration of the test environment. KEPCO has defined the following five criteria to set up a replica controller:

- Dynamic Performance Test (DPT) under large-scale power system: DPT is performed under a sufficiently wide area network to reflect the dynamic characteristics of the power system considering other HVDC and FACTS facilities nearby.
- Preverification: If the controller needs to be modified, improved, or updated during operation stage, such as adding a new function or in/output signal, the adequacy can be verified and de-risk can be achieved using a replica controller, which cannot be achieved with an actual field controller.
- Parameter Tuning: The huge changes in power systems, such as an adjacent network topology or the installation of a new power plant, require adjusting the controller parameters. KEPCO conducts verification and parameter tuning periodically for HVDC and FACTS operations for potential severe events, such as 765 kV line fault and Special Protection Scheme (SPS), by planning winter/summer operation strategies biannually.
- Event Analysis: The replica controller can analyze the root cause of any malfunction or fault due to its detailed internal protection functions. Particularly, KEPCO can simulate not only a single replica controller but also the entire Korean power system, including all HVDC and FACTS manufacturer models with EMT, which facilitates the interaction investigation of the power system and power electronic devices.
- Operator Training: Training to improve the skill level of the field operators can be conducted. Field operators have few opportunities to deal with actual controllers before project completion. Therefore, the replica controller could be a perfect tool for practical training to gain experience in emergency operation, reaction of communication disruption, alarm analysis, and insertion and block.

For this purpose, KEPCO is planning to set up replica controllers for all HVDC and TCSC in the future, starting with the four TCSC replica controllers installed in 2019.

### 2.4. TCSC Replica Controller

The TCSC is installed to compensate for line impedance by the reactance control (boost factor control). The facility configuration consists of a series capacitor, thyristor valve, reactor, and Metal-Oxide Varistor (MOV). TCSC topology and its conceptual control block are shown in Figure 4 [29].

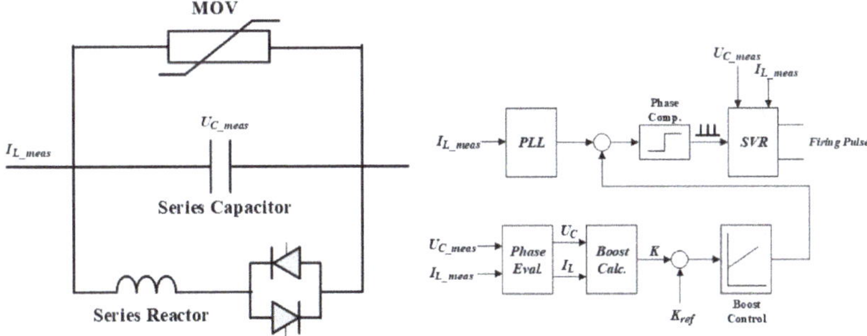


**Figure 4.** Thyristor Controlled Series Compensator (TCSC) topology and conceptual control block.

TCSC topology and the entire network are modeled in a RTS, and the TCSC control and protection systems are implemented on the replica controller. The replica controller receives three analog signals from the real-time simulator: the series capacitor voltage, the thyristor valve current, and the transmission line current. The thyristor firing pulse, the bypass switch, and the boost signal are connected to a digital signal. TCSCs were installed at Shin Youngjoo and Shin Jecheon converter stations. The replica controller is composed of four cubicles: control, protection, operation, and measurement panels. The control, protection, and operation panels have an identical hardware board and embedded software, except for redundancy. Figures 5 and 6 show the actual controller on-site and replica controller in KEPCO laboratory.

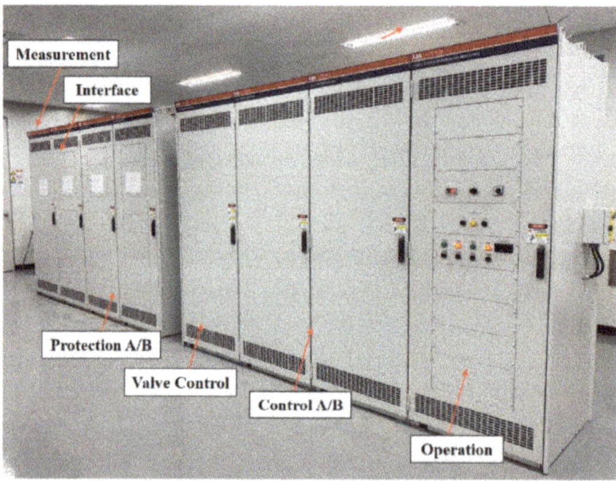


**Figure 5.** TCSC actual controller on-site.

**Figure 6.** TCSC replica controller in Korea Electric Power Corporation (KEPCO) lab.

The TCSC replica controller offers easy playback and on-site problem analysis capability, because users can adjust various control settings, protection settings, and several control block parameters on the operator system.

## 3. Practical Cases

### 3.1. Dynamic Performance Test (DPT) with Adjacent Multiple FACTS

KEPCO recently established a new DPT procedure between Factory Acceptance Test (FAT) and SAT. As the controller DPT is based on large-scale real-time simulator interfaced with a replica controller, it can overcome the barrier of the previous single unit test in FAT. Furthermore, it is a very realistic test because it contains both system dynamics and multiple HVDC and FACTS control characteristics. There are three STATCOM, one SVC, and 765 kV transmission line nearby TCSC. This case covers the verification results of TCSC's boost control scheme to improve the transient stability and the control interaction with nearby parallel FACTS in the event of 765 kV failure, as one of the DPT items.

The 2019 winter operation strategy study data were used as the base case. Large-scale system modeling revealed that RTDS takes 28 racks (based on PB5 processor card), 1399 buses, and 358 generators, and the load is 88,734 MW. There are 22 GW bulk generations over 22 generators on eastern coast of Korea, so two nuclear power plants' trip SPS is applied to maintain transient stability when a 765 kV line fault occurs.

As shown in Figure 7, TCSC maintains a normally capacitive mode (mode: 2) without a bypass or any mode change when a 765 kV line fault occurs, and it follows the boost factor reference appropriately. In addition, both STATCOMs and SVC operate properly to compensate for the low voltage of TCSC and the network near 765 kV transmission line, and to ensure that any control interaction with TCSC does not occur. DPT using the replica controller enables accurate verification of the actual controller's reliability via realistic testing that reflects the network and control characteristics nearby other power electronic devices on-site.

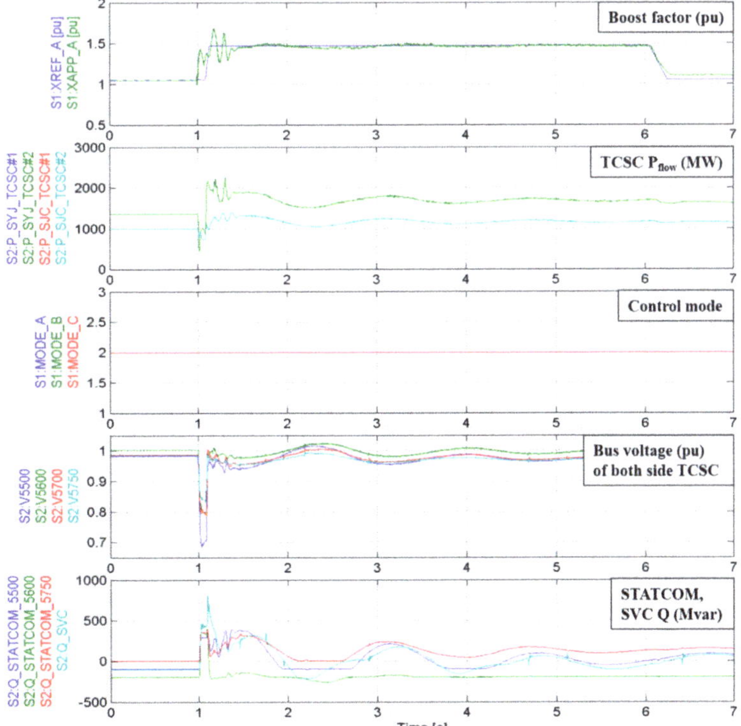

**Figure 7.** TCSC and other FACTS output.

### 3.2. Preverification of a Modified Function for an Actual Controller

In the case of TCSC line internal temporary fault, a reinsertion delay of about 1 s was applied in force for stable synchronization of the internal phase lock loop (PLL) controller in the previous control algorithm. After FAT completion, the manufacturer developed a new function called fast reinsertion, and tried to implement it into the actual controller. However, such implementation directly into the actual controller may incur a risk in the absence of any verification using the actual controller. In this practical case, a replica controller is used to pre-verify a new function or a controller improvement to be updated with an on-site actual controller. Figures 8 and 9 show the verification result using a replica controller before and after improvement.

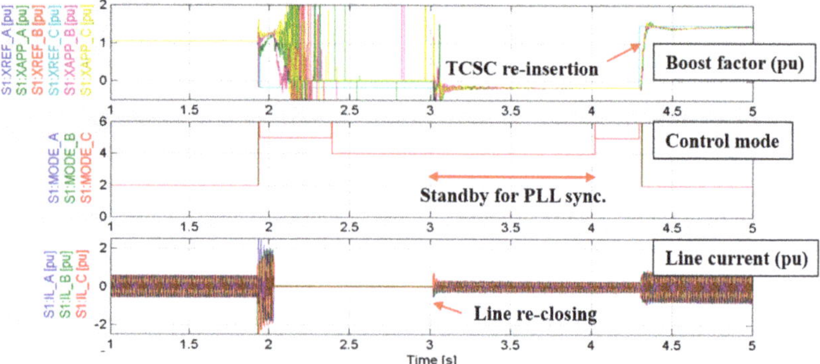


**Figure 8.** TCSC boost factor (before improvement).

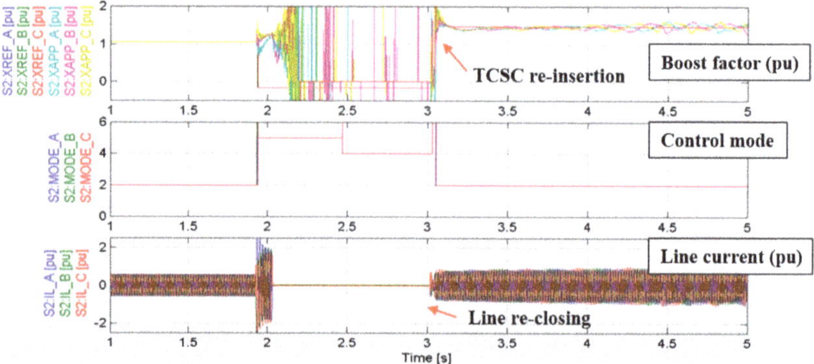


**Figure 9.** TCSC boost factor (after improvement).

As shown in Figure 8, TCSC still maintains a bypass mode (mode: 4, 5) for about 1.3 (at 3–4.3 s), even after successful line reclosing. However, after the improvement, TCSC is reinserted immediately after successful line reclosing, as shown in Figure 9. In addition to this practical case, several other tests were conducted to validate the reliability of the new function, and then on-site actual controller was successfully updated.

### 3.3. Parameter Tuning for Control Interaction

During commissioning for Shin Jecheon TCSC#1 and #2, at 1 AM (light load) on November 2019, TCSC#2 was permanently bypassed by a sudden reactance error protection operation shown below Figure 10. Analysis of the fault recorder revealed that TCSC #1 and #2 were in a state of oscillation and hunting each other while reaching to the light-load condition at dawn. As a result of modeling the network condition at that time with a real-time simulator and analyzing it using a replica controller, the error value of the steady-state PLL controller was found to be continuously increasing rather than decreasing.

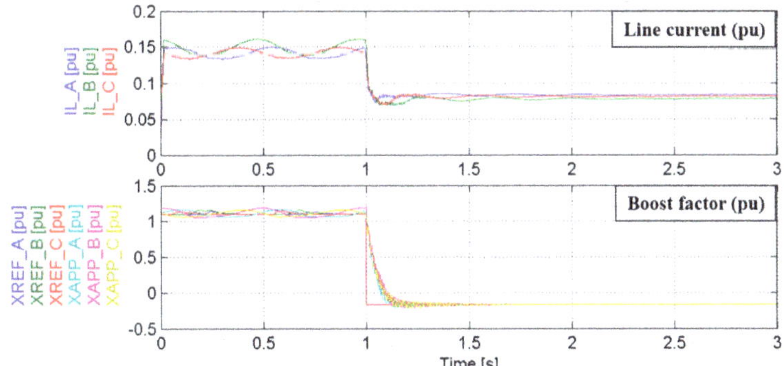

**Figure 10.** TCSC bypassed on site.

To mitigate this control interaction, the sensitivity of the replica controller was tuned so that the steady-state PLL proportional gain was changed from 4 to 3 and integration gain increased from 400 to 1000 via the heuristic method. Figure 11 shows the conceptual PLL control block.

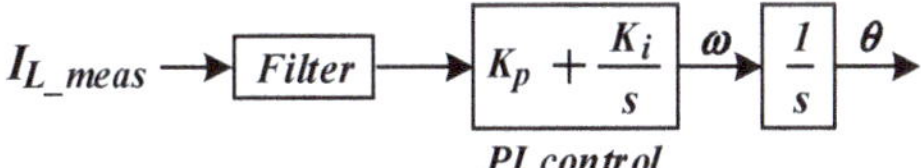


**Figure 11.** Phase lock loop (PLL) conceptual PI control block.

The analysis results of the before and after parameter tuning using the replica controller are presented in Figure 12. The oscillation of TCSC was prevented even if #2 had been inserted during TCSC #1 operation, and the steady-state PLL controller error term was also reduced, which verified that the system was under normal control conditions. Several additional tests among the DPT were conducted to verify whether such parameter tuning affects other control performances, and then the on-site controller parameters were finally tuned successfully. Finally, the steady-state PLL gain was suitable for the current power system conditions.

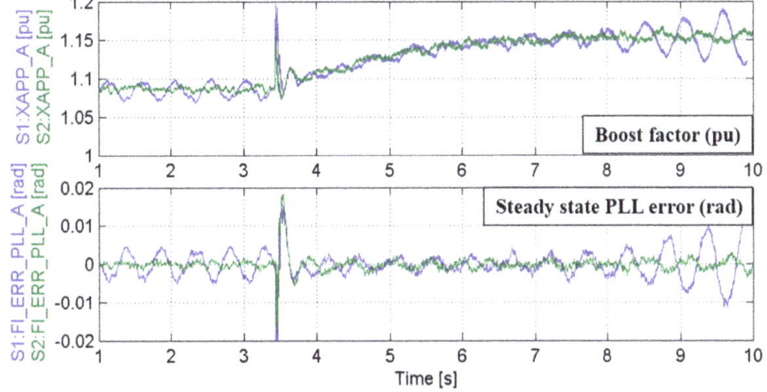

**Figure 12.** Comparison between before/after parameter tuning.

Low-level controller gain tuning does not commonly occur, but it may be needed due to changes in the network condition. Therefore, periodic parameter tuning using the replica controller and the effect on the system should be verified in advance. In June 2020, two 1.4 GW nuclear power plants will be connected to the power system nearby the TCSC sending end, and optimal parameter tuning adapted to the changed network condition will be undertaken again.

### 3.4. Subsynchronous Torsional Interaction (SSTI) Validation under a Large-Scale Power System

During the operation stage, the actual SSTI analysis needs to consider the additional damping provided by the adjacent generator, HVDC, and FACTS. Therefore, KEPCO modeled the entire Korean power system without the equivalent network, interfaced with the TCSC replica controller, and applied damping analysis to investigate the possibility of SSTI occurrence. In the event of SSTI, the target generator is modeled with a multi-mass model and time domain simulation for the entire Korean power system is carried out to identify the problem accurately. In this practical case, three STATCOMs and one SVC are considered in addition to the target generator of SSTI analysis. Figures 13 and 14 show the single line diagram and damping analysis result, respectively, for Hanwool nuclear power plant (NP) #2 and TCSC.

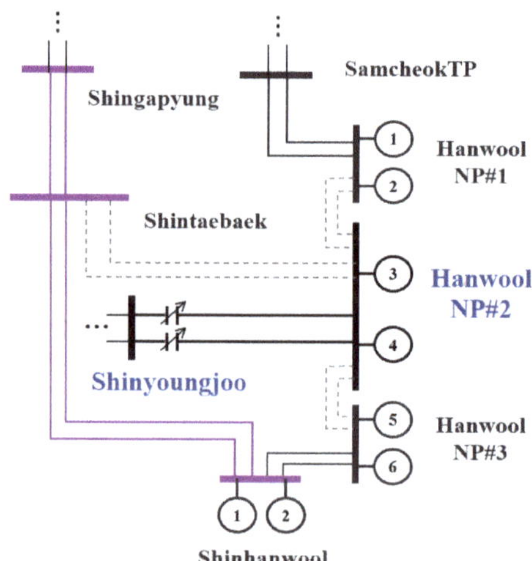


**Figure 13.** Single line diagram near TCSC.

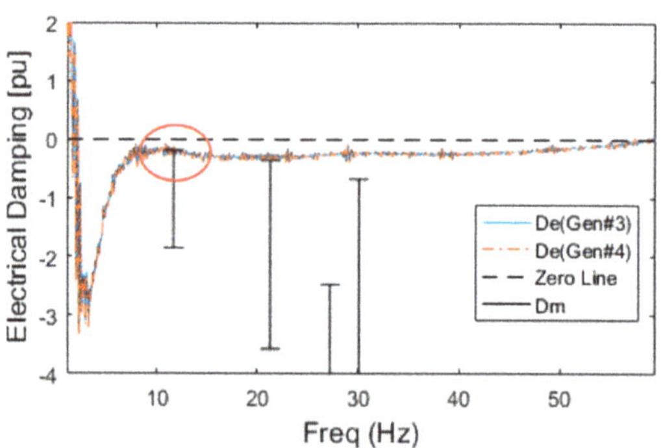


**Figure 14.** Damping analysis result.

The Hanwool #3 and #4, which are the target generators for the analysis, are radially connected to TCSC at the N-6 contingency, which has a little negative damping at the generator's minimum output around 12.3 Hz mode frequency. We finally identified that it had positive damping through the time domain simulation for SSTI possibility, as shown in Figure 15.

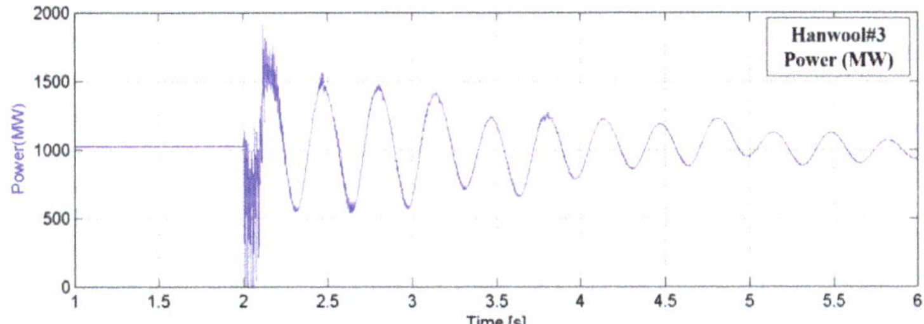


**Figure 15.** Time domain simulation result in light load of generator.

## 4. Conclusions

The importance of replica controllers has recently increased owing to the expansion of HVDC and FACTS facilities. Such replica controllers have been installed mainly in China, Canada, and Europe. In this paper, we described the first replica controller set up in Korea, and presented the more realistic test, analysis, and operation as applicable and necessary under the actual network conditions. In addition, we represented its usefulness for application to the following four practical cases: DPT under a large-scale power system, preverification parameter tuning, event analysis, and SSTI occurrence verification. The replica controllers that demonstrated the usefulness of their on-site application in all four cases could not be performed on-site. The analysis results of these practical experiences offer great value by contributing to the stable future operation of multiple HVDC and FACTS. In particular, since different types and manufacturers of equipment are operated together in the real world, we will prepare and study for interoperability and interaction using multiple replica controllers in the future under an upcoming complicated power system.

Following on from the first TCSC replica controller, KEPCO will continue to setup more HVDC and FACTS replica controllers, and further research and development are being conducted related to its operation and analysis technology for mutual control interaction issues from multivendor replica controllers to adjacent facilities. The integration and comprehensive analysis of multiple replica controllers in the future will require standardization of the replica controllers' technical specifications to optimize the flexibility and interoperability from a perspective of electric utility.

**Author Contributions:** Conceptualization, J.S. (Jiyoung Song) and G.J.; methodology, J.S. (Jiyoung Song) and J.L.; software, J.S. (Jeonghoon Shin). and S.O.; validation, G.L., J.S. (Jeonghoon Shin), and G.J.; investigation, J.S. (Jiyoung Song) and S.O.; resources, S.O. and J.L.; data curation, S.O.; writing—original draft preparation, J.S. (Jiyoung Song); writing—review and editing, J.S. (Jiyoung Song) and G.J.; supervision, G.J.; All authors have read and agreed the published version of the manuscript.

**Funding:** This research received no external funding.

**Conflicts of Interest:** The authors declare no conflict of interest.

## References

1. Barens, M.; Van Hertem, D.; Teeuwsen, S.P.; Callavik, M. HVDC Systems in smart grids. In Proceedings of the IEEE, New York, NY, USA, 11 November 2017; pp. 2082–2098. [CrossRef]
2. Henderson, M.; Bertagnolli, D.; Ramey, D. Planning HVDC and FACTS in New England. In Proceedings of the 2009 IEEE/PES Power Systems Conference and Exposition, Seattle, DC, USA, 24 April 2009; pp. 1–3. [CrossRef]
3. Sun, J.; Li, M.; Zhang, Z.; Xu, T.; He, J.; Wang, H.; Li, H. Renewable energy transmission by HVDC across the continent: System challenges and opportunities. *CSEE J. Power Energy Syst.* **2017**, *3*, 353–364. [CrossRef]

4.	Collados-Rodríguez, C.; Prieto-Araujo, E.; Cheah-Mañé, M.; San José, R.F.; Bellmunt, O.G.; Sanz, S.; Longás, C.; Cordón, A.; Coronado, A. Interaction analysis in islanded power systems with HVDC interconnections. In Proceedings of the 3rd International Hybrid Power Systems Workshop, Tenerife, Spain, 8–9 May 2018; pp. 1–5.

5.	Saad, H.; Dennetière, S.; Clerc, B. Interactions investigations between power electronics devices embedded in HVAC network. In Proceedings of the 13th IET International Conference on AC and DC Power Transmission (ACDC 2017), Manchester, UK, 14–16 February 2017; pp. 1–7. [CrossRef]

6.	Shen, L.; Barnes, M.; Milanovic, J.V.; Bell, K.R.W.; Belivanis, M. Potential interactions between VSC HVDC and STATCOM. In Proceedings of the 2014 Power Systems Computation Conference, Wroclaw, Poland, 18–22 August 2014; pp. 1–7. [CrossRef]

7.	Guillaud, X.; Faruque, M.O.; Teninge, A.; Hariri, A.H.; Vanfretti, L.; Paolone, M.; Dinavahi, V.; Mitra, P.; Lauss, G.; Dufour, C.; et al. Applications of Real-Time simulation technologies in power and energy systems. *IEEE Power Energy Technol. Syst. J.* **2015**, *2*, 103–115. [CrossRef]

8.	Iyoda, I.; Belanger, J. History of power system simulators to analyze and test of power electronics equipment. In Proceedings of the 2017 IEEE HISTory of ELectrotechnolgy CONference (HISTELCON), Kobe, Japan, 7–8 August 2017; pp. 117–120. [CrossRef]

9.	Omar Faruque, M.D.; Strasser, T.; Lauss, G.; Jalili-Marandi, V.; Forsyth, P.; Dufour, C.; Dinavahi, V.; Monti, A.; Kotsampopoulos, P.; Martinez, J.A.; et al. Real-Time simulation technologies for power systems design, testing and analysis. *IEEE Power Energy Technol. Syst. J.* **2015**, *2*, 63–73. [CrossRef]

10.	Guo, Y.; Zhang, J.; Hu, Y.; Han, W.; Ou, K.; Jackson, G.; Zhang, Y. Analysis on the effects of STATCOM on CSG based on RTDS. In Proceedings of the IEEE PES ISGT Europe 2013, Lyngby, Denmark, 6 January 2014; pp. 1–5. [CrossRef]

11.	RTDS Technologies Inc. Large-scale-simulators. Available online: https://legacy.rtds.com/wp-content/uploads/2015/12/Large-scale-simulators.pdf (accessed on 15 January 2020).

12.	Wang, Y.; Syed, M.H.; Guillo-Sansano, E.; Xu, Y.; Burt, G.M. Inverter-Based Voltage Control of Distribution Networks: A Three-Level Coordinated Method and Power Hardware-in-the-Loop Validation. *IEEE Trans. Sustain. Energy* **2019**. [CrossRef]

13.	Song, J.; Nam, S.; Lee, J.; Shin, J.; Kim, T.K. Construction of test bed with RTDS for verifying coordination algorithm for loop distribution system. *J. Int. Counc. Electr. Eng.* **2010**, *2*, 467–472. [CrossRef]

14.	Saran, A.; Kankanala, P.; Srivastava, A.K.; Schulz, N.N. Designing and testing protective overcurrent relay using real time digital simulation. In Proceedings of the 2008 International Simulation Multi-conference & 2008 Summer Simulation Multiconference, Edinburgh, Scotland, 16–19 June 200.

15.	Wang, Z.; Kent, K.; Mantay, C.; Zhou, C.; Wang, P.; Fang, C.; Dhaliwal, N.; Menzies, D. HVDC control replica development for Nelson Reiver Bipole 1&2. In Proceedings of the CIGRE Winnipeg 2017 Colloquium, Winnipeg, MB, Canada, 30 September–6 October 2017; pp. 329–338.

16.	Cowan, I.L.; Marshall, S.C.M. Installation and interfacing HVDC control replicas at the national HVDC centre. In Proceedings of the 15th IET International Conference on AC and DC Power Transmission (ACDC 2019), Coventry, UK, 18 April 2019; pp. 1–6. [CrossRef]

17.	Vernay, Y.; D'Aubigny, A.D.; Benalla, Z.; Dennetière, S. New HVDC LCC replica platform to improve the study and maintenance of the IFA2000 link (Conf, IPST). In Proceedings of the International Conference on Power Systems Transients 2017 (IPST 2017), Seoul, Korea, 26–29 June 2017.

18.	Beijing DC Consultants (BDCC), Zhejiang EPRI and RTDS Technologies Inc. Application of Replica HVDC Controller with Real Time Digital Simulation-Experiences in SGCC. In Proceedings of the 2016 Cigre Session, Paris, France, 22–26 August 2016.

19.	Best Paths. Final Recommendations for Interoperability of Multivendor HVDC. Available online: http://www.bestpaths-project.eu/contents/publications/d93--final-demo2-recommendations--vfinal.pdf (accessed on 12 December 2019).

20.	Kim, S.; Kim, H.; Lee, H.; Lee, J.; Lee, B.J.; Jang, G.; Lan, X.; Kim, T.; Jeon, D.; Kim, Y.; et al. Expanding power systems in the Republic of Korea feasibility studies and future challenges. *IEEE Power Energy Mag.* **2019**, *17*, 61–72. [CrossRef]

21.	Moon, B.; Ko, B.; Choi, J. The study on the efficient HVDC capacity considering extremely low probability of 765 kV double circuit. *J. Electr. Eng. Technol.* **2017**, *12*, 1046–1052. [CrossRef]

22. Kim, S.H.; Oh, H.; Moon, B. A Study on the application of HVDC FACTS System in the Northeast Power System of Korea. *J. Int. Counc. Electr. Eng.* **2011**, *1*, 437–440. [CrossRef]
23. Park, I.K.; Lee, J.; Song, J.; Kim, Y.; Kim, T. Large-scale ACDC EMT level system simulations by a real time digital simulator (RTDS) in KEPRI-KEPCO. *J. Electr. Power Energy* **2017**, *3*, 17–21. [CrossRef]
24. Hur, K.; Lee, J.; Song, J.; Ko, B.; Shin, J. Advanced Real Time Power System Simulator for Korea Electric Power Systems: Challenges and Opportunities. IEEE PES. 2018. Available online: http://site.ieee.org/pes-itst/files/2018/08/2018-Panel-1.pdf (accessed on 23 January 2019).
25. CIGRE Working Group B4.38. *Modeling and Simulation Studies to Be Performed during the Lifecycle of HVDC Systems*; CIGRE Technical Brochure: Paris, France, 2013.
26. Kotsampopoulos, P.; Georgilakis, P.; Lagos, D.T.; Kleftakis, V.; Hatziargyriou, N. FACTS providing grid services: Applications and testing. *Energies* **2019**, *12*, 2554. [CrossRef]
27. Ou, K.; Li, H. Function and dynamic performance test of the updated Tian-Guang HVDC control and protection systems. *Adv. Mater. Res.* **2012**, *516*, 1527–1530. [CrossRef]
28. Guay, F.; Chiasson, P.; Verville, N.; Tremblay, S.; Askvid, P. New hydro Quebec real time Simulation interface HVDC commissioning studies. In Proceedings of the International Conference on Power Systems Transients 2017 (IPST 2017), Seoul, Korea, 25–29 June 2017.
29. Angquist, L. *Synchronous Voltage Reversal Control of Thyristor Controlled Series Capacitor*; TRITA-ETS; Royal Institute of Technology in Stockholm: Stockholm, Sweden, July 2002; ISSN 1650-674X.

*Article*

# Remote Laboratory Testing Demonstration

Luigi Pellegrino [1,*], Carlo Sandroni [1], Enea Bionda [1], Daniele Pala [2], Dimitris T. Lagos [3], Nikos Hatziargyriou [3] and Nabil Akroud [4]

[1]   Ricerca Sistema Energetico, 20134 Milano, Italy; carlo.sandroni@rse-web.it (C.S.); enea.bionda@rse-web.it (E.B.)
[2]   Unareti S.p.A., 25124 Brescia, Italy; daniele.pala@unareti.it
[3]   School of Electrical and Computer Engineering, National Technical University of Athens, 15780 Zografou, Greece; dimitrioslagos@mail.ntua.gr (D.T.L.); nh@power.ece.ntua.gr (N.H.)
[4]   Ormazabal Corporate Technology, 48340 Amorebieta-Etxano, Bizkaia, Spain; nak@ormazabal.com
*   Correspondence: luigi.pellegrino@rse-web.it

Received: 13 March 2020; Accepted: 28 April 2020; Published: 6 May 2020

**Abstract:** The complexity of a smart grid with a high share of renewable energy resources introduces several issues in testing power equipment and controls. In this context, real-time simulation and Hardware in the Loop (HIL) techniques can tackle these problems that are typical for power system testing. However, implementing a convoluted HIL setup in a single infrastructure can be physically impossible or can increase the time required to test a smart grid application in detail. This paper introduces the Joint Test Facility for Smart Energy Networks with Distributed Energy Resources (JaNDER) that allows users to exchange data in real-time between two or more infrastructures. This tool enables the integration of infrastructures, exploiting the synergies between them, and creating a virtual infrastructure that can perform more experiments using a combination of the resources installed in each infrastructure. In particular, JaNDER can extend a HIL setup. In order to validate this new testing concept, a coordinated voltage controller has been tested in a Controller HIL setup where JaNDER was used to interact with an actual On Load Tap Changer (OLTC) controller located in a remote infrastructure. The results show that the latency introduced by JaNDER is not critical; hence, under certain circumstances, it can be used to expand the real-time testing without affecting the stability of the experiment.

**Keywords:** HIL; CHIL; integrated laboratories; real-time communication platform; power system testing

---

## 1. Introduction

In a global warming trend [1], the key enablers to manage the increasing emission of greenhouse gases (GHGs) are energy efficiency and low-carbon technologies. Renewable sources, storage systems, and flexible loads provide enhanced possibilities. However, these new resources introduce several issues that power system operators have to cope with and investigate. Indeed, an infrastructure with a growing amount of heterogeneous components typically has a higher complexity than traditional power plants [2]. Sophisticated component design methods, intelligent information and communication architectures, automation and control concepts, and proper standards are necessary in order to manage the higher complexity of such intelligent power systems (i.e., Smart Grids) [3–5]. Due to the considerable higher complexity of such cyber-physical systems, it is expected that the validation of smart grid configurations will play a major role in future technology developments.

During the last decade, a growing number of various research and technology development activities have already been carried out in this area. This has been brought to form several research infrastructures (RIs) performing experiments on smart grid activities. However, up to now, no integrated approach for analyzing and evaluating smart grid configurations addressing power system, as well as information, communication, and control topics is available. The integration of cyber security and privacy issues are also not sufficiently addressed by existing solutions. In order to guarantee a sustainable and secure supply of electricity in a smart grid system with considerable higher complexity, and to support the expected forthcoming large-scale roll out of new technologies, a proper integrated RI for smart grids is necessary.

In this context, ERIGrid project [6] came up with a tool for integrating laboratories, exchanging data in real-time. This tool, the Joint Test Facility for Smart Energy Networks with Distributed Energy Resources (JaNDER), allows users to exploit the synergies among the RIs in order to increase the number of test cases that can be implemented without any further cost for hardware/software extension. Obviously, some test cases can be performed in an integrated RI, while others, with strict time constraints, cannot. Indeed, the exchange of information between the RIs introduces a delay due to the communication latency that changes as a function of RI locations, the architecture of the communication infrastructure, and the amount of data to exchange.

There are a few research papers that document real-time simulation using geographically distributed resources. Ravikumar et al. [7] integrated two digital real-time simulator (DRTS) setups based on a customized advanced data acquisition system. Similar setups have been used by Faruque et al. [8] and Rentachintala [9]. In recent works [10] and [11] a geographically distributed real-time simulation (GD-RTS) has been performed using VILLASframework [12], a set of tools that can be used for integrating RIs. In particular, VILLASframework includes a lab-to-lab communication mode, thorough the public internet, with an architecture very similar to that of JaNDER. While VILLAS supports various integration modes with different purposes and performances, JaNDER makes lightness and simplicity of use its major strengths.

In Section 2, a description of the main challenges in power system testing are discussed to explain why HIL techniques and the integration of RIs are important to tackle the upcoming issues in the transition of the power system. Section 3 describes the architecture of JaNDER, and Section 4 presents the results of the characterization tests performed. In the end, Section 5 presents a real application involving two RIs to validate the concept of integrated RIs and to evaluate the performance of JaNDER.

## 2. Real-Time Simulation Testing

Real-time simulation and HIL techniques are gaining significant attention as a testing procedure for smart grid applications. These techniques allow the connection of physical devices (e.g., inverters, controllers) to a DRTS that simulates the rest of the power system in real-time. Therefore, instead of using simulated models of hardware equipment, which could be inaccurate compared to the actual device, the actual hardware components are used providing a more realistic view of a smart grid application. Under this setup, exhaustive testing can be achieved in realistic, flexible, controllable, and repeatable conditions leading to fewer problems at the commissioning phase and field operation [13–16].

HIL approaches can be the closest representation that a laboratory test can get to a full hardware application. However, as the HIL setup approaches the actual field implementation, complexity and interfacing challenges increase due to the introduction of various hardware components.

The implementation procedure of such convoluted setup could be quite time consuming. This is derived from the fact that the engineers involved must gain adequate expertise on the different equipment, which could have been developed by various manufacturers, in order to address communication, protection and interface issues between the hardware devices and the DRTS. The physical space required, in the infrastructure hosting the HIL setup, can be another important limitation factor.

To sum up, implementing a convoluted HIL setup in a single infrastructure can be physically impossible or can increase the time required to test a smart grid application in detail. However, platforms like JaNDER, which is described in this work, can be used to overcome the aforementioned issues.

JaNDER allows the interconnection of different infrastructures with very small communication latencies and can be used, as presented in Section 5, to expand the HIL setup. This addresses the space limitation issue and also reduces the time required to interface the different equipment due to the collaboration of different personnel specialized in different components.

## 3. JaNDER Concept

To better understand how an HIL setup can be extended on two or more RIs, it is useful to explain how JaNDER works. At the heart of the JaNDER platform, there is the idea that several RIs, linked with a standardized ICT solution, are integrated into a Virtual Research Infrastructure (VRI), encompassing the simulation/experimental potential made available by each laboratory, and thus being able to participate in tests of complex use cases. The overall idea of VRI is depicted in Figure 1.

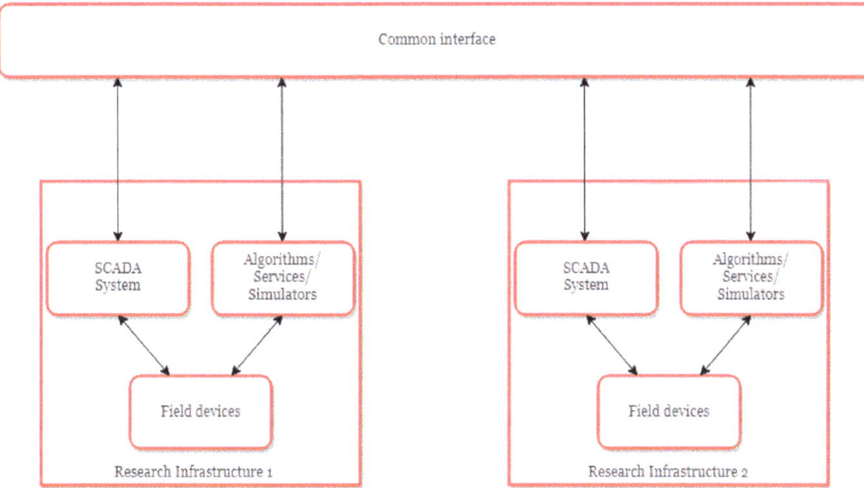


**Figure 1.** Virtual research infrastructure concept.

Figure 1 shows the advantage of the VRI concept: the possibility for one RI to seamlessly access the resources located at a remote site: these resources might be real field devices, typically mediated through a supervisory control and data acquisition (SCADA) system, or resources modelled inside a DRTS. DRTSs are crucial components of smart grid laboratory experiments but are also very expensive and the possibility of sharing them among RIs is a big advantage. Also, software artifacts such as control algorithms (e.g., a voltage control algorithm, operating on both local and remote devices) can be remotely connected. In practice, measurements and control signals transferred through the VRI's common interface can benefit from higher levels of interoperability—at a basic level, there is the need to transfer plain values associated to devices (either real or simulated), while a more semantic level can be placed "above" the basic exchange of field signals. In fact, this "layered" approach is the one used by the JaNDER platform (under the terminology of "Level 0" and "Level 1"), as described in more details in the following subsections.

### 3.1. JaNDER Level 0

JaNDER Level 0 implements the basic mechanism for exchanging live data (i.e., typically measurements and controls) between different RIs. The non-functional constraints associated to this layer are the following:

- Work with a "push" communication model, i.e., do not require any RI to accept incoming TCP connections. This allows for a simpler deployment of the platform.

- Use the HTTPS protocol; the reason for this requirement is the same as the previous one.
- It must be easy to integrate the platform in many different existing laboratory environments, using different communication protocols and programming languages.
- It must be secure. Since RIs are connected over the Internet, proper cyber security mechanisms must be adopted in order to guarantee that only authorized parties can connect to the platform.
- It must be "fast enough"; this actually means fast enough to support the foreseen test cases, both in terms of latency and volume of data.

Given the above requirement, Figure 2 illustrates in a simplified manner the implemented JaNDER Level 0 architecture, which is then explained in details.

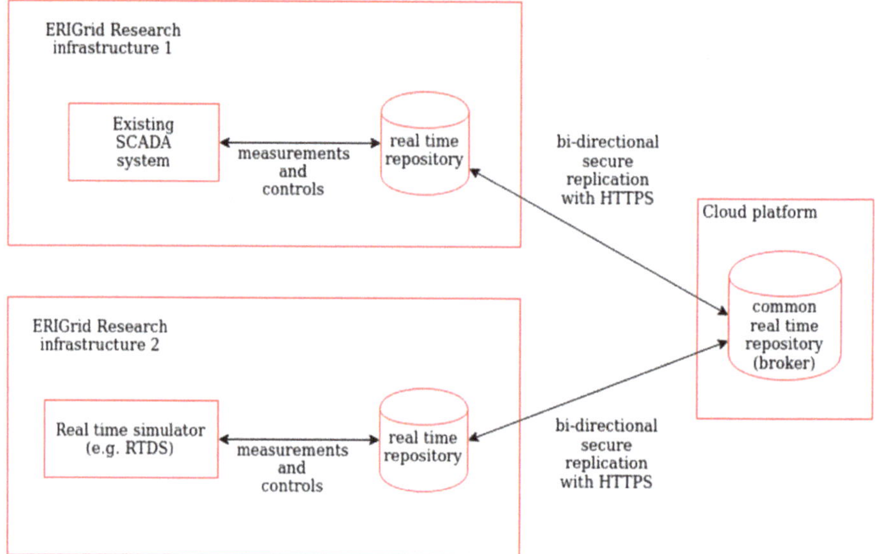


**Figure 2.** Joint Test Facility for Smart Energy Networks with Distributed Energy Resources (JaNDER) Level 0 architecture.

The starting point for each RI is a real-time repository used to collect measurements and controls from the field (or more often, as shown in Figure 2, from an already existing SCADA system or DRTS). The reason for this repository is the decoupling of the JaNDER platform from any specific automation solution already installed in the infrastructure. The idea is to have data points from each partner available in the same basic format using a simple key-value repository. The remote connection of remote infrastructures is then implemented by deploying a common real-time repository (which can be hosted in a cloud environment, for example) that is automatically synchronized with all the local real-time databases of the partners. In other words, the common repository acts as a central broker for connecting the different local repositories of the partners, and can be thought as a "virtual bus" connecting all authorized facilities. The real-time repository is an existing open source product, called Redis [17], which is a widely tested, supported, and documented solution with client libraries already developed for many programming languages and environments. However, the HTTPS replication logic is not part of the Redis product and has been developed originally within the project and then released as open-source software.

A crucial implementation aspect of JaNDER Level 0 is the connection between Redis and the existing SCADA system (or individual field devices, depending on each laboratory architecture, such as a DRTS); this link is of course different for each partner, and therefore, the associated software development effort can vary from a really straightforward task, in case the SCADA system provides an open and well-documented interface for connection with external systems to a difficult and time-demanding task, in case there is no central SCADA system and each device must be integrated separately (possibly with different communication protocols). Given the impossibility to address each of the partner's ICT infrastructures in a unified way, it should be noted that the extremely simple data model of the Redis database and the fact that client libraries are available in almost any programming language and environment (including LabVIEW and Matlab, which are highly popular in laboratory environments) helps in keeping the integration effort minimal. In conclusion, all of the requirements mentioned at the beginning of this section have been addressed with the described implementation: the performance argument in particular is detailed in the next section.

The fully open source nature of JaNDER Level 0 [18] makes it easy to extend the VRI community in the future with new participants, however external users will also typically be interested in having a standardized protocol for interfacing: this is handled by the higher JaNDER levels, as described in the next subsection.

*3.2. JaNDER Level 1*

JaNDER Level 1 is a software abstraction build on top of Level 0, and its purpose is to provide an IEC 61850 interface on top of the very simple data structures defined in Redis. The reason for adding this level is of course to provide access to the VRI by means of an internationally accepted standard where applicable, such as IEC 61850, which is a standard of primary importance in the field of Smart Grids. The high level architecture of JaNDER Level 1, as an extension of the picture presented for JaNDER level 0, is the following.

In Figure 3, one RI wants to access some contents of its Redis database using the IEC 61850 protocol. It should be stressed that the data points which are in its repository can also come from another RI by means of the JaNDER Level 0 replication mechanism. This means that the IEC 61850 server in the picture can give access to devices in infrastructure 2 but also in infrastructure 1, in a seamless way; at the same time, since the IEC 61850 connection is local to infrastructure 2, there is no need to setup sophisticated cyber security mechanisms for this connection, i.e., the basic MMS protocol without any encryption can be safely used. Of course this is possible because strong cyber security mechanisms are already implemented at Level 0, and are therefore completely transparent for Level 1. The "Mapping" and "CID" files shown as inputs in the above figure are the fundamental inputs needed by the IEC 61850 server in order to work. More in detail, the configured IED description (CID) is the standard IEC 61850 file used for configuring a device (an IED) and contains a data model representing (a subset of) the contents of the Redis repository in terms of IEC 61850 logical nodes. Apart from this file, it is of course necessary to link the data attributes defined inside it with the live values stored in Redis: this is done by means of a mapping file, which is a text file where each line contains an IEC 61850 data attribute name and a corresponding Redis data point name. The server will use this file in order to connect the IEC 61850 data model specified in the CID to Redis.

The IEC 61850 server software is completely open source [19] and based on the OpenIEC61850 Java library provided by Fraunhofer ISE [20] and others in the context of the wider OpenMUC framework. The library has been integrated with a Redis interfacing mechanism which allows the implementation of the behavior discussed above. The interfacing is of course bidirectional, so that measurements can be retrieved and control commands can be issued. (It should be highlighted that the more advanced IEC 61850 control services (enhanced security and select-before-operate) have not been implemented since the direct control with normal security is adequate for the testing scenarios.) In conclusion, the Level 1 of the JaNDER platform provides a simplified way of adding an IEC 61850 standard interface to a RI, built on top of the Level 0 basic solution.

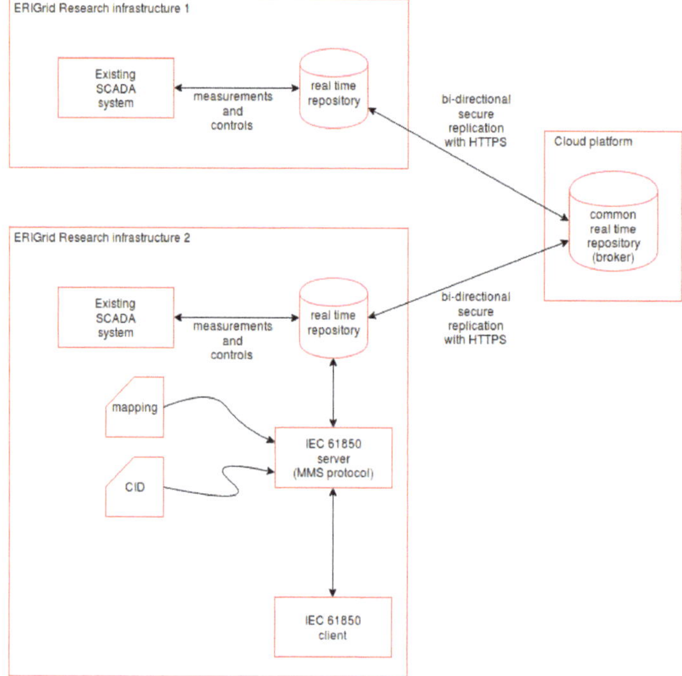


**Figure 3.** JaNDER Level 1 architecture.

## 4. JaNDER Characterization

### 4.1. JaNDER Characterization Introduction

To better understand the behavior of JaNDER in different contexts a characterization of the different JaNDER levels is necessary. First of all, two of the principal features that are usually taken into account in a distributed system have been considered: latency and response time. In this work only one level has been tested: JaNDER Level 0. The difference between JaNDER Level 0 and Level 1, in terms of latency, is negligible; the delay introduced by JaNDER Level 1 is only few milliseconds due to the mapping between the two layers. For this reason, only a single test has been performed. For this type of architecture, it is important to evaluate the latency to exchange measurements between a single RI and the cloud platform. The next subsection introduces a useful platform to log the measurements used during the tests in order to characterize JaNDER and the test procedure adopted for testing JaNDER Level 0.

### 4.2. Logging Data Platform: a Big Data Solution

Since the characterization of JaNDER needs a large number of experiments, to simplify the management of the test, a Big Data solution has been developed. The idea is to automate the analysis of test results as much as possible and to make it easy to manage the collection and analysis of data with a Big Data platform. The platform for data analysis is deployed using the Amazon AWS [21] infrastructure. Databricks [22] was used for the Big Data analysis. Every single RI records the files coming from tests, and at the end of the test, these logs are sent to the single cloud repository (AWS S3 in Figure 4). Using the framework Apache Spark [23] in Databricks, it is possible to aggregate and explore the data coming from all the tests performed by the RI in a single place with powerful Big Data tools and save the results again in AWS S3 to be eventually visualized or processed with other tools.

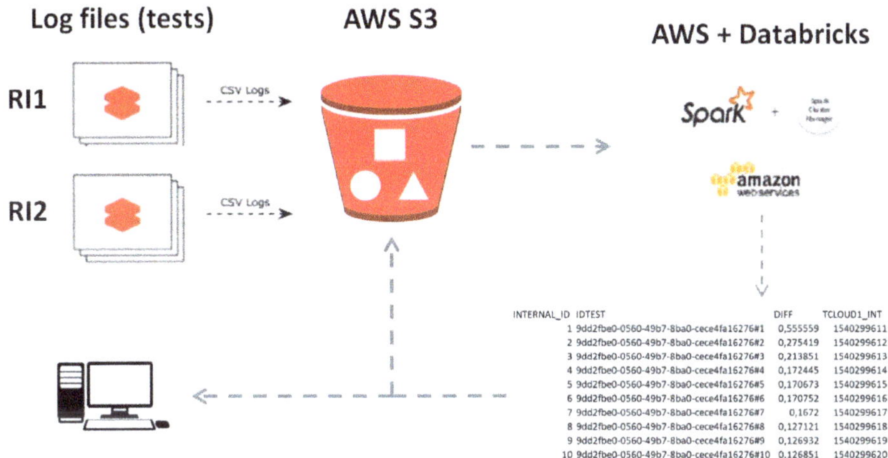



| | INTERNAL_ID IDTEST | DIFF | TCLOUD1_INT |
|---|---|---|---|
| 1 | 9dd2fbe0-0560-49b7-8ba0-cece4fa16276#1 | 0,555559 | 1540299611 |
| 2 | 9dd2fbe0-0560-49b7-8ba0-cece4fa16276#2 | 0,275419 | 1540299612 |
| 3 | 9dd2fbe0-0560-49b7-8ba0-cece4fa16276#3 | 0,213851 | 1540299613 |
| 4 | 9dd2fbe0-0560-49b7-8ba0-cece4fa16276#4 | 0,172445 | 1540299614 |
| 5 | 9dd2fbe0-0560-49b7-8ba0-cece4fa16276#5 | 0,170673 | 1540299615 |
| 6 | 9dd2fbe0-0560-49b7-8ba0-cece4fa16276#6 | 0,170752 | 1540299616 |
| 7 | 9dd2fbe0-0560-49b7-8ba0-cece4fa16276#7 | 0,1672 | 1540299617 |
| 8 | 9dd2fbe0-0560-49b7-8ba0-cece4fa16276#8 | 0,127121 | 1540299618 |
| 9 | 9dd2fbe0-0560-49b7-8ba0-cece4fa16276#9 | 0,126932 | 1540299619 |
| 10 | 9dd2fbe0-0560-49b7-8ba0-cece4fa16276#10 | 0,126851 | 1540299620 |

**Figure 4.** Logging data platform.

To permit that every single test is realized without the overlap of other tests, scheduling the tests is recommended in order to automatically run the test for every RI. In this way, the latency of every single measurement is not compromised by other tests. In the next paragraph, the JaNDER Level 0 test is described.

### 4.3. Test JaNDER Level 0: Test Description

The layer considered for this test is JaNDER Level 0. Figure 5 shows the conceptual model of this test. It aims to measure the time required for the data synchronization from an RI to the Cloud Platform: $t_2 - t_1$. In practice this time should be similar to the latency value of a public network such as the Internet. Since the Internet is not a deterministic network and does not present Quality of Service (QoS), but it is based on best-effort service, obviously the results must always consider this characteristic of the Internet, so it means that sometimes the measured latency may contain outliers with very high values (also in the order of tens of seconds).

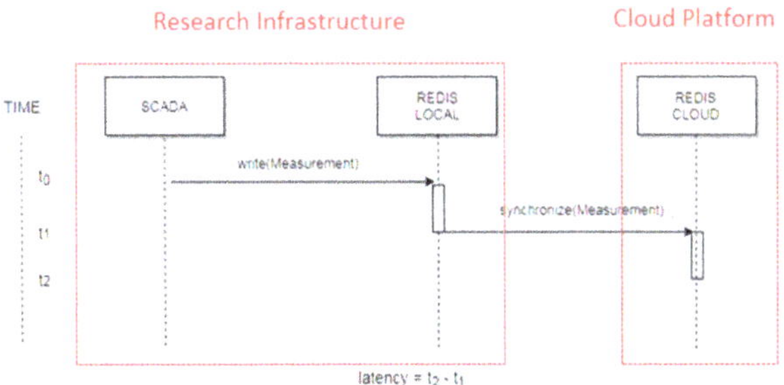



**Figure 5.** Conceptual model to test latency for JaNDER Level 0.

Every single RI, at the scheduled time, will send 1000 times a group of measurements. The test is repeated for four times with groups of 1, 10, 100, and 1000 measurements every time. The tests have been executed for a week, in slots of 30 min from 09:00 to 17:00, for each single RI. As already mentioned these experiments were performed using a scheduler in such a way as to have at any moment only a laboratory connected for the tests. The test parameters for JaNDER Level 0 can be summarized as follows:

- Target measure: latency of updating data on cloud platform;
- N° of measurements: 1, 10, 100, and 1000 repetitions for every single test (one hour per day for each RI);
- The synchronization between the machines is obtained using NTP.

For this type of test, a lot of data was collected—16 tests a day, four for every partners for four number of measurements. Every test consists of 1000 repetitions, for a total of 16,000 single repetitions for day. This was possible thanks to the automatic script system. Here a test summarizations (after a data cleaning):

- n° of considered tests:     67
- n° of measurement synchronization (local to cloud):   66,580
- n° of 1 measurement sync:   16,820 (hours 7–8)
- n° of 10 measurement sync:   18,985 (hours 9–10)
- n° of 100 measurement sync:   17,775 (hours 11–12)
- n° of 1000 measurement sync:   13,000 (hours 13–14)

Figure 6 shows that during the test the server has no Central Processing Unit (CPU), memory, network problems (of course, there is only one active client). This means that the latency measurement is not influenced by the machine (an Amazon AWS EC2 c3.large).

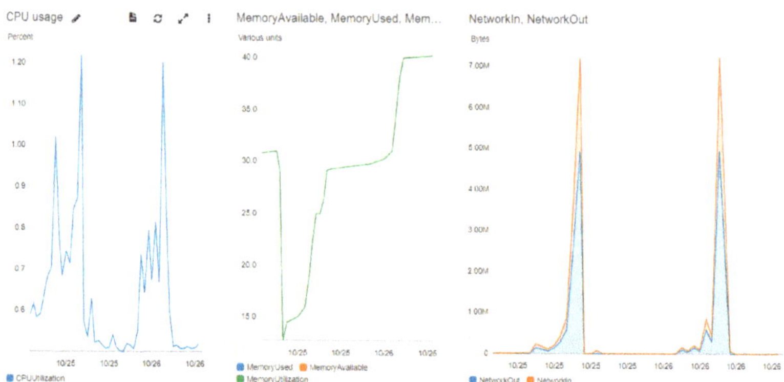

**Figure 6.** CPU, memory, network measurements during the tests.

*4.4. Test JaNDER Level 0: Test Results*

This paragraph shows some results of the JaNDER Level 0 characterization. Considering one of the RI, Figures 7–10 show the latency as function of the number of measurements sent from the RI to the cloud for every single repetitions.

The first set of tests considers one single measurement for repetition. Figure 7 shows the results coming from tests related to a single RI for JaNDER Level 0 test.

In this case the picture on the right shows that there are 3997 measurements exchange between the RI and the cloud platform. The average latency is 32.7 ms, and the third quartile is equal to 29.3 ms.

The second set of experiments considers 10 measurements for time. Figure 8 displays the results. In this case, the third quartile is equal to 31 ms.

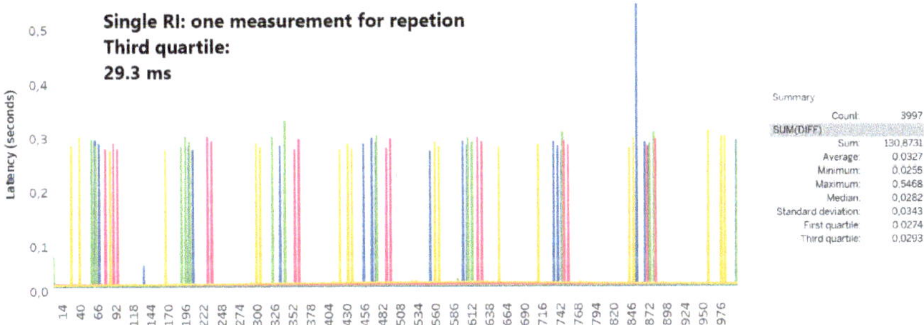

**Figure 7.** Single research infrastructure (RI) results for JaNDER Level 0 test: one single measurement for every repetition.

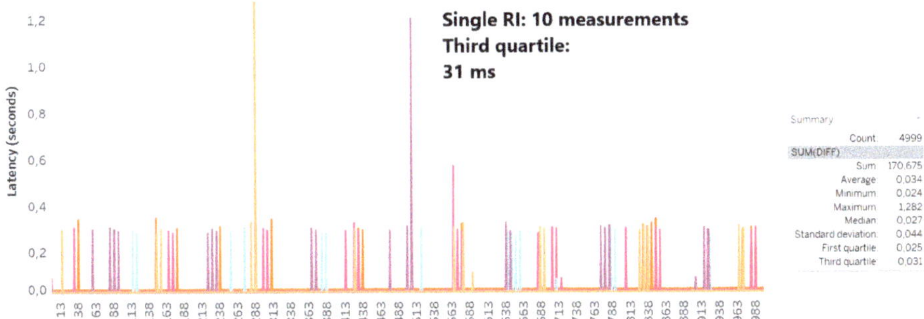

**Figure 8.** Single RI results for JaNDER Level 0 test: 10 measurements for every repetition.

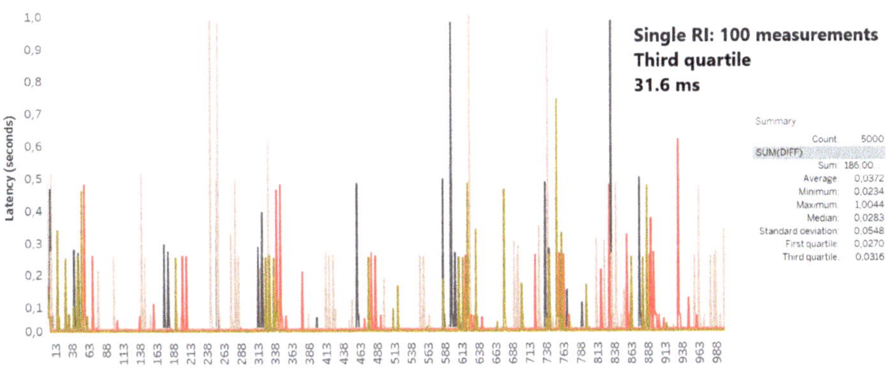

**Figure 9.** Single RI results for JaNDER Level 0 test: 100 measurements for every repetition.

The third experiment considers 100 measurements for time. Figure 9 shows the results. The third quartile is equal to 31.6 ms. The latency is practically in line with previous tests.

The fourth and last test is with 1000 measurements sent from the RI to the cloud platform. Figure 10 shows that the third quartile is equal to 132 ms. This means that with 1000 measurements at a time, the latency increases from about 30 ms to 130 ms.

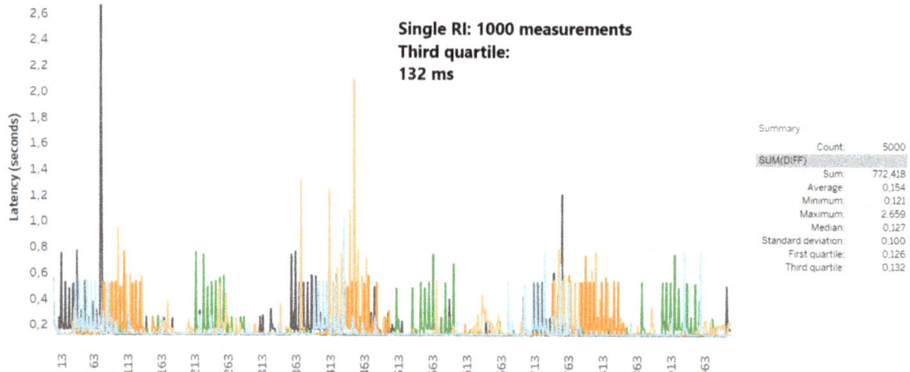



**Figure 10.** Single RI results for JaNDER Level 0 test: 1000 measurements for every repetition.

Figure 11 presents a summary of the tests. Figure 11 compares all the tests for different number of measurements (1, 10, 100, 1000) and showing the total statistics (grand total). Blue points represent the outlier of every test. The Internet latency, and thus the JaNDER Level latency, is sometime more than one second.

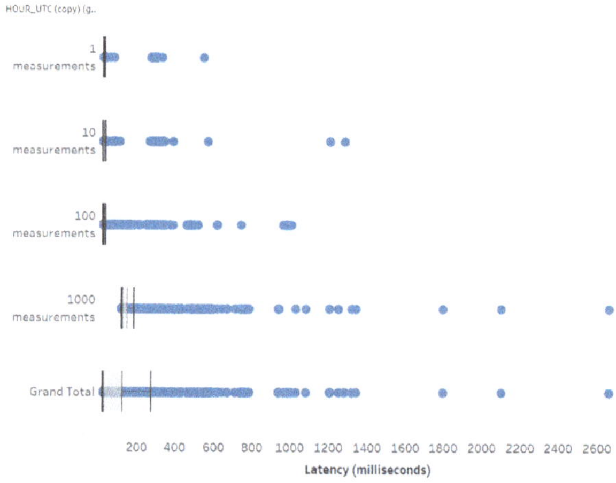



**Figure 11.** RI results for JaNDER Level 0 test.

The JaNDER Level 0 test results confirms that applications with time constant in the range of seconds are feasible with a number of measurements lower than 1000 measurements.

## 5. Example of Application: Integration of a Remote OLTC Controller

Real-time testing, using HIL techniques, can evaluate equipment performance in realistic conditions. This could lead in a decrease of their integration time while also reducing potential risks in their field deployment [14]. Components such as controllers and power equipment that are part of actual applications make the HIL setup more realistic and close to the actual conditions.

However, there might be a limitation on how many devices a research infrastructure can host, due to space limitation or due to the fact that different components probably are developed by different manufacturers. Therefore, it might be difficult to investigate if any integration issue occurs between the devices in a safe laboratory environment prior to field implementations. For example, possible issues might exist due to communication delays between the different IED's control devices.

In this section the adequacy of JaNDER platform to serve as the interface between equipment located at remote RIs coupling them in the same HIL setup is investigated.

In order to illustrate that, a Coordinated Voltage Controller, which aims to reduce voltage deviations utilizing different inverters and an On Load Tap Changer (OLTC) located in a Low Voltage (LV) benchmark grid, will be tested in a Control Hardware in the Loop (CHIL) setup. At the same time, it will utilize JaNDER to interact with an actual OLTC controller located in a remote infrastructure.

Finally, in actual field implementations, an interface with an industrial communication protocol is established between a centralized controller and remote IEDs. Therefore, using the JaNDER Level 1 platform, which integrates different components located in remote infrastructures with the industrial protocol IEC 61850, a more realistic environment can be achieved.

*5.1. Test Description*

In the aforementioned setup, a centralized voltage controller (CVC) operates in the premises of Electrical Energy Systems Laboratory (EESL) of the National Technical University of Athens (NTUA). The aim of this controller is to minimize the voltage deviations, power losses and the required OLTC operations [14]. In order to do that, the CVC receives measurements such as the load demand, the PVs' production, the batteries state of charge (SoC), and the OLTC's tap position. Then, an optimization problem is solved providing the required setpoints for the PV and battery inverters as well as the OLTC's required tap changes in order to achieve the aforementioned goals. The aim of this optimization is to minimize the voltage deviation from the nominal value while at the same time to ensure that the power losses are not increased considerably, due to the additional reactive power injection, and the OLTC operations are restricted to avoid its wear. This optimization problem is solved with a commercial solver that is able to find a local minimum for non-linear problems. The timestep of each iteration is around 1 min [8].

In this setup, a DRTS located in the EESL, is the backbone of a CHIL setup. Under this setup, the centralized controller can receive and send measurements in real-time. Therefore, both its behavior and its impact on the power system can be investigated under realistic conditions. In the DRTS, the LV Cigre benchmark network is simulated in a similar way to previous works [14,15]. The main goal of the test-case described in this work is to demonstrate the ability of JaNDER to further enhance the setup including an OLTC controller located in a remote installation. Therefore, possible interaction issues between the CVC controller located in EESL in Athens, Greece, and an OLTC controller located in the UDEX laboratory at the Ormazabal premises in Bilbao, Spain, have been investigated. It should be highlighted that the CVC controller has been integrated through the JaNDER Level 1 platform with the OLTC controller. To meet the test setup requirements, an OLTC controller emulator has been used instead of the actual transformer with the OLTC. This controller emulator includes the control of the OLTC and also serves as a real-time simulator of the whole OLTC system. This is achieved by including in the emulator the expected behavior of the OLTC hardware, such as delays in changing taps [24,25]. Therefore, through the JaNDER platform, a HIL setup is interconnected with a custom-made real-time simulator.

JaNDER-Level 1 was used to receive measurements (tap position) and send commands (tap changes) to the OLTC controller. Finally, the tap position measured by the OLTC controller is forced to a simulated OLTC in the DRTS at NTUA's premises in order to include the impact of the actual OLTC's changes to the simulated system. The described advanced setup is presented in Figure 12.

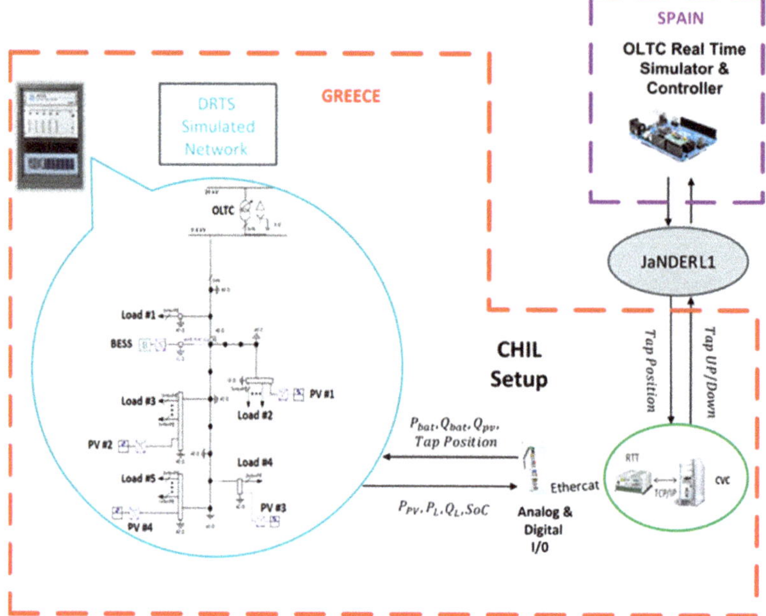



**Figure 12.** Test setup using JaNDER.

### 5.2. Results

The operation of CVC controller in parallel with the OLTC controller was tested for a daily scenario with increased PV production. Under this scenario, the CVC is expected to utilize the OLTC in order to mitigate voltage rise issues.

In Figures 13 and 14, the voltage profiles with and without the CVC operation are presented. In Figure 15, the Tap position forced to the OLTC controller by the CVC controller is presented.

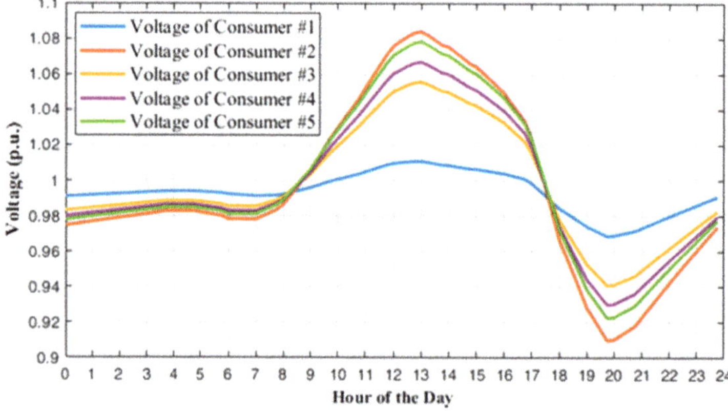



**Figure 13.** Voltage profile at consumer premises without the centralized voltage controller (CVC).

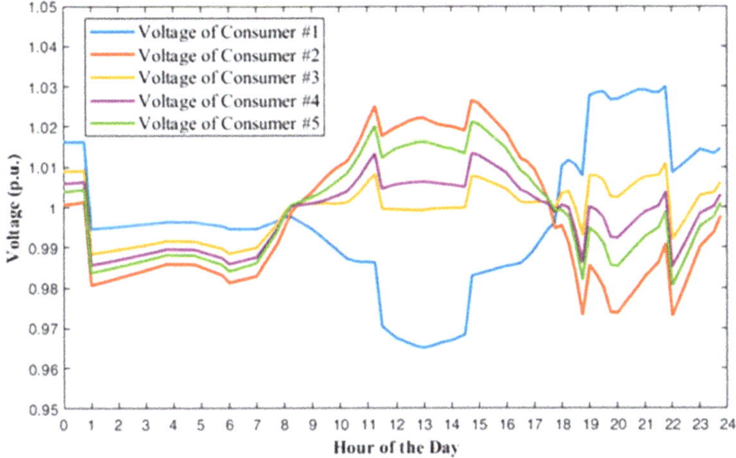

**Figure 14.** Voltage profile at consumer premises with the CVC.

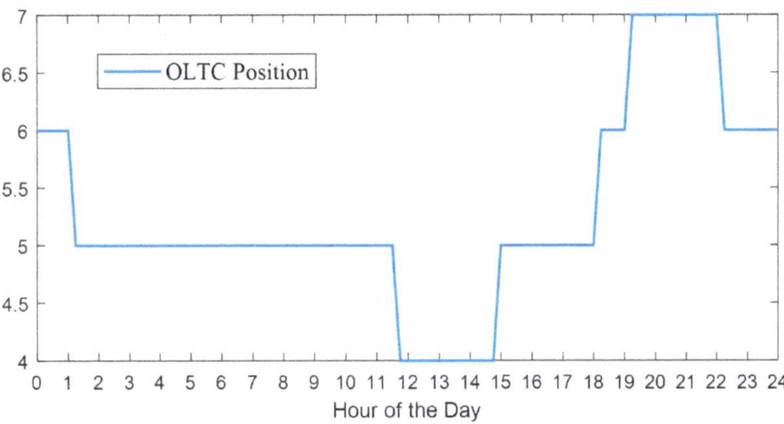

**Figure 15.** On Load Tap Changer (OLTC) position changes.

Under this setup, the ability of JaNDER platform to enhance real-time testing was investigated. The CVC was tested and validated against an OLTC controller and real-time simulator designed by specialized personnel in a different location. This test case clearly showed that JaNDER can be used for testing applications, like the CVC controller with equipment in different RIs without affecting its operation due to the time delays introduced by the interface. In Figure 16, the delays measured during a request for a tap position by the CVC to the OLTC are presented. The most significant delay is introduced by the OLTC controller/OLTC real-time simulator which emulates the actual performance of an OLTC in a transformer. The inverters, which are simulated in the DRTS located in the same infrastructure as the CVC controller, receive their setpoint faster, which results in faster acknowledgement of the setpoint implementation by the CVC controller. Therefore, both the distributed DERs implement their setpoints in different time frames and the CVC controller acknowledges those setpoints at different time stamps.

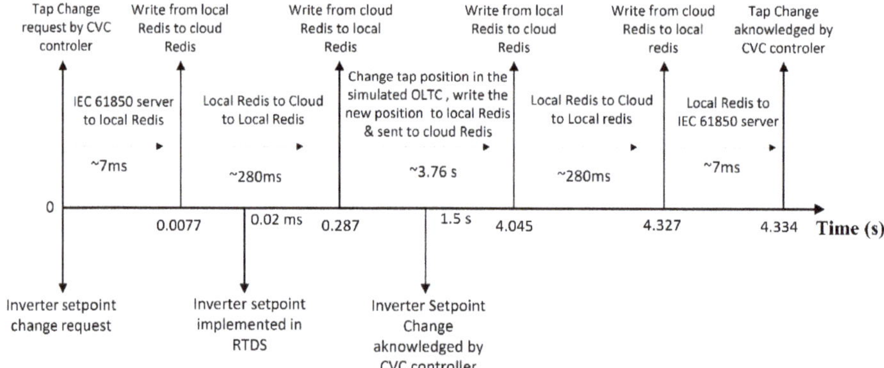


**Figure 16.** Time delays introduced in the setup.

However, the CVC time step (~1 min) is significantly larger than the time required from the CVC controller to send the setpoints to the DERs and OLTC and acknowledge their implementation (~5 s). Therefore, the overall behavior of this application is not affected by the different time frames that the devices react to the CVC controller setpoints.

As shown in Figure 16, the most significant delay is introduced by the OLTC operation according to the CVC request. This delay could differ according to the OLTC manufacturer. It is clear of course that if the OLTC controller's response is comparable to the centralized controller's timestep, the overall behavior of the system would be affected. For example, during online operation the centralized controller initiates a sequence by sending setpoints to all the local devices. If the OLTC controller response is comparable to the centralized controller timestep, then it is possible that the tap position will remain the same when the central controller begins measurements in the next timestep. The central controller would then send an additional request (e.g., tap increase) to the OLTC controller. If the OLTC controller implements those request in series, the resulting tap position would be wrong leading to undesired operation of the OLTC. To avoid this, the CVC controller should request, for example acknowledgement from the local controllers that the ordered setpoints have been implemented. The setup presented can effectively reveal such weaknesses of the controller design in a safe laboratory environment.

In this implementation, the most significant factor of failure is the OLTC response compared to the CVC timestep, since the platform's delays are insignificant compared to the CVC timestep. Nevertheless, the platform delays can affect the overall stability of the setup and the accuracy of the results, if applications requiring faster response are considered. The stability of a control system considering the sample rate of the controller for systems with communication network delays has been investigated in [26,27]. A controller with slower response or slower sample rate is usually a way to increase the robustness of the control against communication delays. This is not applicable in setups with several controllers located in remote locations, since altering their response might lead to a behaviour different than anticipated in the field deployment. A general rule of thumb when using the JaNDER platform to create a testing setup is to compare the network delays introduced by JaNDER, with the actual network delays expected in the field deployment or to ensure that the controller has a sample rate that is not affected by the communication delays. Otherwise, the testing setup is not suitable, as shown in the example of decentralized secondary voltage and frequency control of an islanded microgrid under varying time delays [27].

## 6. Discussion

From the point of view of the CVC users, this test case using JaNDER was more realistic by interfacing their DRTS with the industrial product. At the same time, the collaboration with the experienced industrial personnel has assured a safe and fast interfacing of both devices, since every team was able to prepare the equipment in their area of expertise, thus boosting the overall implementation time.

From the point of view of the industrial partner, the JaNDER implementation has provided an insight on state-of-the-art approaches on smart grid applications. At the same time, the introduction of their equipment in such application presented a new opportunity for future research and investigation area.

Since most of commercial real-time simulators have a considerable capital cost, from the overall project perspective, JaNDER can promote collaboration between different infrastructures in order to expand their real-time testing capabilities without spending significant resources in the upgrade of their equipment and can provide considerable benefits for both a commercial and academic field by connecting geographically distributed real-time simulators.

On the other hand, since the industrial protocol (IEC 61850) is widely supported by JaNDER platform, studying the impact of this communication alongside the electrical grid in real-time, can provide a valuable insight on the impact of the delays and interfacing issues of both networks.

Finally, connecting power components (e.g., inverters) to different infrastructures can also be utilized through this platform to study the latency impact and to test slow varying phenomena of these components. Knowing that this platform was proven as a reliable and valuable tool with no critical latency, it is interesting to expand it in real-time testing without affecting the stability of the experiment.

## 7. Conclusions

The integration of distributed renewable energy resources to the grid is increasing its complexity and, therefore, new flexible system test methods are needed to study their impact and to ensure their efficiency and security. A novel HIL setup, based on remote hardware integration, is presented in this paper. This new testing setup is available thanks to the real-time communication tool developed in the context of ERIGrid project called JaNDER.

To demonstrate this new testing setup, a real test case, interfacing an industrial device like an OLTC controller with a remote DRTS with a LV network model, has been implemented. The test results shows that the latency for data exchanging due to JaNDER is lower than 300 ms. Hence, this platform is suitable for testing remote hardware/software and analyzing slow varying phenomena. Indeed the latency does not affect the stability of the experiments.

**Author Contributions:** Conceptualization, C.S. and D.P.; Data curation, E.B. and D.T.L.; Project administration, L.P.; Resources, D.T.L., N.H. and N.A.; Software, E.B. and D.P.; Supervision, L.P.; Writing—original draft, L.P., E.B., D.P., D.T.L. and N.A.; Writing—review and editing, L.P., C.S., D.T.L. and N.H. All authors have read and agreed to the published version of the manuscript.

**Funding:** This work is supported by the European Community's Horizon 2020 Program (H2020/2014–2020) under project "ERIGrid" (Grant Agreement No. 654113). Further information is available at the corresponding website www.erigrid.eu.

**Conflicts of Interest:** The authors declare no conflict of interest.

## References

1. IPCC. *Climate Change 2014: Mitigation of Climate Change. Contribution of Working Group III to the Fifth Assessment Report of the Intergovernmental Panel on Climate Change*; Edenhofer, O., Pichs-Madruga, R., Sokona, Y., Farahani, E., Kadner, S., Seyboth, K., Adler, A., Baum, I., Brunner, S., Eickemeier, P., et al., Eds.; Cambridge University Press: Cambridge, UK; New York, NY, USA, 2014.
2. *Strategic Research Agenda for Europe's Electricity Networks of the Future*; European Commission, Directorate-General for Research: Brussels, Belgium, 2007.
3. *Smart Grid Mandate—Standardization Mandate to European Standardisation Organisations (ESOs) to Support European Smart Grid Deployment*; European Commission: Brussels, Belgium, 2011.
4. Farhangi, H. The path of the smart grid. *IEEE Power Energy Mag.* **2010**, *8*, 18–28. [CrossRef]

5.	IEA. *Technology Roadmap—Smart Grids*; IEA: Paris, France, 2011; Available online: https://www.iea.org/reports/technology-roadmap-smart-grids (accessed on 5 May 2020).

6.	Strasser, T.I.; Pröstl Andrén, F.; Widl, E.; Lauss, G.; De Jong, E.C.W.; Calin, M.; Sosnina, M.; Khavari, A.; Rodriguez, J.E.; Kotsampopoulos, P.; et al. Elektrotech. *Inftech* **2018**, *135*, 616. [CrossRef]

7.	Ravikumar, K.; Schulz, N.; Srivastava, A. Distributed simulation of power systems using real-time digital simulator. In Proceedings of the IEEE/PES Power Systems Conference and Exposition, Washington, DC, USA, 15–18 March 2009; pp. 1–6.

8.	Faruque, M.; Sloderbeck, M.; Steurer, M.; Dinavahi, V. Thermo-electric co-simulation on geographically distributed real-time simulators. In Proceedings of the IEEE Power Energy Society General Meeting, Calgary, AB, Canada, 26–30 July 2009; pp. 1–7.

9.	Rentachintala, K.S. *Multi-Rate Co-Simulation Interfaces between the RTDS and the Opal-RT*; Florida State University: Tallahassee, FL, USA, 2012.

10.	Stevic, M.; Vogel, S.; Monti, A. From Monolithic to Geographically Distributed Simulation of HVdc Systems. In Proceedings of the 2018 IEEE 19th Workshop on Control and Modeling for Power Electronics (COMPEL), Padova, Italy, 25–28 June 2018; pp. 1–5.

11.	Monti, A.; Stevic, M.; Vogel, S.; De Doncker, R.W.; Bompard, E.; Estebsari, A.; Gevorgian, V. A Global Real-Time Superlab: Enabling High Penetration of Power Electronics in the Electric Grid. *IEEE Power Electron. Mag.* **2018**, *5*, 35–44. [CrossRef]

12.	Available online: https://www.fein-aachen.org/projects/villas-framework/ (accessed on 5 May 2020).

13.	Henderson, M.I.; Novosel, D.; Crow, M.L. Electric Power Grid Modernization Trends, Challenges, and Opportunities. *IEEE Power Energy* **2017**.

14.	Maniatopoulos, M.; Lagos, D.; Kotsampopoulos, P.; Hatziargyriou, N. Combined control and power hardware in-the-loop simulation for testing smart grid control algorithms. *IET Gener. Transm. Distrib.* **2017**, *11*, 3009–3018. [CrossRef]

15.	Kotsampopoulos, P.; Lagos, D.; Hatziargyriou, N.; Faruque, M.O.; Lauss, G.; Nzimako, O.; Dinavahi, V. A Benchmark System for Hardware-in-the-Loop Testing of Distributed Energy Resources. *IEEE Power Energy Technol. Syst. J.* **2018**, *5*, 94–103. [CrossRef]

16.	Kotsampopoulos, P.; Georgilakis, P.; Lagos, D.; Kleftakis, V.; Hatziargyriou, N. FACTS Providing Grid Services: Applications and Testing. *Energies* **2019**, *12*, 2554. [CrossRef]

17.	Available online: https://redis.io/ (accessed on 5 May 2020).

18.	Available online: https://bitbucket.org/danielePala/redisrepl (accessed on 5 May 2020).

19.	Available online: https://bitbucket.org/danielePala/openiec61850-redis (accessed on 5 May 2020).

20.	Available online: https://github.com/beanit/openiec61850 (accessed on 5 May 2020).

21.	Available online: https://aws.amazon.com/ (accessed on 5 May 2020).

22.	Available online: https://databricks.com/ (accessed on 5 May 2020).

23.	Available online: https://spark.apache.org/ (accessed on 5 May 2020).

24.	Brandl, R.; Arnold, G.; Cirujano, P.; Pérez de Nanclares, G.; Del Rio Etayo, L.; Soto, A.; Larracoechea, I. Assesement of the Low Voltage Ride Through capability of a smart distribution transformer with On-Load Tap Changer (OLTC) for renewable applications. In Proceedings of the CIRED 2019 Ljubljana Workshop, Ljubljana, Slovenia, 7–8 June 2018.

25.	Ulasenka, A.; Del Rio Etayo, L.U.I.S.; Cirujano, P.; Ortiz, A.; Brandl, R.; Montoya, J. Holistic Coordination of Smart Technologies for efficiently LV operation, increasing hosting capacity and reducing grid losses. In Proceedings of the 25th International Conference on Electricity Distribution, Madrid, Slovenia, 3–6 June 2019.

26.	Zhang, W.; Branicky, M.; Phillips, S. Stability of networked control systems. *IEEE Control Syst.* **2001**, *21*, 84–99.

27.	Zhao, C.; Sun, W.; Wang, J.; Li, Q.; Mu, D.; Xu, X. Distributed Cooperative Secondary Control for Islanded Microgrid With Markov Time-Varying Delays. *IEEE Trans. Energy Convers.* **2019**, *34*, 2235–2247. [CrossRef]

*Article*

# A Comparison of DER Voltage Regulation Technologies Using Real-Time Simulations

Adam Summers [1],*, Jay Johnson [1],*, Rachid Darbali-Zamora [1], Clifford Hansen [1], Jithendar Anandan [2] and Chad Showalter [3]

[1]  Sandia National Laboratories, Albuquerque, NM 87123, USA; rdarbal@sandia.gov (R.D.-Z.); cwhanse@sandia.gov (C.H.)
[2]  Electric Power Research Institute, Palo Alto, CA 94304, USA; janandan@epri.com
[3]  Connected Energy, Pittsburgh, PA 15220, USA; chad.showalter@connectedenergy.com
*  Correspondence: asummer@sandia.gov (A.S.); jjohns2@sandia.gov (J.J.); Tel.: +1-505-284-9586 (J.J.)

Received: 18 May 2020; Accepted: 30 June 2020; Published: 10 July 2020

**Abstract:** Grid operators are now considering using distributed energy resources (DERs) to provide distribution voltage regulation rather than installing costly voltage regulation hardware. DER devices include multiple adjustable reactive power control functions, so grid operators have the difficult decision of selecting the best operating mode and settings for the DER. In this work, we develop a novel state estimation-based particle swarm optimization (PSO) for distribution voltage regulation using DER-reactive power setpoints and establish a methodology to validate and compare it against alternative DER control technologies (volt–VAR (VV), extremum seeking control (ESC)) in increasingly higher fidelity environments. Distribution system real-time simulations with virtualized and power hardware-in-the-loop (PHIL)-interfaced DER equipment were run to evaluate the implementations and select the best voltage regulation technique. Each method improved the distribution system voltage profile; VV did not reach the global optimum but the PSO and ESC methods optimized the reactive power contributions of multiple DER devices to approach the optimal solution.

**Keywords:** voltage regulation; distribution system; power hardware-in-the-loop; distributed energy resources; extremum seeking control; particle swarm optimization; state estimation; reactive power support; volt–VAR

---

## 1. Introduction

Installed variable distributed renewable energy is continuing to grow, with much of this growth being related to distribution systems [1]. In the past, utilities could assume unidirectional power flow and regulation of grid voltage within ANSI Standard C84.1 [2] limits. The penetration of distributed energy resources (DER) presents challenges for distribution circuit voltage regulation [3] because fluctuating DER current injection on the feeder from renewable energy resources leads to voltage perturbations [4]. The voltage swings are currently corrected by load tap changing (LTC) transformers, capacitor banks, and other voltage regulation equipment—often designed or placed prior to the addition of the DER systems. Many distribution system operators (DSOs) are concerned about increasing DER deployments, and often create screening criteria to limit the deployments [5–7], because they cannot guarantee power quality in high DER penetration scenarios [8].

New grid code requirements in Hawaii, California, and at the national level (i.e., IEEE 1547-2018 [9]) are requiring newly installed DER grid support functions, including volt–watt, fixed power factor, and volt–VAR. When these functions are configured correctly, distribution hosting capacity can be drastically increased by mitigating thermal and voltage excursions, minimizing losses, and maintaining ANSI limits [10,11].

Extensive research has been conducted to mitigate voltage deviations on distribution systems and can be categorized into three broad areas. The first area uses standardized DER grid support functions and can be further separated into supervised or unsupervised methods. The second area uses machine learning algorithms to mitigate the voltage deviations. The third area uses OLTCs and other distribution equipment to solve the voltage deviation problem, however, this is out of the scope of this paper.

Unsupervised functionality relies on the autonomous functionality in DER equipment and is characterized by speed and simplicity. Many of the original unsupervised voltage regulation studies were conducted by EPRI, Sandia National Laboratories (Sandia), National Renewable Energy Laboratory (NREL), and Georgia Institute of Technology in the 2010s, and typically focused on the autonomous voltage-reactive power (volt–VAR) function which adjusts DER-reactive power based on local grid measurements [12–15]. There has also been extensive research into the optimal placement of DER assets to avoid reaching voltage and thermal constraints [16,17].

Unfortunately, unsupervised voltage regulation algorithms rarely achieve the global optimal operation for a DER fleet because each DER does not have visibility into the rest of the circuit. Supervised control algorithms pull in data from multiple sources of telemetry in order to achieve greater control accuracy, accounting for the voltages at multiple points on the distribution circuit. These optimal DER setpoints are achieved with more sophisticated techniques combining distribution state estimation [18,19], and centralized or decentralized controls [20,21]. Other researchers have investigated the coordination of on-load tap changing (OLTC) transformers and wind systems reactive power control [22] and DER active power curtailment for voltage regulation [23]; while others have studied control of demand response loads [24] and community energy management systems on voltage regulation [25]. These supervised methods depend on bidirectional communications and novel optimization algorithms which require knowledge of the feeder states, topology of the circuit, and DER locations. Unfortunately, these approaches are currently infeasible because DSOs often do not have in-depth knowledge of their distribution system designs, locations where customer-owned DER equipment is interconnected, or the communications to end devices.

The two primary contributions of this work are (a) the development of a novel DER management system (DERMS), called the Programmable Distribution Resource Open Management Optimization System (ProDROMOS) [26], and (b) developing a standardized methodology to tune, evaluate, and compare different distribution voltage regulation approaches in increasing levels of fidelity. ProDROMOS was designed to ingest real-time feeder data to provide extensive visibility into distribution circuits using the Georgia Tech-developed Integrated Grounding System Analysis program for Windows (WinIGS) distribution system state estimation tool [27] and then optimize the DER-reactive power settings to meet voltage regulation and economic objectives. In the case of supervised voltage regulation methods, there are varying degrees of visibility into the power system. In some of the approaches (e.g., [28,29]), the DER are only aware of the other DER interconnection voltages and they adjust their reactive power to improve the visible voltage levels. In more sophisticated control methods, like that developed in ProDROMOS, distribution system state estimation [26,30–32] is used to expand the visibility into the distribution system, and therefore improve the ability of the DER to optimize their reactive power contributions to optimize the voltage profile across the entire feeder.

The second contribution is a standardized evaluation methodology comparing three DER-reactive power technologies using two different levels of fidelity: real-time (RT) power simulations and RT power hardware-in-the-loop (PHIL) simulations. For this project, the following voltage regulation methods were compared:

1. an unsupervised, volt–VAR (VV) function, defined by the IEEE 1547-2018 standard;
2. a real-time communication-based optimization technique called extremum seeking control (ESC) [33–36];
3. the ProDROMOS particle swarm optimization optimal power flow (PSO OPF) method.

In each of the fidelity levels, timing, control, and communication parameters were adjusted to enact the desired voltage regulation controls. A comparative analysis of the three methods was then performed to evaluate the effectiveness of each voltage regulation method for two reduced-order utility distribution circuits from Albuquerque, NM and Grafton, MA. The results showed that ESC and ProDROMOS provided the best voltage regulation and maintained ANSI C84.1-2011 Range A limits and reached near-optimal reactive power setpoints when controlling multiple DER. While not presented here, the results in this paper provided sufficient confidence in the control techniques that they were then executed on a live feeder in Grafton, MA—the results of the field demo are included in [26].

The reminder of the paper is organized as follows: Section 2 describes the voltage regulation problem, the DER control methods and the RT simulation platform; Section 3 describes the simulation tests and results; Section 4 provides a discussion of the results; Section 5 provides a conclusion to the study. The Supplementary Materials section provide access to portions of the non-proprietary models and code used for these experiments.

## 2. Voltage Regulation Methods

In a distribution circuit without generation, power flows from a substation to customers and the voltage generally drops from source to load. As previously stated, the utility will regulate the voltage with LTC transformers, capacitor banks, and other voltage regulation equipment. However, a distribution circuit with DER may create bidirectional power flows and increase voltages at the point of common coupling (PCC). The increasing voltage operating envelope and variable power output of the DER create voltage perturbations and may violate ANSI C84.1 Range A limits. The voltage perturbations increase the mechanical operations of the utility-owned voltage regulation equipment [8,35]. This equipment was not designed for such rapid switching, which increase operations and maintenance (O&M) costs and decreases the DSOs ability to provide safe and reliably power [8]. An alternative solution is to use the equipment that is causing the voltage rise to mitigate the problem by using the DER grid support functions with reactive power capabilities. In this paper, we investigate volt–VAR, ESC, and ProDROMOS controls. Volt-var is a common function required by many European and North American grid codes and interconnection standards [37–39] included in most DER devices. ESC and ProDROMOS are more sophisticated controllers which calculate the target reactive power of DER and then set a power factor function to produce the reactive power level. It should be noted that in new DER interconnection standards, such as IEEE Std 1547-2018, there are constant reactive power modes—but these functions were not available in the equipment at the time of these experiments.

### 2.1. Autonomous Volt–VAR Control

The volt–VAR (VV) function adjusts the reactive power of the DER based on the PCC voltage measurements according to an adjustable curve of (voltage and reactive power) points. This distributed autonomous control function was the first reactive power factor evaluated. In this project, the VV curve for the RT experiments was defined with non-aggressive voltage points = {95%, 99%, 101%, 105%} and Var points = {25%, 0%, 0%, −25%} because there were oscillations between the simulated DER device and the power simulation.

### 2.2. Extremum Seeking Control

Extremum seeking control (ESC) is a real-time optimization technique for multi-agent, nonlinear, and infinite-dimensional systems [33–36]. The decentralized, model free control algorithm operates by adjusting the reactive power to optimize measured outputs, in this case, feeder voltage. The system adjusts inputs; ($u$); via a sinusoidal injection, demodulates system outputs; ($J(u)$); to extract approximate gradients, and finally performs a gradient descent. A block diagram of this approach is shown in Figure 1. The designer sets the parameters $k$, $l$, $h$, $a$ and $\omega$. Additionally, unique probing frequencies for each controllable DER are chosen, but one must ensure that $l$ and $h << \omega$, ensuring proper operation of the high and low pass filters in each ESC loop. The reader is referred to [36] for more details.

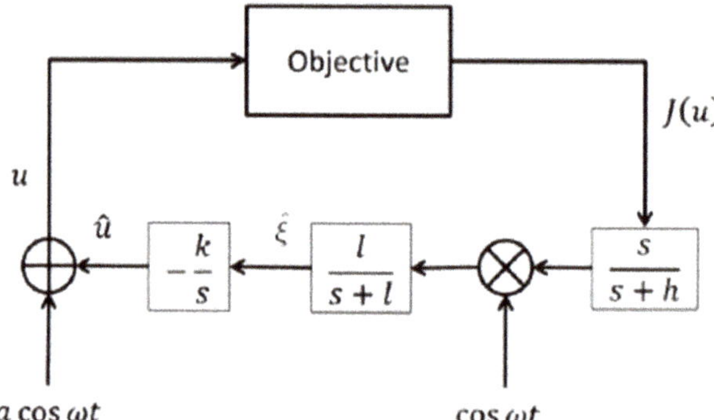


**Figure 1.** Block diagram of extremum seeking control (ESC). Refer to [36] for description.

The optimization function for the simulations is given in Equation (1):

$$J = \frac{1}{n} \sum_{i=1}^{n} \sum_{ph=1}^{3} \left( V_{i,ph} - V_{nom,i,ph} \right)^2 \tag{1}$$

where $V_i$, is the voltage measurement at Bus $i$ and $V_{non}$ is the nominal voltage. This is a proposed new grid support function that can achieve the global optimum. We assume that each bus in the circuit feeder, can provide measurements in these experiments and more details will be provided in the RT simulation platform section.

Extremum Seeking Control Parameter Selection

These parameters were chosen based on prior experience with ESC simulations and laboratory demonstrations [33–36], and a systematic debugging process that considered communication latencies, DER power factor (PF) accuracies, and prior lessons learned. The sequence to implement ESC control follows:

1. The ProDROMOS manager constructs the objective function, $J$.
2. The ProDROMOS manager configures the DER with unique $\omega$s to avoid controller conflict while producing 10 or more data points per cycle.
3. The parameters $l$ and $h$ are set significantly less than $\omega$, such that the perturbation is passed through the washout filter $s/(s + h)$ but is removed by the lowpass filter $l/(s + l)$.
4. Parameter $a$ is selected to produce the smallest reactive power oscillation that is observable in the objective function, $J$.
5. Parameter $k$ is set based on designer experience, the stability of ESC simulation, and desired time to reach local minimum of $J$.
6. The ProDROMOS monitors the objective function and makes PF changes to each DER to improve the performance of the system.

The optimization of ESC parameters is a future research interest due to the time requirements necessary to tune the parameters for the application. The use of the ProDROMOS platform and the RT simulator allowed the ESC designer confidence in the parameters before real world deployment.

### 2.3. ProDROMOS

The ProDROMOS system completed the DER-reactive power setpoint optimization in multiple stages. Initially, intelligent electronic devices (IEDs) from the power system would collect data from the RT, PHIL, or live feeder and send those to the WinIGS state estimation system tool. For the RT and PHIL experiments, these data were collected from virtualized micro-phasor measurement units (µPMUs) and formatted in IEEE C37.118 format using a physical SEL-3373 Phasor Data Concentrator (PDC) in the Distributed Energy Technologies Laboratory (DETL) at Sandia National Laboratories. These data were then parsed into different sections of the WinIGS feeder. By segmenting the state estimation problem into smaller feeder pieces, WinIGS was able to solve the state estimation at 30 Hz. Details of the construction of the WinIGS models for the Public Service Company of New Mexico (PNM) and National Grid (NG) feeders in NM and MA are presented in [26]. The solution to the state estimation included active and reactive power values for different loads in the model. These values were passed to the optimization routine as another IEEE C37.118 data stream.

The Particle Swarm Optimization (PSO) was used to determine the Optimal Power Factor (OPF) settings for the DER devices because of the nonconvex fitness landscape [26]. PSO OPF uses a time-series OpenDSS simulation wrapped in a PSO loop to calculate the OPF values. The load data for OpenDSS were populated with state estimation solutions before each run of the optimization routine, which occurred every 1 or 2 min. A solar production persistence forecast was created to estimate the solar energy over the next time horizon (5 min). This forecast was used as the photovoltaic (PV) power profile in the OpenDSS simulation. The WinIGS software and the python-based PSO code were implemented in a portable Connected Energy cloud-based Windows virtual machine (VM) that allowed geographically distributed developers to access the environment. Figure 2 provides a simplified approach of the PSO OPF method.

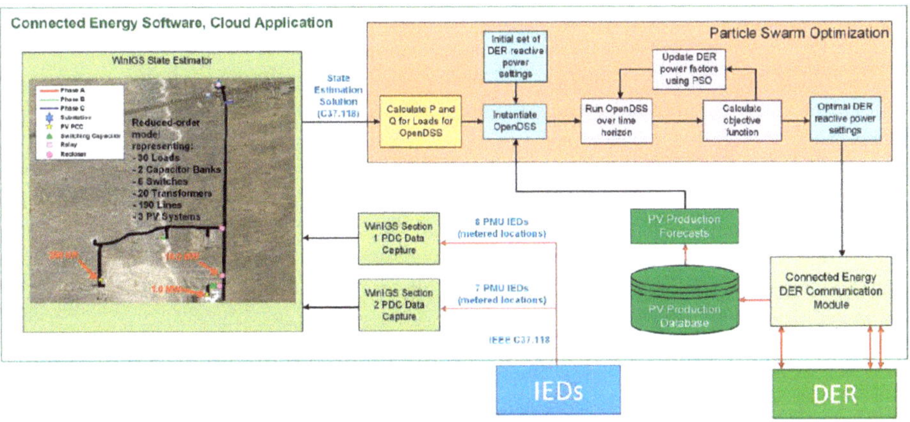


**Figure 2.** The particle swarm optimization optimal power flow (PSO OPF) optimization method and information flows.

Once populated with the anticipated PV production and the current—and assumed static—active and reactive power levels for the loads in the OpenDSS simulations, the PSO ran the OpenDSS model with different DER PF setpoints to calculate the optimal setpoints for each of the DER devices. The optimization formulation was designed to capture the voltage regulation and economic considerations of operating PV systems with a non-unity power factor (PF):

$$\text{Minimize}_{PF}\left[w_0\delta_{violation}(\boldsymbol{V}) + w_1\sigma(\boldsymbol{V} - \boldsymbol{V}_{base}) + w_2C(\boldsymbol{PF})\right] \qquad (2)$$

where

$$\delta_{violation}(\mathbf{V}) = 1 \tag{3}$$

if any

$$|V| > V_{lim} \tag{4}$$

$$\sigma(\mathbf{V} - \mathbf{V}_{base}) \tag{5}$$

is the standard deviation of $\mathbf{V} - \mathbf{V}_{base}$

$$C(PF) = \sum 1 - |\mathbf{PF}| \tag{6}$$

$V$ is a vector of bus voltages, $\mathbf{V}_{base}$ is a vector of the nominal voltages for each bus, and $\mathbf{PF}$ is a vector of the DER PFs. The objective function is minimized when the bus voltages are at $\mathbf{V}_{base}$ and PF = 1. $V_{lim}$ was selected to be the ANSI C84.1 Range A limits of ±0.05 per unit (pu), so any solutions outside the limits would be highly penalized. The third term was a simplified method to discourage solutions that moved away from unity power factor, because these solutions would curtail active power (and expense the PV owner through net metering, power purchase agreements, etc.) at high irradiance times. More sophisticated methods for determining the curtailment magnitude were considered, but the simple approach shown here was implemented. For the experiments conducted in this project, $w_0 = 1.0$, $w_1 = 2.0$, and $w_2 = 0.05$. The optimization was configured so that if all the bus voltages were within an acceptance threshold (set to 0.2% of nominal voltage) the PSO would not run. If any of the voltages were outside ANSI Range A the PSO would run. Furthermore, if the new PF values did not change the objective function by an objective threshold (set to $1 \times 10^{-7}$), the new PFs would not be sent to the DER devices to minimize communications and DER memory writes.

## 3. Evaluation Environments

### 3.1. Distribution Systems of Study

Several distribution systems models were acquired from PNM and NG for evaluation. The PNM models were in the SynerGEE software format and NG models were in the CYME software format. To run the power hardware-in-the-loop (PHIL) experiments, a series of conversions were needed to convert the models to OPAL RT-Lab-compliant formats (MATLAB/Simulink). The CYME and SynerGEE models were first converted to OpenDSS. The objective of the project was to show voltage regulation with 50% PV penetration level (defined by feeder PV nameplate/peak feeder load). For those feeders that did not meet this threshold, additional residential PV systems were randomly added to the buses in the model to meet this goal. Time-domain RT-Lab simulations are not possible with feeder models with thousands of buses, so the full OpenDSS models were reduced in complexity to be incorporated in the RT-Lab and WinIGS platform. The circuit reduction algorithms reduced the number of buses in a model while maintaining an equivalent voltage profile throughout the model at feeder locations of interest [40,41]. Ultimately, a commercial 12.47-kV PNM feeder and a suburban 13.80 kV NG feeder were used for the evaluation of the different voltage regulation algorithms. The PNM reduced order feeder that was used in these experiments is shown in Figure 3 and the NG feeder is shown in Figure 4.

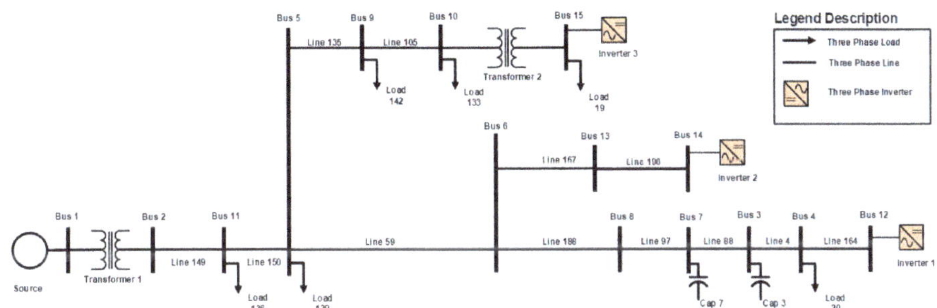

**Figure 3.** One Line Diagram of PNM Feeder.

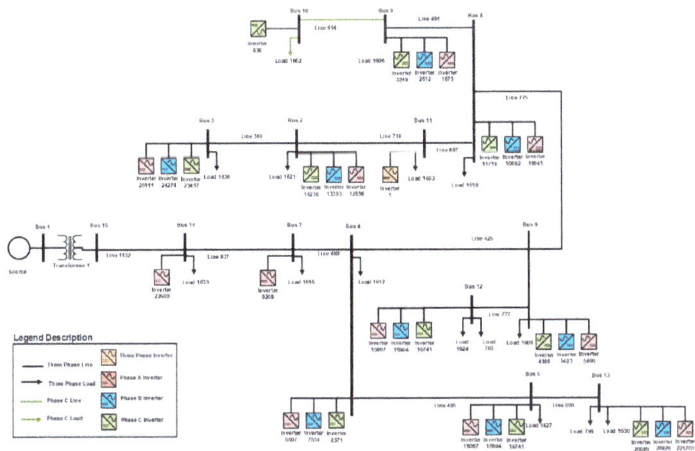

**Figure 4.** One-line diagram of National Grid (NG) feeder.

## 3.2. Real-Time Evaluation Platform

The evaluation platform could be run in a software-in-the-loop (SIL), a power-hardware-in-the-loop (PHIL), or a combination of SIL and PHIL setup. The SIL and PHIL simulations were conducted with an Opal-RT OP5600 real-time digital simulator running distribution feeder models in RT-LAB with a 100 us timestep. To complete the PHIL experiments, a physical 3.0-kW inverter was interfaced to the power simulation via a 180 kVA Ametek RS180 grid simulator. The current waveform was measured by a Person CT110 and returned to the power simulation via an analog input channel. The DC side of the inverter was connected to an Ametek TerraSAS PV simulator configured to represent a 3.0-kWp silicon PV system.

In the RT simulations, the EPRI-developed DER Simulator [42] software, emulated smart solar inverters with several grid support functionalities and communication interfaces. Any number of DER devices, with different nameplate capacities and phasing configurations, were instantiated with independent irradiance profiles. The EPRI-developed TCP/IP protocol called Data Bus (DBus) was used as the co-simulation interface to exchange data between the OPAL-RT simulator and the DER simulator. The DER simulator determined the active and reactive power of the DER devices based on irradiance profiles, voltage, frequency, and DER settings. The active and reactive power outputs are exchanged via DBus with the RT simulation. The RT simulation passed the voltage and frequency of the points of common coupling (PCCs) where the DER were interconnected to the feeder.

The PHIL simulations used the RT simulation environment but replaced one of the EPRI PV systems with a 3.0-kW residential inverter that was then scaled to a device in the PNM model (inverter 3, 258 kW) and NG feeder (inverter 1, 658 kW). The 258-kW and 684-kW inverters were three-phase balanced output inverters where each phase produced the same current magnitude and phase angle. The 3-kW inverter was scaled to match its per phase output to the 684-kW inverter for the NG experiments and the 258-kW inverter for the PNM experiments. The advantage of incorporating a physical inverter into the simulations was twofold: the control algorithms had to be built with communication interfaces to interact with real equipment over the public internet networks and the abnormal operations of physical equipment (e.g., startup times, ramp rates, non-optimal maximum power point tracking, converter losses, etc.) were represented in the power simulation and needed to be managed by the control algorithms. Scaling a residential PV system to represent a utility-owned site likely produced higher reactive power errors because of the lower-cost control and power electronics components in the 3-kW inverter, but this helped approximate inverter interoperability and control functionality, with the expectation that the fielded system would perform better.

## 4. Results for PNM Model

### 4.1. Simulations with the PNM Model

The PNM model included two scenarios. In the first scenario, there were three simulated EPRI PV inverters, and in the second scenario, the 258-kW inverter at bus 15 was replaced with a physical single phase 3.0 kW that was then scaled to the 258-kW rating in the simulation. The EPRI simulator provided three correlated, highly variable irradiance profiles from a PV site on the US east coast to feed the simulated inverters. These profiles were generated from a single Global Horizontal Irradiance (GHI) measurement and then shifted and scaled to account for geographical separation of the PV sites. The PV irradiance profiles were then smoothed based on the plant sizes using the Wavelet Variability Model [43–46] to account for spatial averaging with large PV systems. To understand the impact of spatial averaging on the irradiance profiles the reader is invited to review [47,48]. The final PV profiles are shown in Figure 5. The smoothed irradiance profiles are used because clouds do not instantaneously shade larger PV systems and it takes time for them to pass over the entire array. Additionally, variable load profiles were used on loads connected to bus 6 and bus 13 to create transient voltage variability.

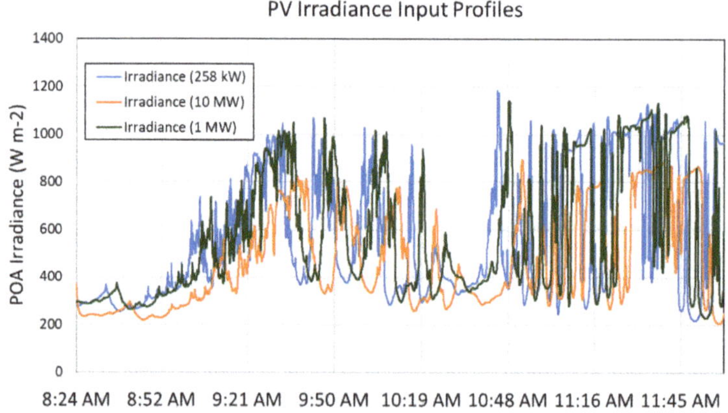


**Figure 5.** PV irradiance profiles for the Public Service Company of New Mexico (PNM) model.

## 4.2. Baseline Simulation

The PNM model was simulated using P/Q data passed from the EPRI DER simulator to the RT power simulation using the Data Bus (DBus) exchange. The hardware PV inverter was interfaced using the Ideal Transformer Method (ITM) [48]. The simulated PV inverters connected to bus 12 and bus 14 were simulated using a P/Q controlled PV inverter model [49]. In this feeder model, controllable capacitor banks were removed. At each PV site, irradiance is converted to AC power using pvlib-python [50]; an example of the power simulation is shown in Figure 6. The voltage profile for each of the 15 buses is shown in Figure 7. As can be seen from these results, there is relatively little voltage rise when the PV systems inject power at their points of common coupling (PCCs). Overall, the feeder maximum voltage is driven by the 10-MW PV system on this feeder.

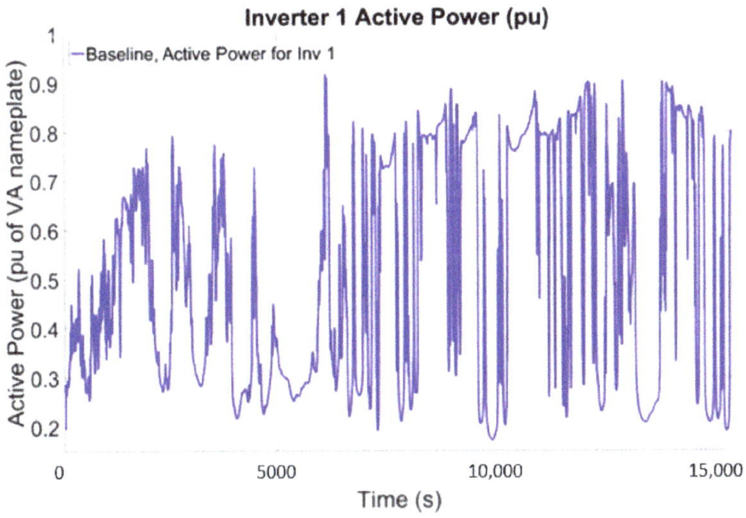


**Figure 6.** PV power levels for 4- hour simulation.

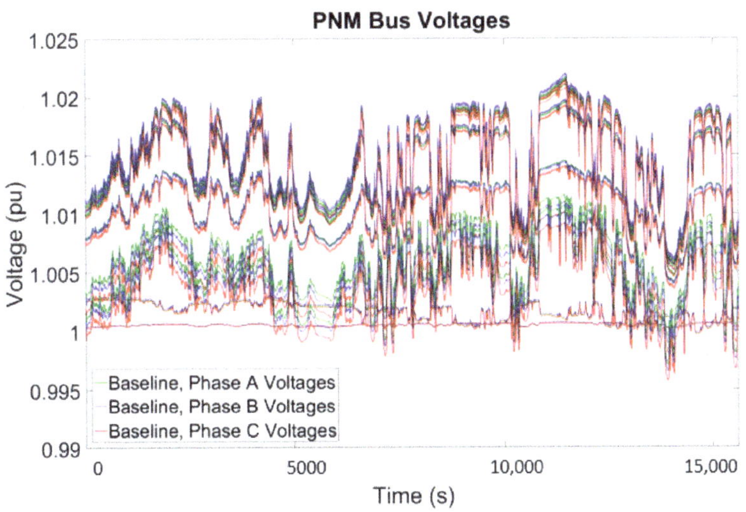


**Figure 7.** Bus voltages for baseline simulation.

### 4.3. Volt–VAR Simulation

The volt–VAR function was implemented by programming the VV function in the EPRI DER devices using a DNP3 command. The deadband was set to ±0.01 pu so the DER would support the feeder voltage. Aggressive VV parameters produced reactive power oscillations because the EPRI simulator-to-Opal-RT communication rates were limited to once/second—e.g., DBus read the DER bus voltage once a second and then DBus updated the Opal-RT DER-reactive power set point on the next DBus write. As a result, a stable VV curve defined as V = {92, 99, 101, 108} pu, Q = {25, 0, 0, −25}% nameplate VA capacity were used for the simulations. The reactive power contribution from the 10-MW plant was modest, but it reduced the maximum and average feeder voltages, as shown in Figure 8—where the minimum, maximum, and average voltages were plotted using for all buses and all phases. The line represents the average voltage for all the buses and the colored patch represents the range of voltages over time for the feeder. Notice from the simulation results that the VV function can bring the average bus voltage closer to the desired nominal value. These effects can also be observed from the maximum voltage, reducing the voltage band created from the maximum and minimum values.

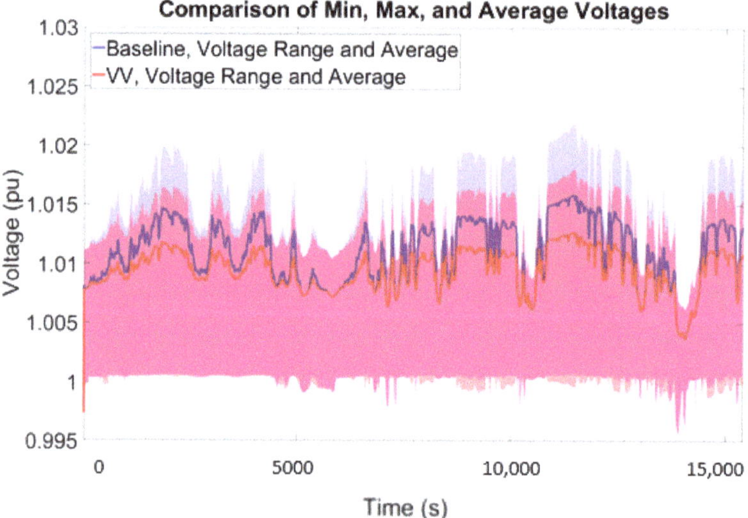


**Figure 8.** Volt–Var (VV) vs baseline feeder voltage comparison.

### 4.4. Extremum Seeking Control Simulation

ESC was implemented with a RT model of the PNM feeder using multiple parameter sets. When selecting ESC parameters, there are multiple tradeoffs that make it challenging. For instance, large probing magnitudes help the objective function signal stand out from feeder noise, but it causes larger voltage deviations once a solution is reached. The normalized voltage deviation objective function was used with probing signal frequencies that could be independently demodulated. Ultimately, the parameters shown in Table 1 were used. As shown in Figure 9, the reactive power from the DER devices to move the average bus voltage toward the nominal set point, i.e., 1 pu as compared to the baseline simulation. The maximum and minimum bus voltage values were adjusted closer to nominal as well. The reactive power probing signal's impact on the feeder voltage is clearly seen in the results.

**Table 1.** ESC Parameters.

| J Function | $l$ | $h$ | $r_{comm}$ | Inverter 1 (258 kW) | | | | Inverter 2 (10 MW) | | | | Inverter 3 (1 MW) | | | |
|---|---|---|---|---|---|---|---|---|---|---|---|---|---|---|---|
| | | | | $P$ | $f$ | $a$ | $k$ | $P$ | $f$ | $a$ | $k$ | $P$ | $f$ | $a$ | $k$ |
| $\frac{1}{n}\sum_{i=1}^{n}\left(\frac{V_i - V_n}{V_n}\right)^2$ | $\frac{\sqrt{5}}{800}$ | $\frac{\sqrt{5}}{800}$ | 2 s | 258 kW | $\frac{\sqrt{2}}{40}$ | 51.6 kVar | $-2.58 \times 10^7$ | 10 MW | $\frac{\sqrt{3}}{40}$ | 50 kVar | $-1 \times 10^9$ | 1 MW | $\frac{\sqrt{5}}{40}$ | 50 kVar | $-1 \times 10^8$ |

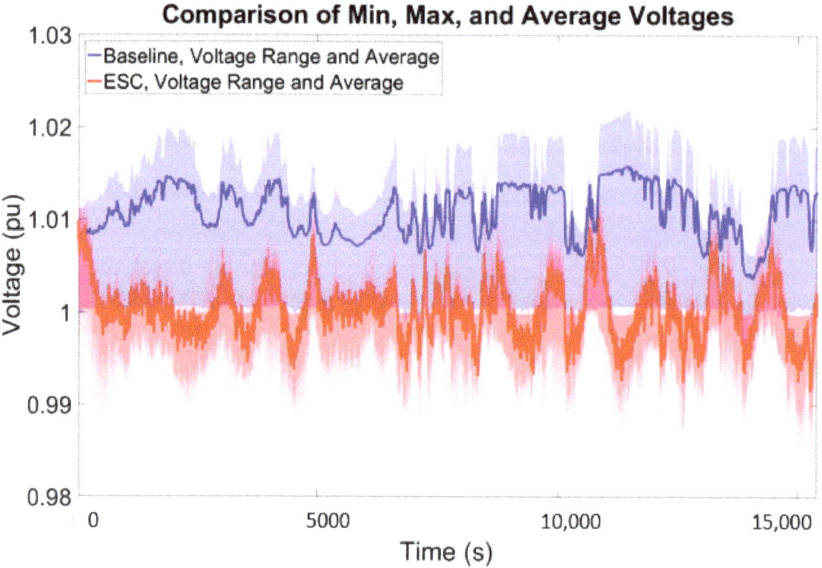

**Figure 9.** ESC results for the PNM model with parameter set 2 for a period of 4 hours.

*4.5. State-Estimation-Based Particle Swarm Optimization*

The setup for the PSO simulations is shown in Figure 2. In this case, the state estimation was used to update the load data in the OpenDSS time series simulation and forecasts of three-minute average DER power were used to update the OpenDSS power levels. Figure 10 illustrates the average power forecasts. The forecasts track the average irradiance reasonably well with some lag as is common for persistence forecasts. Note that it takes 1 h for the forecast to begin because it uses prior production data and the clear sky index to generate the power prediction.

The OpenDSS time series simulation was run multiple times using PSO to determine the optimal PF set points for each of the DER devices. It was found that absolute care must be taken to use the same component and settings for the Simulink/RT-Lab and OpenDSS models; otherwise erroneous solutions were found. Binary variables (e.g., capacitor banks) and discrete variables (e.g., tap changes) must also be re-initialized (i.e., reset) with each new run of the OpenDSS time series simulation to prevent initial conditions from being set by the prior simulation—an unfortunate byproduct of using the OpenDSS communication interface implementation. As shown in Figure 11, the reactive power from the DER devices to move the average bus voltage toward the nominal set point, i.e., 1 pu as compared to the baseline simulation. Interestingly, with the constraints placed on updating a new PF level (e.g., the solution must be a certain amount better than the previous one), there are only 3 times the DER devices had their PF setting updated as shown in Figure 12. This threshold can be tuned, but it is desirable because it reduces the number of PF write commands and shows the solution is robust to changing PV irradiance. As seen in Figure 12, the initial PF setpoints for DER 1 and 2 absorbed reactive power while DER 3 injected reactive power. For this feeder topology and DER locations, the global optimum required reactive power absorption or injection to minimize the voltage deviation from nominal at all buses in the feeder.

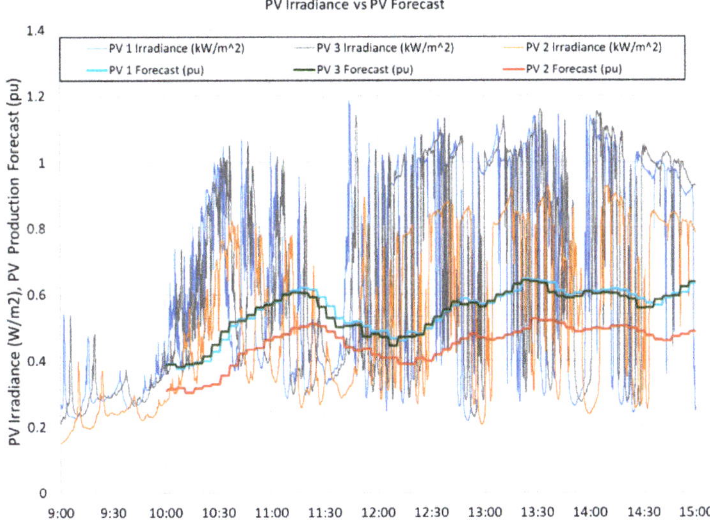



**Figure 10.** Irradiance and PV Energy Forecast Profiles. 1000 W/m$^2$ produces 1pu of Active Power.

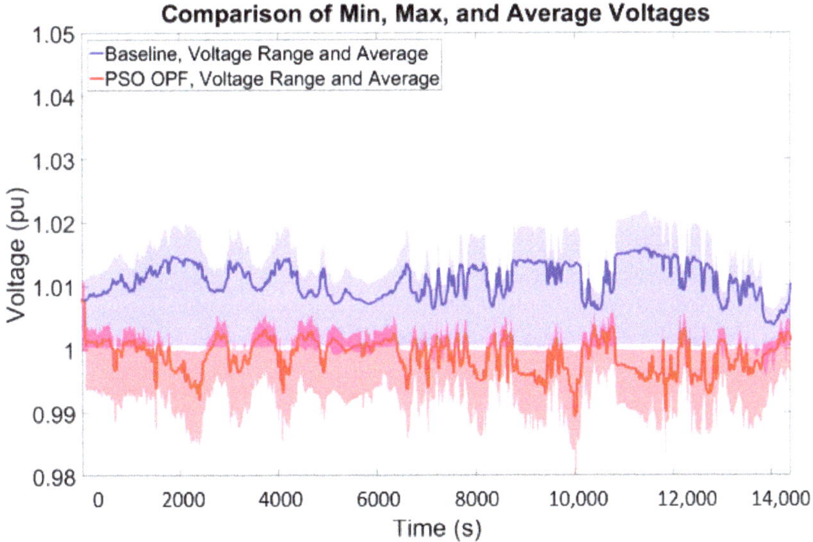



**Figure 11.** PSO results for the OpenDSS PNM model without Switching Capacitor Banks.

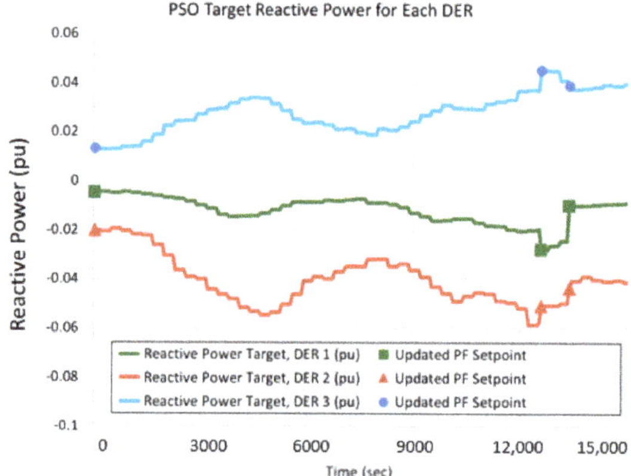

**Figure 12.** PSO PNM target reactive power levels for each distributed energy resources (DER). The target reactive power is calculated from the PSO power factor (PF) set point and the PV forecast at that time.

## 5. Results for National Grid Model

### 5.1. Simulations with the National Grid Model

Like the PNM model, VV, ESC, and PSO voltage regulation approaches were compared on the NG feeder model. In this model, there is a 684-kW utility-owned three-phase PV site and 30 single phase PV systems. The NG system was unbalanced, with phase B above nominal voltage, phase C below nominal voltage, and phase A operating approximately at nominal voltage. This prevented many of the voltage regulation methods from making significant improvements to the feeder voltage profile when only the three-phase utility-owned inverter was controlled. Therefore, we proposed controlling all 31 DER on the NG model for further testing.

### 5.2. Baseline Simulation with National Grid Model

Four-hour irradiance profiles, at one second resolution were created for the 31 DER devices in this model. Figure 13 shows the voltage profiles for each of the phases.

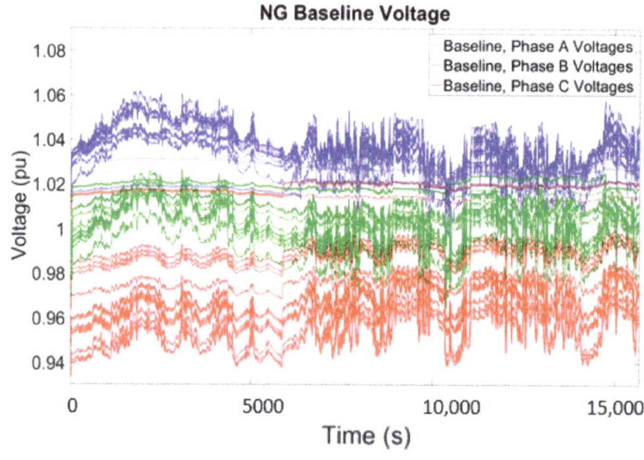

**Figure 13.** Voltage profiles for the NG simulations with each phase colored.

### 5.3. Volt–VAR

There were negligible reactive power contributions from the utility-owned 684-kW PV site because the average voltage at the PCC was within the VV dead band. However, the single-phase inverters were able to shrink the feeder voltage envelope, as shown in Figure 14.

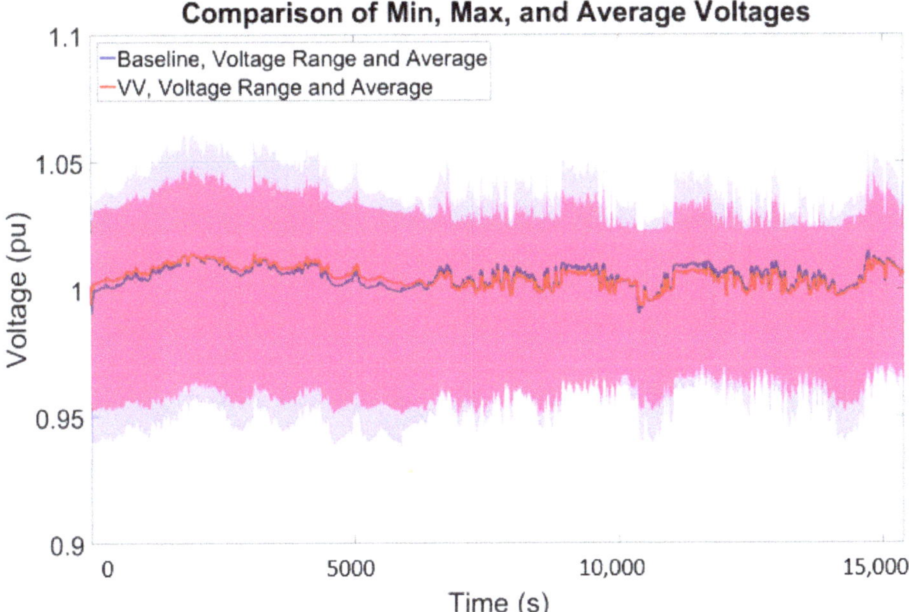



**Figure 14.** Minimum, maximum, and average bus voltages vs time for the VV test compared to the baseline data controlling all PV inverters.

### 5.4. Extremum Seeking Control

ESC was conducted on the NG feeder model. In this feeder, there was significant phase imbalance and at the PV system: phase A voltages are close to nominal (~0.99–1.01 pu), phase B voltages are significantly above nominal (at times >1.05 pu), and phase C voltages are <0.95 pu—as shown in Figure 15. This presented a challenge for the ESC approach (like all the other voltage regulation methods) because the DER can only inject symmetric (positive sequence) reactive power. This means all the phase voltages are shifted in the same direction with a change in the DER power factor. Grouping the DER based on what phase they were interconnected was a good method of limiting the number of probing frequencies but allowed each of the phases to adjust their reactive power contributions independently. Unfortunately, this also produced a sizable voltage ripple on the power system because all the inverters on each phase had the same probing frequency. Ultimately, the parameters shown in Table 2 were used for ESC testing on the NG model, and the feeder voltage profile is shown in Figure 16.

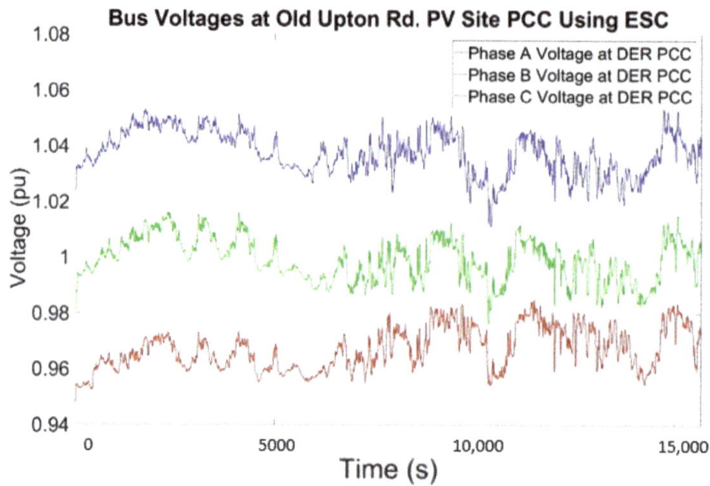


**Figure 15.** Imbalanced phase voltages; DER device during ESC simulation using parameter set 1.

**Table 2.** ESC parameters for all inverters in the NG feeder model.

| $J$ function | $l$ | $h$ | $r_{comm}$ | Three-Phase Inverter | | | Inverters on Phase A | | | Inverters on Phase B | | | Inverters on Phase C | | |
|---|---|---|---|---|---|---|---|---|---|---|---|---|---|---|---|
| | | | | $f$ | $a$ | $k$ | $f$ | $a$ | $k$ | $f$ | $a$ | $k$ | $f$ | $a$ | $k$ |
| $\frac{2}{n}\sum_{i=1}^{n}\left(\frac{\overline{V_i}-V_n}{V_n}\right)^2$ | $\frac{\sqrt{5}}{800}$ | $\frac{\sqrt{5}}{800}$ | 2 s | $\frac{1}{40}$ | $\frac{S}{15}$ | $\frac{-S}{100}$ | $\frac{\sqrt{2}}{40}$ | $\frac{S}{10}$ | $\frac{-S}{100}$ | $\frac{\sqrt{3}}{40}$ | $\frac{S}{10}$ | $\frac{-S}{100}$ | $\frac{\sqrt{5}}{40}$ | $\frac{S}{10}$ | $\frac{-S}{100}$ |

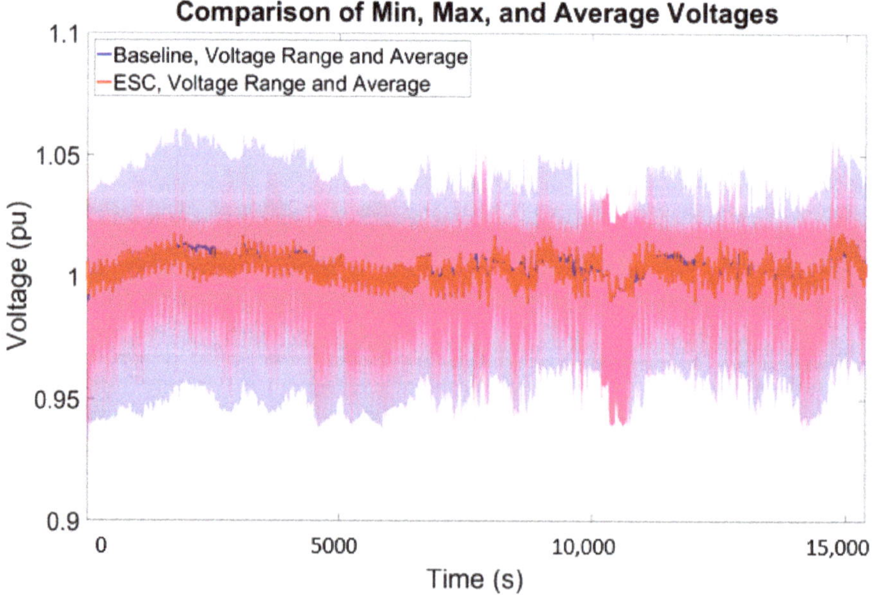


**Figure 16.** Minimum, maximum, and average bus voltages vs time for the ESC test, compared to baseline results controlling all PV inverters.

## 5.5. State-Estimation-Based Particle Swarm Optimization

The DER PF PSO optimization technique wrapped NG OpenDSS time-series simulations. The code was configured to execute once a minute to determine the optimal set point for all the DER devices. Using the same swarm size of 60, the solution often took 10 iterations, where it was capped. In those cases, it took longer than 60 s, so the optimization was configured to run every two minutes. As shown in Figure 17, optimizing 31 DER devices were difficult with a swarm size of 60 because the first two solutions produced results that pushed the minimum or maximum voltages outside the baseline envelope. However, after a few more solutions, the PSO significantly improved the voltage profile of the feeder for the remainder of the simulation. It should be noted that we did not attempt to optimize the PSO code to minimize run time.

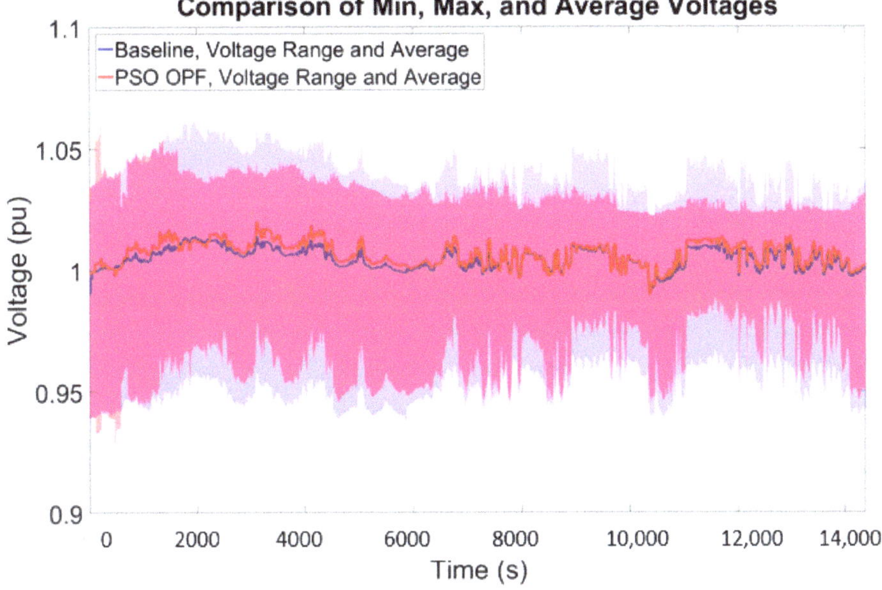


**Figure 17.** Minimum, maximum, and average bus voltages vs time for the PSO test compared to the baseline results controlling all PV inverters.

## 6. PHIL Results

To further validate the operational effectiveness of the PSO voltage regulation method, realistic DER power hardware-in-the-loop (PHIL) simulations were conducted with both the PNM and NG feeder models. These simulations provide better fidelity because they show ProDROMOS can work with real PV communications systems and ramp rates.

### 6.1. PNM Baseline

To validate the PHIL setup for the PNM feeder model, a comparison was made between active power, reactive power, and bus voltage for the RT and PHIL simulations. Figure 18 shows the voltage profile for the RT and PHIL simulations. The positive and negative reactive power changes around 1000 s in the baseline case were due to setting the PF setpoint on the 10-MW system to +0.85 and −0.85 to verify communications and that the PHIL simulation was closed-loop. Overall, the RT and PHIL simulations are closely matched.

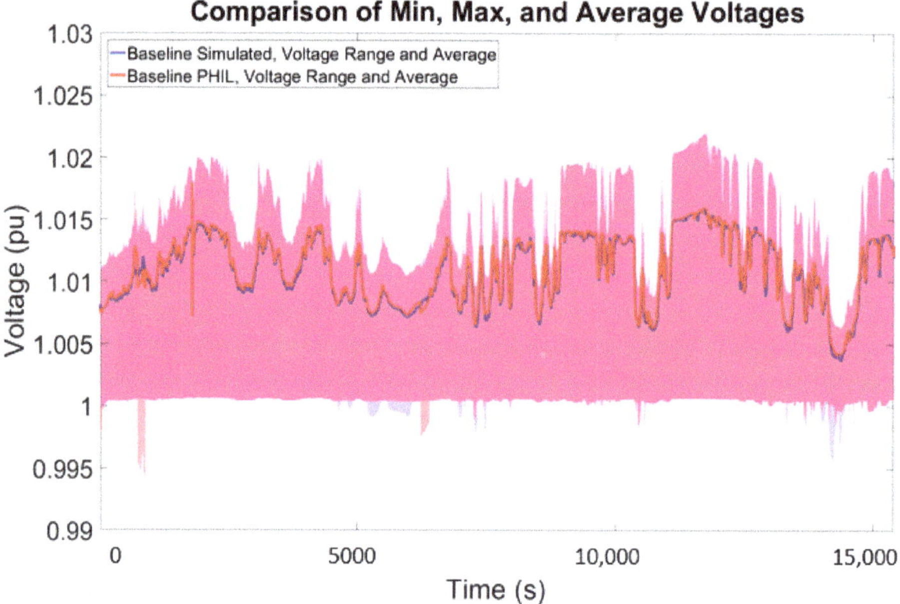


**Figure 18.** Comparison between simulated and power hardware-in-the-loop (PHIL) baseline minimum, maximum, and average bus voltages for the PNM model.

*6.2. PNM Particle Swarm Optimixation PHIL*

The only change from the RT simulations was that commands were communicated with the physical device to change the power factor and the DER was connected to the power simulation using a PHIL interface, as opposed to DBUS. The 258-kW system was recreated by scaling a 3-kW DER device. It was found that the physical equipment recreated the power profile reasonably well. Slight differences in the active power are due to efficiencies of the devices and slightly oversizing the simulated PV system in the Ametek PV simulator. The PSO solutions were repeatable for multiple runs. There were some deviations in the reactive power contributions from the PSO solutions, but overall, they matched well, as shown in Figure 19. Do note the PHIL PSO simulation was not started for approximately 10 min at the beginning of the RT-Lab simulation, which is why it takes some time for the reactive power on the physical DER to change. Similar results for the NG RT and PHIL simulations are described in [26].

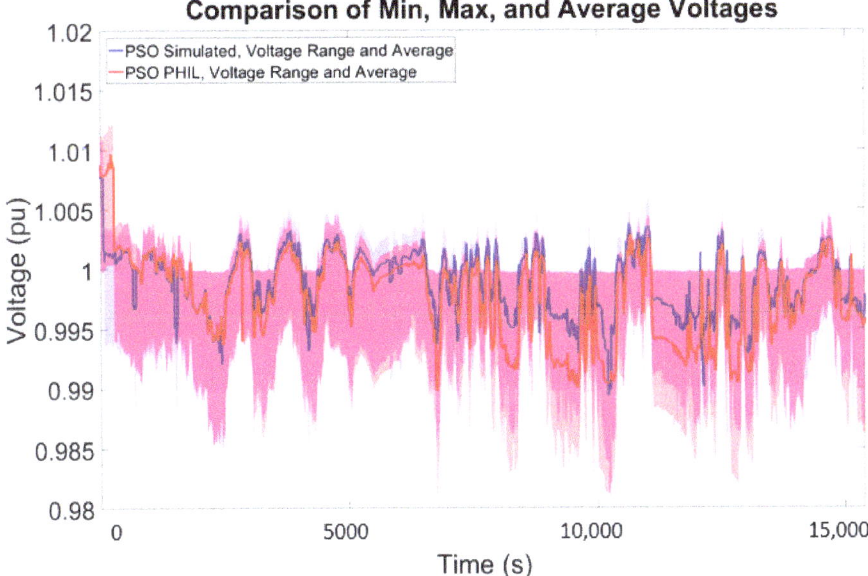


**Figure 19.** Comparison between simulated and PHIL baseline minimum, maximum, and average bus voltages for the PNM model.

## 7. Discussion

Each of the voltage regulation methods have their advantages and disadvantages in terms of communications and computational overhead, implementation complexity, and availability in DER equipment. Overall, each of them can reduce the voltage deviation from nominal and maintain the feeder well within ANSI Range A voltage limits. For the PNM model, the DER PCC voltages are barely outside the dead-band, so the DER does not absorb much reactive power. The ESC method produces a DER-reactive power probing signal that causes a voltage ripple on the feeder, but it allows the DER to track the optimal reactive power setpoint well. The PSO method is the most complicated to implement, but issues optimal set points to each DER every minute. A comparison of each of the voltage regulation methods is shown in Figure 20. The simulation results indicate ESC and PSO can regulate the voltage closer to nominal, compared to VV.

To better understand the differences in these approaches, an analytical score was developed to summarize the effectiveness of each voltage regulation method, and a best score was calculated where the voltage regulation approach drove the solution to 1 pu:

$$score = \frac{1}{T} \int\limits_{t=0}^{t_{end}} \frac{1}{N} \sum_{b=1}^{N} \left( \left|v_{bl}(t) - v_{nom}\right| - \left|v_{reg}(t) - v_{nom}\right| \right) dt \tag{7}$$

$$best\ score = \frac{1}{T} \int\limits_{t=0}^{t_{end}} \frac{1}{N} \sum_{b=1}^{N} \left( \left|v_{bl}(t) - v_{nom}\right| \right) dt \tag{8}$$

where $v_{bl}$ is the baseline voltage, $v_{nom}$ is the nominal voltage (1 pu), $v_{reg}$ is the voltage from the voltage regulation method, $T$ is the time period of the simulation, $b$ is the bus, and $t$ is the simulation time. The scores representing the average voltage improvement for all buses averaged over a four-hour simulation period in units of pu. Table 3 summarizes the effectiveness for the PNM model of each approach per phase as well as the average of each phase, calculated with:

$$Impact = \frac{score}{best\ score} \tag{9}$$

VV slightly improved the average bus voltage, with an impact percentage of 12.3%. Implementing ESC and PSO demonstrated a larger impact on the feeder, with an average impact percentage of 74.5% and 73.7% respectively. A more detailed look at the impact of this regulation approaches per phase, shows that, since the system is relatively balanced, each phase is affected equally.

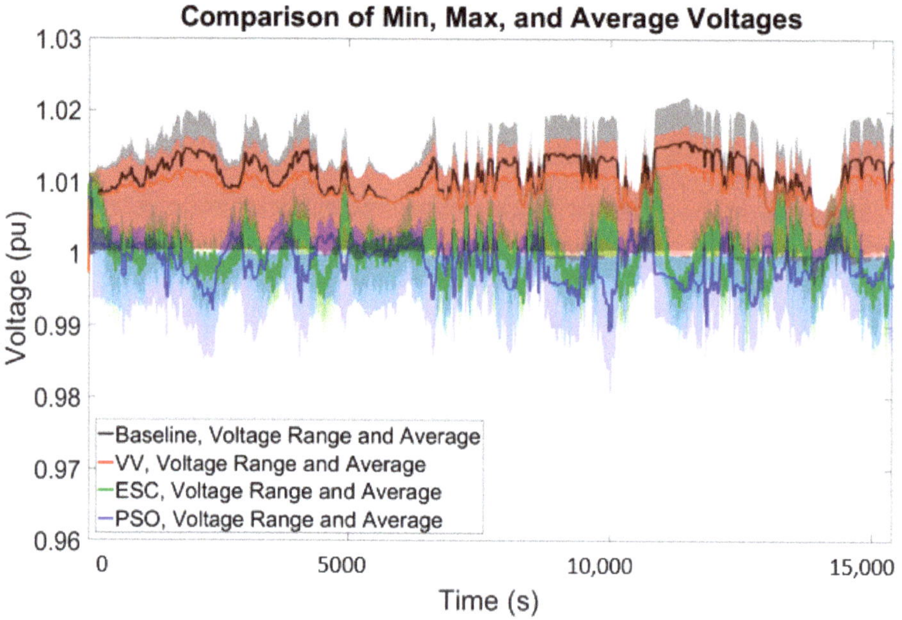


**Figure 20.** Comparison of voltage regulation approaches for the PNM feeder.

**Table 3.** PNM voltage regulation scores.

| | Phase A (×1000) | Phase B (×1000) | Phase C (×1000) | Average (×1000) | Average Impact (%) |
|---|---|---|---|---|---|
| | | | **PNM Feeder Score** | | |
| **VV** | 0.467 | 0.468 | 0.466 | 1.401 | 12.9% |
| **ESC** | 2.745 | 2.748 | 2.591 | 8.084 | 74.5% |
| **PSO** | 2.727 | 2.731 | 2.541 | 7.999 | 73.7% |
| **Best Score** | 3.650 | 3.681 | 3.519 | 10.850 | |

A comparison of the bus voltages for the NG model is illustrated in Figure 21 for each of the control methods. VV, ESC, and PSO all collapse the voltage envelope toward nominal voltage. In this case, ESC slightly outperforms the other two methods, shown in Figure 21. Table 4 summarizes the effectiveness for the NG model of each method. Interestingly, all the methods caused phase A (that was close to nominal to start) to deviate from the nominal voltage. The phases of the distribution system are coupled; in order to reach the global minimum for the system, and substantially improve the voltages on phases B and C, there is an adverse impact on phase A. This results in phase A deviating further from the nominal voltage.

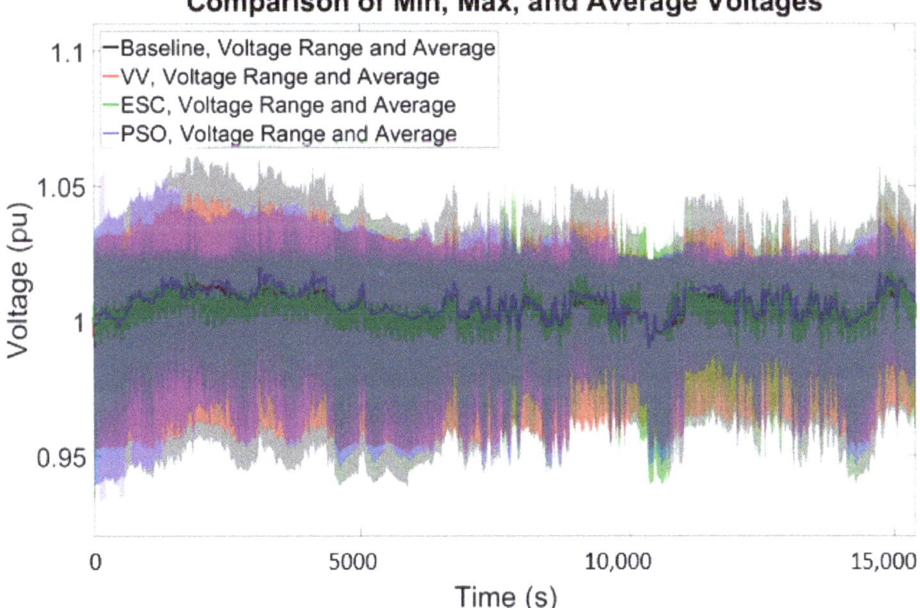



**Figure 21.** Comparison of minimum, maximum and average voltage regulation approaches for the NG feeder controlling all PV inverters.

**Table 4.** NG feeder score results controlling all PV inverters.

| | NG Feeder Score Controlling All PV | | | | |
|---|---|---|---|---|---|
| | Phase A (×1000) | Phase B (×1000) | Phase C (×1000) | Average (×1000) | Average Impact (%) |
| **VV** | −0.058 | 1.855 | 1.281 | 3.078 | 15.2% |
| **ESC** | −0.345 | 4.971 | 3.068 | 7.694 | 38.0% |
| **PSO** | −0.345 | 1.878 | 2.079 | 3.612 | 17.8% |
| **Best Score** | 2.937 | 9.624 | 7.678 | 20.238 | |

## 8. Conclusions

The novel RT and PHIL platform and scoring criterion introduced in this paper allow for the development of a standardized methodology to tune, evaluate, and compare different inverter-based distribution voltage regulation approaches on different feeder models. Utilities and distribution system operators generally do not have the sensor infrastructure or DERMS communication network to execute centralized control of DER. However, as the number of measurement devices and interoperable DER increases, it will become possible to calculate power system states and calculate optimal DER setpoints to provide voltage regulation and provide protection assurance on the system. These capabilities were demonstrated in this project with both SIL and PHIL simulations.

This project investigated and compared three voltage regulation approaches: volt–VAR, extremum seeking control, and particle swarm optimization. Each of the approaches were shown to help provide voltage support on a feeder with symmetrically elevated phase voltages. In the case of the imbalanced NG feeder voltages, the approaches were ineffective when employed with only the three-phase inverter, but an aggregate of single-phase devices were shown to improve the NG feeder voltage profiles. This was demonstrated using RT simulations with simulated PV devices and power hardware-in-the-loop simulations with a 3-kW PV inverter. The extremum seeking control voltage regulation technique was shown to be effective at controlling groups of DER devices to improve feeder voltages, even in cases of phase imbalance. This is the first reported demonstration of this ESC application. The particle swarm optimization approach worked well when the OpenDSS feeder matched the RT/PHIL simulation environment. The technique incorporated live telemetry and PV forecast data to generate the optimal PF setpoints for a collection of PV systems over a fixed time horizon.

## 9. Patents

A U.S. Provisional Patent titled "Digital Twin Advanced Distribution Management System (ADMS) and Methods" was filed on 9 March 2020 based on this work.

**Supplementary Materials:** The following are available online at http://www.mdpi.com/1996-1073/13/14/3562/s1, The anonymized, reduced-order OpenDSS and Opal-RT feeder models, and all portions of the non-proprietary ProDROMOS and ESC codebases are included in the project GitHub repository: https://github.com/sunspec/prodromos.

**Author Contributions:** Conceptualization, J.J.; methodology, J.J.; software, J.J., C.H., J.A. and C.S.; model preparation, R.D.-Z., A.S. and C.H; validation, A.S., R.D.-Z, J.J., and C.H.; formal analysis, J.J., A.S., and R.D.-Z.; resources, J.J and C.S.; data curation, A.S. and R.D.-Z.; writing—original draft preparation, A.S.; writing—review and editing, J.J. and C.H.; visualization, C.H. and C.S; supervision, J.J.; project administration, J.J.; funding acquisition, J.J. All authors have read and agreed to the published version of the manuscript.

**Funding:** This work was sponsored by the US Department of Energy Solar Energy Technologies Office, grant DE-EE0001495-1593, under the "Voltage Regulation and Protection Assurance using DER Advanced Grid Functions" project in the Enabling Extreme Real-time Grid Integration of Solar Energy (ENERGISE) program. The DER ESC implementation was initially developed in the "Community Control of Distributed Resources for Wide Area Reserve Provision" project supported by the U.S. Department of Energy Grid Modernization Laboratory Consortium (GMLC), grant number GM0187.

**Acknowledgments:** The authors would like to thank the ENERGISE team that completed this work, including National Grid, Georgia Institute of Technology, Connected Energy, SunSpec Alliance, SunPower Corporation, and Electric Power Research Institute. The team would also like to thank Trimark for working with the team to establish specific DNP3 points to control and measure the NG PV system. Special thanks to Matthew Reno who validated a proof-of-concept PSO using a simulated feeder in OpenDSS. Sandia National Laboratories is a multi-mission laboratory managed and operated by National Technology and Engineering Solutions of Sandia, LLC., a wholly owned subsidiary of Honeywell International, Inc., for the U.S. Department of Energy's National Nuclear Security Administration under contract DE-NA-0003525. SAND2020-5161 J.

**Conflicts of Interest:** This paper describes objective technical results and analysis. Any subjective views or opinions that might be expressed in the paper do not necessarily represent the views of the U.S. Department of Energy or the United States Government. The funders had no role in the design of the study; in the collection, analyses, or interpretation of data; in the writing of the manuscript, or in the decision to publish the results.

## Acronyms and Definitions

| Abbreviation | Definition |
| --- | --- |
| AC | Alternating Current |
| ANSI | American National Standards Institute |
| Dbus | Data Bus |
| DER | Distributed Energy Resource(s) |
| DERMS | Distributed Energy Resource Management System |
| DETL | Distributed Energy Technology Laboratory |
| DNP3 | Distributed Network Protocol 3 |
| DSO | Distribution System Operator |
| EPRI | Electric Power Research Institute |
| ESC | Extremum Seeking Control |
| GHI | Global Horizontal Irradiance |
| IED | Intelligent Electronic Device |
| IEEE | Institute of Electrical and Electronics Engineers |
| ITM | Ideal Transformer Method |
| LTC | Load Tap Changer |
| MA | Maine |

| NG | National Grid |
| NM | New Mexico |
| NREL | National Renewable Energy Laboratories |
| O&M | Operations and Maintenance |
| OLTC | On-Load Tap Changer |
| OPF | Optimal Power Flow |
| PCC | Point of Common Coupling |
| PF | Power Factor |
| PHIL | Power Hardware-in-the-Loop |
| PNM | Public Service Company of New Mexico |
| PRoDROMOS | Programmable Distribution Resource Open Management |
| PSO | Particle Swarm Optimization |
| PSO PF | Particle Swarm Optimization Optimal Power Factor |
| pu | Per unit |
| PV | Photovoltaic |
| RT | Real-Time |
| SANDIA | Sandia National Laboratories |
| SIL | Software in-the Loop |
| VV | Volt–VAR |
| WinIGS | Integrated Grounding System Analysis program for Windows |

## References


1. International Renewable Energy Agency. *Future of Solar Photovoltaic: Deployment, Investment, Technology, Grid Integration and Socio-Economic Aspects (A Global Energy Transformation: Paper)*; International Renewable Energy Agency: Abu Dhab, UAE, 2019.

2. National Electrical Manufactures Association. *American National Standard for Electric Power Systems and Equipment-Voltage Rating (60 Hz)*; National Electrical Manufactures Association: Rosslyn, VA, USA, 2016.

3. Denholm, P.; Clark, K.; O'Connell, M. *On the Path to SunShot: Emerging Issues and Challenges in Integrating High Levels of Solar into the Electrical Generation and Transmission System*; EERE Publication and Product Library: Washington, DC, USA, 2016.

4. Sadeghian, H.; Wang, Z. Photovoltaic generation in distribution networks: Optimal vs. random installation. In Proceedings of the 2018 IEEE Power Energy Society Innovative Smart Grid Technologies Conference (ISGT), Washington, DC, USA, 19–22 February 2018; pp. 1–5. [CrossRef]

5. Coddington, M.; Kroposki, B.; Mather, B.; Ellis, A.; Hill, R.; Lynn, K.; Razon, A.; Key, T.; Nicole, K.; Smith, J. Updating technical screens for PV interconnection. In Proceedings of the 2012 IEEE Photovoltaic Specialists Conference, Austin, TX, USA, 3–8 June 2012.

6. Reno, M.J.; Broderick, R.J. Technical evaluation of the 15% of peak load PV interconnection screen. In Proceedings of the 2015 IEEE 42nd Photovoltaic Specialist Conference (PVSC), New Orleans, LA, USA, 14–19 June 2015; pp. 1–6.

7. Rylander, M.; Broderick, R.J.; Reno, M.J.; Munoz-Ramos, K.; Quiroz, J.E.; Smith, J.; Rogers, L.; Dugan, R.; Mather, B.; Gotseff, P.; et al. *Alternatives to the 15% Rule*; Sandia Technical Report SAND2015-10099; Sandia National Laboratories: Albuquerque, NM, USA, 2015.

8. Seguin, R.; Woyak, J.; Costyk, D.; Hambrick, J.; Mather, B. *High-Penetration PV Integration Handbook for Distribution Engineers Technical Report*; NREL/TP-5D00-63114; National Renewable Energy Lab: Golden, CO, USA, 2016.

9. IEEE. *IEEE Standard for Interconnection and Interoperability of Distributed Energy Resources with Associated Electric Power Systems Interfaces*; IEEE Std 1547-2018; IEEE: Piscataway, NJ, USA, 2018.

10. Seuss, J.; Reno, M.J.; Broderick, R.J.; Grijalva, S. Improving distribution network PV hosting capacity via smart inverter reactive power support. In Proceedings of the 2015 IEEE Power Energy Society General Meeting, Denver, CO, USA, 26–30 July 2015; pp. 1–5. [CrossRef]

11. Farivar, M.; Neal, R.; Clarke, C.; Low, S. Optimal inverter VAR control in distribution systems with high PV penetration. In Proceedings of the 2012 IEEE Power and Energy Society General Meeting, San Diego, CA, USA, 22–26 July 2012; pp. 1–7. [CrossRef]

12. Ding, F.; Mather, B.; Gotseff, P. Technologies to increase PV hosting capacity in distribution feeders. In Proceedings of the 2016 IEEE Power and Energy Society General Meeting (PESGM), Boston, MA, USA, 17–21 July 2016; pp. 1–5.

13. Rylander, M.; Reno, M.J.; Quiroz, J.E.; Ding, F.; Li, H.; Broderick, R.J. Methods to determine recommended feeder-wide inverter settings for improving distribution performance. In Proceedings of the 2016 IEEE 43rd Photovoltaic Specialists Conference (PVSC), Portland, OR, USA, 5–10 June 2016; pp. 1393–1398. [CrossRef]

14. Haghi, H.V.; Pecenak, Z.; Kleissl, J.; Peppanen, J.; Rylander, M.; Renjit, A.; Coley, S. Feeder impact assessment of smart inverter settings to support high PV penetration in California. In Proceedings of the 2019 IEEE Power & Energy Society General Meeting (PESGM), Atlanta, GA, USA, 4–8 August 2019; pp. 1–5. [CrossRef]

15. Reno, M.J.; Lave, M.; Quiroz, J.E.; Broderick, R. PV ramp rate smoothing using energy storage to mitigate increased voltage regulator tapping. In Proceedings of the IEEE Photovoltaic Specialists Conference (PVSC), Portland, OR, USA, 5–10 May 2016.

16. Georgilakis, P.S.; Hatziargyriou, N.D. Optimal distributed generation placement in power distribution networks: Models, methods, and future research. *IEEE Trans. Power Syst.* **2013**, *28*, 3420–3428. [CrossRef]

17. Reno, M.J.; Broderick, R.J. Optimal siting of PV on the distribution system with smart inverters. In Proceedings of the 2018 IEEE 7th World Conference on Photovoltaic Energy Conversion (WCPEC) (A Joint Conference of 45th IEEE PVSC, 28th PVSEC & 34th EU PVSEC), Waikoloa Village, HI, USA, 10–15 June 2018; pp. 1468–1470.

18. Antoniadou-Plytaria, K.E.; Kouveliotis-Lysikatos, I.N.; Georgilakis, P.S.; Hatziargyriou, N.D. Distributed and decentralized voltage control of smart distribution networks: Models, methods, and future research. *IEEE Trans. Smart Grid* **2017**, *8*, 2999–3008. [CrossRef]

19. Primadianto, A.; Lu, C. A review on distribution system state estimation. *IEEE Trans. Power Syst.* **2017**, *32*, 3875–3883. [CrossRef]

20. Dehghanpour, K.; Wang, Z.; Wang, J.; Yuan, Y.; Bu, F. A survey on state estimation techniques and challenges in smart distribution systems. *IEEE Trans. Smart Grid* **2019**, *10*, 2312–2322. [CrossRef]

21. Elkhatib, M.E.; El-Shatshat, R.; Salama, M.M.A. Novel coordinated voltage control for smart distribution networks with DG. *IEEE Trans. Smart Grid* **2011**, *2*, 598–605. [CrossRef]

22. Salih, S.N.; Chen, P. On coordinated control of OLTC and reactive power compensation for voltage regulation in distribution systems with wind power. *IEEE Trans. Power Syst.* **2016**, *31*, 4026–4035. [CrossRef]

23. Xie, Q.; Hara, R.; Kita, H.; Tanaka, E. Coordinated control of OLTC and multi-CEMSs for overvoltage prevention in power distribution system. *IEEJ Trans. Electr. Electron. Eng.* **2017**, *12*, 692–701. [CrossRef]

24. Zhu, X.; Zhang, Y. Coordinative voltage control strategy with multiple resources for distribution systems of high PV penetration. In Proceedings of the 2018 IEEE 7th World Conference on Photovoltaic Energy Conversion (WCPEC) (A Joint Conference of 45th IEEE PVSC, 28th PVSEC & 34th EU PVSEC), Waikoloa Village, HI, USA, 10–15 June 2018; pp. 1497–1502.

25. Tonkoski, R.; Lopes, L.A.C.; El-Fouly, T.H.M. Coordinated active power curtailment of grid connected PV inverters for overvoltage prevention. *IEEE Trans. Sustain. Energy* **2011**, *2*, 139–147. [CrossRef]

26. Johnson, J.; Summers, A.; Darbali-Zamora, R.; Hansen, C.; Reno, M.; Castillo, A.; Gonzalez, S.; Hernandez-Alvidrez, J.; Gurule, N.; Xie, B.; et al. *Optimal Distribution System Voltage Regulation using State Estimation and DER Grid-Support Functions*; Sandia Technical Report, SAND2020-2331; Sandia National Laboratories: Albuquerque, NM, USA, 2020.

27. Meliopoulos, A.P.S. Windows Based Integrated Grounding System Design Program. Training Guide. 2017. Available online: http://www.ap-concepts.com/_downloads/IGS_TrainingGuide.pdf (accessed on 18 January 2017).

28. Quiroz, J.E.; Reno, M.J.; Lavrova, O.; Byrne, R.H. Communication requirements for hierarchical control of volt-VAr function for steady-state voltage. In Proceedings of the 2017 IEEE Power & Energy Society Innovative Smart Grid Technologies Conference (ISGT), Washington, DC, USA, 23–26 April 2017; pp. 1–5. [CrossRef]

29. Reno, M.J.; Quiroz, J.E.; Lavrova, O.; Byrne, R.H. Evaluation of communication requirements for voltage regulation control with advanced inverters. In Proceedings of the 2016 North American Power Symposium (NAPS), Denver, CO, USA, 18–20 September 2016; pp. 1–6.

30. Xie, B.; Meliopoulos, A.P.S.; Zhong, C.; Liu, Y.; Sun, L.; Xie, J. Distributed quasi-dynamic state estimation incorporating distributed energy resources. In Proceedings of the 2018 North American Power Symposium (NAPS), Fargo, ND, USA, 9–11 September 2018.

31. Choi, S.; Kim, B.; Cokkinides, G.J.; Meliopoulos, A.P.S. Feasibility study: Autonomous state estimation in distribution systems. *IEEE Trans. Power Syst.* **2011**, *26*, 2109–2117. [CrossRef]

32. Xie, B.; Meliopoulos, A.P.S.; Liu, Y.; Sun, L. Distributed quasi-dynamic state estimation with both GPS-synchronized and non-synchronized data. In Proceedings of the 2017 North American Power Symposium (NAPS), Morgantown, WV, USA, 17–19 September 2017; pp. 1–6.

33. Johnson, J.; Summers, A.; Darbali-Zamora, R.; Hernandez-Alvidrez, J.; Quiroz, J.; Arnold, D.; Anandan, J. Distribution voltage regulation using extremum seeking control with power hardware-in-the-loop. *IEEE J. Photovolt.* **2018**, *8*, 1824–1832. [CrossRef]

34. Johnson, J.; Gonzalez, S.; Arnold, D.B. Experimental distribution circuit voltage regulation using DER power factor, volt-var, and extremum seeking control methods. In Proceedings of the 2017 IEEE 44th Photovoltaic Specialist Conference (PVSC), Washington, DC, USA, 25–30 June 2017; pp. 3002–3007. [CrossRef]

35. Arnold, D.B.; Sankur, M.D.; Negrete-Pincetic, M.; Callaway, D.S. Model-free optimal coordination of distributed energy resources for provisioning transmission-level services. *IEEE Trans. Power Syst.* **2018**, *33*, 817–828. [CrossRef]

36. Arnold, D.B.; Negrete-Pincetic, M.; Sankur, M.D.; Auslander, D.M.; Callaway, D.S. Model-free optimal control of VAR resources in distribution systems: An extremum seeking approach. *IEEE Trans. Power Syst.* **2016**, *31*, 3583–3593. [CrossRef]

37. Distributed Energy Resources Connection Modeling and Reliability Considerations. 2017. Available online: http://www.nerc.com/comm/Other/essntlrlbltysrvcstskfrcDL/Distributed_Energy_Resource_Report.pdf. (accessed on 5 March 2020).

38. Bründlinger, R. Grid codes in Europe for low and medium voltage. In Proceedings of the 6th International Conference on Integration of Renewable and Distributed Energy Resources, Kyoto, Japan, 17–20 November 2014.

39. Reno, M.J.; Coogan, K.; Broderick, R.; Grijalva, S. Reduction of distribution feeders for simplified PV impact studies. In Proceedings of the 2013 IEEE 39th Photovoltaic Specialists Conference (PVSC), Tampa, FL, USA, 16–21 June 2013; pp. 2337–2342. [CrossRef]

40. Pecenak, Z.K.; Disfani, V.R.; Reno, M.J.; Kleissl, J. Multiphase distribution feeder reduction. *IEEE Trans. Power Syst.* **2018**, *33*, 1320–1328. [CrossRef]

41. EPRI. *Overview of EPRI's DER Simulation Tool for Emulating Smart Solar Inverters and Energy Storage Systems on Communication Networks: An Overview of EPRI's Distributed Energy Resource Simulator*; EPRI Report 3002009851; EPRI: Palo Alto, CA, USA, 2018.

42. PV Performance Modeling Collaborative. Wavelet Variability Model. Available online: https://pvpmc.sandia.gov/applications/wavelet-variability-model/ (accessed on 5 March 2020).

43. Lave, M.; Kleissl, J.; Stein, J.S. A wavelet-based variability model (WVM) for solar PV power plants. *IEEE Trans. Sustain. Energy* **2013**, *4*, 501–509. [CrossRef]

44. Lave, M.; Kleissl, J.; Arias-Castro, E. High-frequency irradiance fluctuations and geographic smoothing. *Sol. Energy* **2012**, *86*, 2190–2199. [CrossRef]

45. Lave, M.; Kleissl, J. Cloud speed impact on solar variability scaling—Application to the wavelet variability model. *Sol. Energy* **2013**, *91*, 11–21. [CrossRef]

46. Lohmann, G.M. Irradiance variability quantification and small-scale averaging in space and time: A short review. *Atmosphere* **2018**, *9*, 264. [CrossRef]

47. Bosch, J.L.; Zheng, Y.; Kleissl, J. Deriving cloud velocity from an array of solar radiation measurements. *Sol. Energy* **2013**, *87*, 196–203. [CrossRef]

48. Summers, A.; Hernandez-Alvidrez, J.; Darbali-Zamora, R.; Reno, M.J.; Johnson, J.; Gurule, N.S. Comparison of ideal transformer method and damping impedance method for PV power-hardware-in-the-loop experiments. In Proceedings of the 2019 IEEE 46th Photovoltaic Specialists Conference (PVSC), Chicago, IL, USA, 16–21 June 2019; pp. 2989–2996. [CrossRef]

49. Hernandez-Alvidrez, J.; Summers, A.; Pragallapati, N.; Reno, M.J.; Ranade, S.; Johnson, J.; Brahma, S.; Quiroz, J. PV-inverter dynamic model validation and comparison under fault scenarios using a power hardware-in-the-loop testbed. In Proceedings of the 2018 IEEE 7th World Conference on Photovoltaic Energy Conversion (WCPEC) (A Joint Conference of 45th IEEE PVSC, 28th PVSEC & 34th EU PVSEC), Waikoloa Village, HI, USA, 10–15 June 2018; pp. 1412–1417.
50. Holmgren, W.F.; Hansen, C.W.; Mikofski, M.A. pvlib python: A python package for modeling solar energy systems. *J. Open Source Softw.* **2018**, *3*, 884. [CrossRef]

 *energies*

*Article*

# Power System Hardware in the Loop (PSHIL): A Holistic Testing Approach for Smart Grid Technologies

Manuel Barragán-Villarejo *, Francisco de Paula García-López, Alejandro Marano-Marcolini and José María Maza-Ortega

Department of Electrical Engineering, Universidad de Sevilla, Camino de los Descubrimientos s/n, 41092 Sevilla, Spain; fdpgarcia@us.es (F.d.P.G.-L.); alejandromm@us.es (A.M.-M.); jmmaza@us.es (J.M.M.-O.)
* Correspondence: manuelbarragan@us.es

Received: 20 June 2020; Accepted: 24 July 2020; Published: 28 July 2020

**Abstract:** The smart-grid era is characterized by a progressive penetration of distributed energy resources into the power systems. To ensure the safe operation of the system, it is necessary to evaluate the interactions that those devices and their associated control algorithms have between themselves and the pre-existing network. In this regard, Hardware-in-the-Loop (HIL) testing approaches are a necessary step before integrating new devices into the actual network. However, HIL is a device-oriented testing approach with some limitations, particularly considering the possible impact that the device under test may have in the power system. This paper proposes the Power System Hardware-in-the-Loop (PSHIL) concept, which widens the focus from a device- to a system-oriented testing approach. Under this perspective, it is possible to evaluate holistically the impact of a given technology over the power system, considering all of its power and control components. This paper describes in detail the PSHIL architecture and its main hardware and software components. Three application examples, using the infrastructure available in the electrical engineering laboratory of the University of Sevilla, are included, remarking the new possibilities and benefits of using PSHIL with respect to previous approaches.

**Keywords:** Hardware-in-the-Loop (HIL); Control HIL (CHIL); Power HIL (PHIL); testing of smart grid technologies

---

## 1. Introduction

The power system is undergoing a revolution nowadays, with the fast evolution of new hardware and software technologies. Power electronics, automation, cloud computing, big data, information, and communication technologies are some of the key components of the so-called smart grid. These new technologies are the enablers of the electricity sector decarbonization pursued by a society worried about energy dependence and the environmental impact of the consumed energy [1]. Without any doubt, the fulfillment of the Kyoto protocol [2] for restricting $CO_2$ emissions and other greenhouse gases is resulting in drastic changes in terms of the generation mix of the power system, where a more active participation of renewable energies is required, but also the electrification of the transportation sector [3].

As a result, the operation and planning of power systems both in the transmission and distribution sides turn out to be even more complex than before. In fact, the uncertainty associated with renewable power sources and electrical vehicle charging, the security of supply and power quality expected by the final users with competitive energy prices and the constrained investment on network assets are stressing the power system [4]. At the same time, however, the technologies required for reducing this stress and also improving network operation are ready, but their network deployment has to be carefully done. This is of utmost importance for those components that interact with the power system in an active manner, i.e., generators, protection devices, capacitor and reactor banks, On-Load Tap Changers (OLTCs), Flexible AC Transmission Systems (FACTS), High-Voltage DC links (HVDC), etc. Given the current automation level of the power system, where different control devices widespread along the network react to either local conditions or set points determined by a control center, it is mandatory to be sure that any new component does not interfere with the already existing ones.

For this purpose, simulation tools have brought a cost-effective solution to figure out the performance of a device or a system from its initial design stages, reducing the development costs and the time to market. The current available simulation tools cover almost all of the electrical engineering fields from electromagnetic transients to steady-state analysis. These design-support tools provide invaluable information, especially in the case of complex devices and systems. Simulation, however, is just one of the design process steps, experimental validation through a prototype or proof of concept always being required before any commercial product or even a pilot project. In fact, not all physical phenomena can be reproduced on a computer simulation tool because the reality is quite complex. In this regard, a recent branch of research, namely Hardware-in-the-Loop (HIL), focuses on the use of hardware and software tools to test devices and systems in the most realistic possible way and with a low investment. The main idea is to take advantage of the real-time simulation capabilities to test controllers or devices in a situation as close as possible to the reality. Two different and complementary options have been proposed in the HIL paradigm: Controller-Hardware-in-the-Loop (CHIL) and Power-Hardware-in-the-Loop (PHIL). Figure 1 describes these two HIL options and compares them with simulation tools. This figure highlights the simulated and physical devices (green and blue colored parts, respectively), the information flow, the power flows, and the interfaces between the actual and simulated environments. Let's consider that the objective of the simulation is to validate the performance of a device, i.e., the Device Under Test (DUT), connected to a power system. For this purpose, it is required to model all of the system components with the required detail level, but also the DUT including its control algorithms. The simulation tool can be executed on a conventional computer or a Real-Time Control System (RTCS) [5–9]. A step forward on the technology validation consists of applying CHIL [10–13]. In this case, all of the power components (power system and power components of the device) are still simulated in a RTCS, but the device control algorithm is embedded into a physical controller, e.g., a digital signal processor, with the objective of its real-time testing. The interface between the DUT (physical controller) and the RTCS is done using adequate analog/digital input/output ports as shown in Figure 1. It is possible, however, to move a step forward towards the reality by incorporating power components in the tests using a PHIL approach. This requires a power amplifier, which acts as a front end between the simulated and physical parts. The power amplifier is in charge of imposing the voltage at the point of common coupling with the DUT, which is computed by the RTCS considering its reaction. For this purpose, it is required to feedback the DUT injected current in the real-time simulated power system by using adequate analog/digital input/output ports [14–19].

In spite of the benefits provided by HIL testing, it is also important to point out that these techniques, and especially PHIL, are devoted to analyzing the behavior of a single DUT connected to a network node. Therefore, some aspects related to the introduction of a new technology in the power system can be masked. Interactions between different DUTs, the impact on the power system of a massive deployment of a new smart grid technology or the interaction between physical devices (DUTs) and the algorithms in charge of determining their setpoints are just some of these. In order to overcome these shortcomings, this paper proposes to extend the PHIL approach by: (i) introducing the algorithms determining the DUT setpoints, i.e., Algorithm Under Test (AUT); (ii) expanding the test focus from the current device-oriented approach to a system-oriented concept including the impact of the tested technology on the power system. This new approach, namely Power System Hardware-in-the-Loop (PSHIL), may allow the evaluation of the impact of a given technology over the power system in a holistic manner, considering all its power and control components.

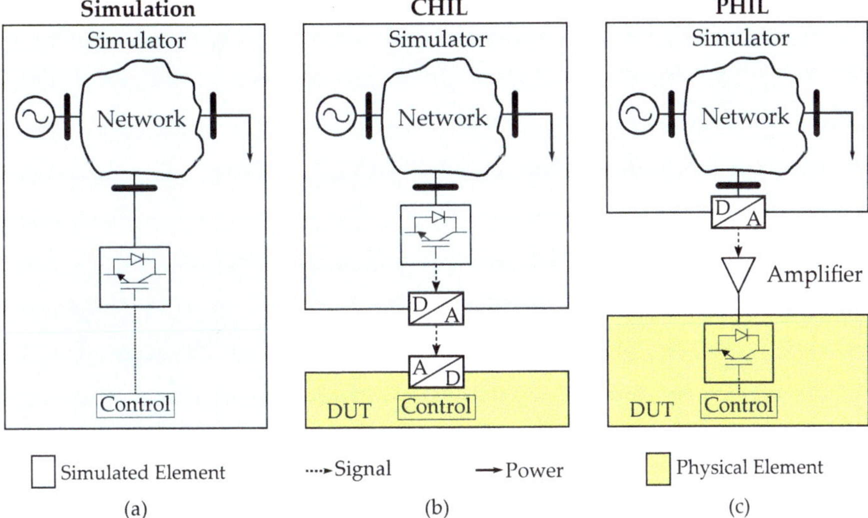



**Figure 1.** Possible testing setups for assessing the effect of a new smart grid technology in the power system: (**a**) Simulation, (**b**) Control Hardware-in-the-Loop (CHIL), (**c**) Power-Hardware-in-the-Loop (PHIL).

The structure of this document is as follows. Section 2 elaborates on the PSHIL architecture and components. Then, the PSHIL functionalities and benefits with respect to simulation and conventional HIL approaches are outlined. Subsequently, Section 4 describes the PSHIL environment of the Department of Electrical Engineering at the University of Sevilla. Section 5 shows some case studies to evidence the wide range of possibilities that PSHIL testing may offer to the power community. Finally, the paper closes with the main conclusions and future lines of research.

## 2. PSHIL Architecture

This section describes the different PSHIL components and how they interact each other. Basically, a general PSHIL architecture is shown in Figure 2, where the simulated and physical environments are differentiated. On the simulation side, all the components, including network, generators, loads, and AUTs, are represented using adequate models which interact on a real-time simulation platform. In contrast, the physical side comprises a set of actual elements which may involve networks, generators, loads, DUTs, and AUTs implemented in different ways. The interaction between the simulated and the physical environments must be done through adequate interfaces. Finally, note that a management layer is required to coordinate the simultaneous testing in both environments. The next subsections are devoted to detailing each of these components.

### 2.1. Real-Time Simulated Power System

In this case, the power system used for testing purposes is simulated using an RTCS. Buses, branches, loads, generators, etc. are fully simulated. The physical devices (DUTs) are connected to this simulated network through a controllable voltage source, which acts as a power amplifier [20] as shown in Figure 2. The purpose of this amplifier is to physically reproduce the instantaneous voltage on a given bus of the real-time simulated network. Then, the DUT connected to the amplifier reacts to this voltage according to its control algorithm. This reaction is normally reflected in a current absorption/injection from the power amplifier, which is measured and sent back to the RTCS to be considered in the real-time simulation. In this way, it is possible also to evaluate the DUT's local impact in the simulated network. Note that this is quite convenient for testing local controllers (LCs) and hardware components of a single DUT (PHIL). However, it is also possible to envision configurations with multiple power amplifiers for simultaneously testing different DUTs. In this case, it should be possible to evaluate the interactions between the DUTs through the real-time simulated network, but at a cost of requiring one power amplifier for each tested DUT.

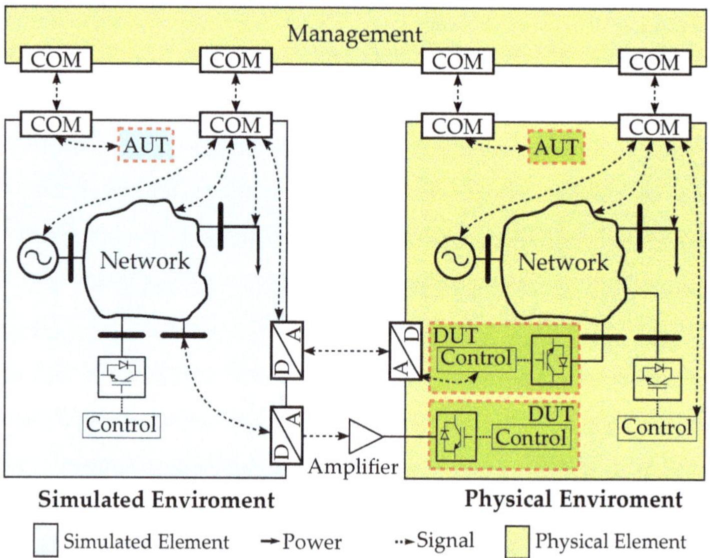


**Figure 2.** Generic scheme describing the Power System Hardware-in-the-Loop (PSHIL) configuration.

*2.2. Physical Power System*

Two alternatives can be presented in this case: full-scaled networks and scaled-down networks. In the first case, different buses are interconnected through electrical lines and the DUTs are full-scale prototypes that interact with the different network components. Without any doubt, this testing environment gives to the test results the highest credibility because it is quite close to an actual field deployment. The space requirements and cost for high-voltage an high-power applications have to be considered, however. On the other hand, it is possible to use scaled-down networks where a change of the system base (rated voltage and rated power) is applied to scale down a given original network to an adequate value to be used on a laboratory environment. In this way, the main scaled-down magnitudes (voltage, current, and power) expressed in per-unit values are exactly the same as the ones in the original system. The buses of the scaled-down systems are interconnected through resistors and inductors with adequate values for emulating the original electrical lines. This network requires physical generators and loads that will be analyzed later. The main advantage of these physical testing environments is that some phenomena difficult to reproduce in simulations, such as heating and electromagnetic interference, always appear. The main disadvantage, however, is the reduced flexibility to modify either the topology or the parameters of the electrical network. This may hamper the impact of the DUTs on networks with different characteristics.

It is important to highlight that the PSHIL approach also considers that simulated and physical networks may coexist in a testing environment, as shown in Figure 2. This may provide an extended flexibility degree, which will be later exemplified in Section 5.1.

### 2.3. Power Supply

The objective of the power supply used in the physical network is to energize the testing environment providing the required power for the loads and DUTs. For this purpose, it is possible to consider the following not mutually exclusive options:

- External network. This is the most cost-effective option to supply the PSHIL platform since only a connection with an available external network is required. This usually corresponds to a node of the laboratory network in case of scaled-down distribution systems. The rated power and voltage depend on the characteristic of this connection node. This, however, has limited flexibility, as the PSHIL voltage and frequency are mainly imposed by the laboratory network. Therefore, intentional events such as voltage and frequency disturbances required for testing the provision of ancillary services [21–23] cannot be reproduced in a straightforward manner. This option, however, is adequate for analyzing the interaction of different DUTs and the power system in quasi-steady-state conditions [24–27].
- Synchronous generator. In this case, a synchronous generator driven by a primary energy source (steam/hydraulic turbine, diesel motor, DC machine or induction motor fed with a variable speed drive) provides the power to the network [28]. Note that without a connection to an external network, the physical environment is operated in islanded mode. This endows the system with greater flexibility because it should be possible to reproduce frequency and voltage disturbances. Moreover, with the adequate control actions on those prime movers based on electrical machines (DC machine or induction motor fed with a variable speed drive), it should be possible to reproduce the dynamic behavior of any actual generator driven by a hydraulic/steam turbine or diesel motor without any complex auxiliary systems [29].
- Controllable voltage source. This is the power-electronic counterpart of the synchronous generator where the mechanically coupled rotating machines are substituted by two voltage-source converters (VSCs) in a back-to-back configuration. This allows a total controlability of the output voltage with really fast dynamics. Note also that controllable voltage sources are an indispensable component for interfacing the real-time simulated with the physical network if required. The main drawback of these devices is the high cost, especially for those that are based on linear technology.
- Actual renewable generators. This type of power supply, including photovoltaic and wind generators, provides realism to the testing. Its main drawback, however, refers to the impossibility of controlling the primary energy source, which considerably limits the replicability of the testing conditions.

### 2.4. Loads

The loads are elements that absorb power from the buses of the physical power system under study. These can be classified into passive or active elements depending on the control capability:

- Passive elements. These are elements with no control capability to change their operating states according to setpoints. Resistors, inductors, and capacitors are within this group. The main advantage of these elements is that they are inexpensive, but at the cost of providing a limited flexibility, which may limit the testing scenarios.
- Active elements. This kind of load allows controlling the power demand according to setpoints, with adapting the testing scenarios according to the user needs being possible. Moreover, this capability allows reproducing the daily load profiles of different customer types (domestic, commercial, and industrial loads). Commercial electronic loads or VSCs supplied from the DC side with an external DC source may act as a controllable active element. The main drawback, however, is that they are complex devices involving different technologies like power electronics and control and communication systems.

### 2.5. Devices and Algorithms under Test—DUTs and AUTs

DUTs correspond to an actual device whose functionality and impact on the power system is tested using a PHIL approach. The possible list of devices that can be found in the specialized literature is endless, ranging from photovoltaic (PV) inverters [30], wind generators [31], HVDC [32], FACTS [14], energy storage systems [33], electric vehicle charging stations [26], OLTCs for power transformers [34], to digital protective relays [35]. Similarly, the functionalities that can be tested are numerous: primary control of voltage and frequency of Distributed Energy Resources (DERs) [36], current control in VSCs, inertia emulation in DERs [37,38], protection of VSCs during short-circuit faults [20], high-frequency power smoothing of renewable energy resources [39], MPPT in PV systems [40,41], etc. The simultaneous implementation of these strategies in several DUTs within a PSHIL environment allows evaluating the global impact on the power system under study, but also, mutual interactions between different DUTs can be analyzed, such as the elimination of zero-sequence current flow between transformerless grid-connected VSCs with a common DC bus [42] or the resonance frequencies originating between VSCs [43].

The PSHIL approach, in addition, considers the possibility of also testing algorithms (AUTs) for optimizing real-time system operation and which may interact with DUTs through adequate setpoints. Therefore, the PSHIL concept extends the testing capabilities associated with the PHIL approach. Several objectives can be implemented for centralized or decentralized control structures, such as voltage control [44–46], minimization of power losses [47], congestion release [48], or the optimal allocation of ancillary services provided by different distributed control resources [49], among others. Basically, these algorithms are implemented on a computer that receives all of the network information (static and real-time data) following the general scheme outlined in Figure 2. The AUT computes and sends the DUT setpoints according to the available information and the intended objective. Furthermore, the PSHIL approach allows including in the loop the validation of other components such as the communication infrastructure, e.g., communication latencies, which are key for a comprehensive assessment of the real-time control strategies.

Therefore, the PSHIL concept simultaneously combines device- and system-oriented testing, which provides a new holistic approach especially suited to complex smart grid technologies.

### 2.6. PSHIL Management

The objective of the management layer is to coordinate the different components involved in the PSHIL testing platform. Basically, it consists of a hierarchical control architecture comprising two layers:

- Local controllers. The first control layer consists of the different LCs associated with each element, i.e., loads, generators, or DUTs. The objective of each LC is to guarantee that the corresponding device follows specific setpoints during the tests, which are sent by the centralized PSHIL management. Additionally, LCs are in charge of monitoring and protecting the controlled device.
- Centralized PSHIL management. This layer is in charge of several tasks, which can be broadly classified into two main categories [24]: off-line and on-line tasks. On the one hand, off-line tasks are devoted to configuring the testing scenario by: (i) adjusting the topology and parameters of the networks; (ii) defining the operation of loads and generators; and (iii) setting the functionalities of DUTs and AUTs. On the other hand, online tasks are committed to controlling and supervising the tests. Control tasks are required to send to the different LCs the adequate setpoints according to the defined testing scenario. Note that some of these setpoints are defined by the offline tasks, e.g., active power demand of a load, but other ones can be determined in real-time by the AUT, e.g., optimal reactive power of a distributed generator. In any case, all of these setpoints are managed by this centralized management layer. In turn, supervision tasks are in charge of monitoring the relevant electrical magnitudes with a twofold objective: guaranteeing the safe operation of all of the physical components (network, loads, generators, and DUTs) and gathering

all of the required measurements. These measurements are provided to the AUTs, which may use them to compute the DUT setpoints, but are also stored for later analysis.

The information flow between the central management and LCs of generators, loads, and DUTs, as well as the interfaces between the simulated and physical networks, are key for proper PSHIL operation. Following a top-down description, the centralized PSHIL management layer is connected with each LC by using digital signals according to a communication protocol (CAN, Modbus, UDP, etc.) over a communication channel (RS-232, RS-485, Ethernet, etc.). Communication latencies depend on the performed tests, but usual values are in the order of seconds. Then, LCs are in charge of following these setpoints commanded by the centralized management layer, usually based on closed-loop algorithms to achieve a good dynamic response. For this purpose, it is required to manipulate analog signals related to the physical controlled device. This control loop latency depends on the required control bandwidth, but typical values are around dozens of microseconds. Regarding the interfaces between the simulated and physical environments, an information exchange as rapid as possible is required to avoid the problems generated by the time delays [50]. For this reason, the information exchange from the real-time simulation to the physical environment, and vice versa, is done by means of analog measurements that are regularly sampled at dozens of microseconds.

*2.7. Energy Management*

The energy management between the components of the PSHIL depends on whether the system under study is connected to a main electrical network or operates in islanded mode. In the first case, the electrical network is responsible for supplying or absorbing the energy necessary to maintain the power balance between loads and generators in the system. In islanded mode, however, it is required to provide adequate energy management to assure this active and reactive power balance. In this regard, two operation modes are clearly distinguished: communication-based and communication-less schemes [51]. The former schemes collect data from the PSHIL components (voltage, current, power, etc.) using a communication infrastructure in order to set the output power of the generators. Two alternative energy management algorithms within this group are usually implemented: centralized and decentralized energy management schemes [52]. The centralized algorithms receive all of the data from the PSHIL components in the Centralized PSHIL management layer and sets the operating points of the generators according to an objective, such as minimizing system operation and maintenance costs, environmental impact, or power losses [53,54]. In decentralized energy management schemes, local controllers of generation units exchange information through a communication bus. This information is used by the local controllers to determine an adequate setpoint by means of an optimization algorithm based on the shared information. Artificial-intelligence-based methods such as neural network or fuzzy systems [55] along with genetic algorithms have been used for this purpose [56]. Conversely, in communication-less schemes each generator operates independently without a communication infrastructure to exchange information with a control center or other generators. Droop-based strategies for frequency and voltage control implemented in each generation local controller are the most common techniques for energy management in this case [57].

## 3. PSHIL Benefits

The flexibility provided by the PSHIL approach clearly enlarges the testing capability with respect to the HIL concept. Basically, PSHIL may simultaneously test device and system functionalities provided by the DUTs and AUTs, as shown in Table 1. In this regard, it is interesting to note that the PSHIL approach is as powerful as a simulation tool but within an actual hardware environment close to the reality. Some examples of the functionalities tested using a PSHIL approach have been reported in [25,26]. In the first work, an Optimal Power Flow (OPF) for losses minimization (AUT) using an OLTC, a DC link, and distributed generators (DUTs) have been tested in a scaled-down MV distribution system. The second work deals with the integration of an electric vehicle charging station in an LV distribution network (DUT) to balance the MV/LV transformer loading (AUT). Both works reveal that the PSHIL approach allows us to obtain specific conclusions about the DUT and AUT's performance but also the global impact of the tested technology on the network.

It has to be said, however, that this proximity to the reality is at the cost of relying on a physical network, which is difficult to reconfigure or adapt to other scenarios. In spite of this limitation, the PSHIL concept has several advantages with respect to simulation and conventional HIL strategies:

- Simultaneous testing of different DUTs considering their mutual interaction through the real-time simulated or actual power system.
- Simultaneous testing of DUTs and AUTs, which analyzing the performance of closed control loop actions in real conditions.
- Simultaneous testing of different technologies such as power electronics, control software, communication infrastructure, etc., which may interact each other.
- Power-system-oriented testing capability in addition to the classical device-oriented approach.

**Table 1.** Testing approaches for DUTs and AUTs.

| Tested Element | Simulation | CHIL | PHIL | PSHIL |
|:---:|:---:|:---:|:---:|:---:|
| DUT | • | • | • | • |
| AUT | • | • | | • |
| AUT & DUT | • | | | • |

## 4. PSHIL Infrastructure at the University of Sevilla

This section is devoted to describing the PSHIL infrastructure implemented in the Department of Electrical Engineering at the University of Sevilla. Figures 3 and 4 show a scheme and a view of this laboratory, respectively, whose main components are the following.

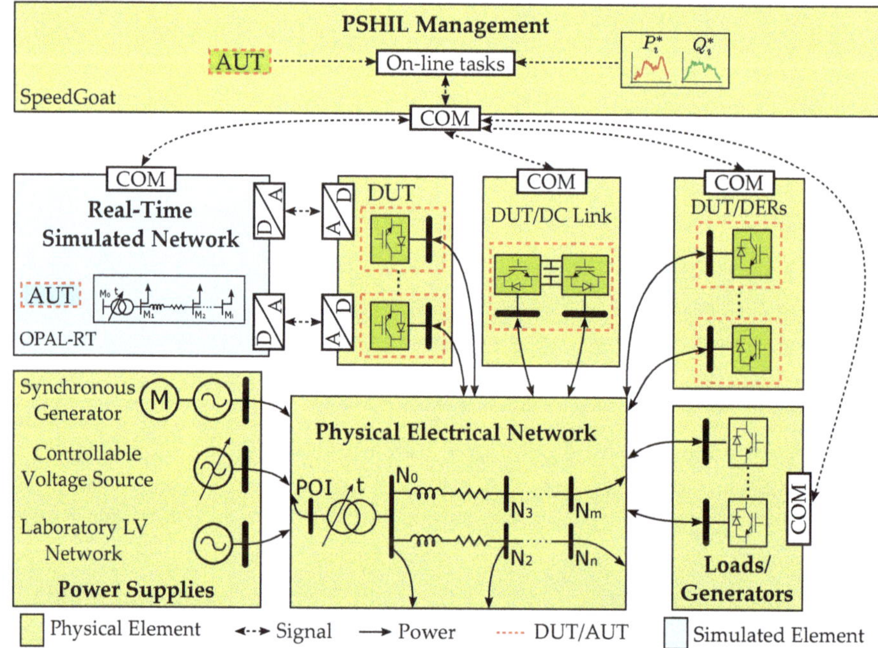


**Figure 3.** PSHIL platform of the Department of Electrical Engineering at the University of Sevilla.

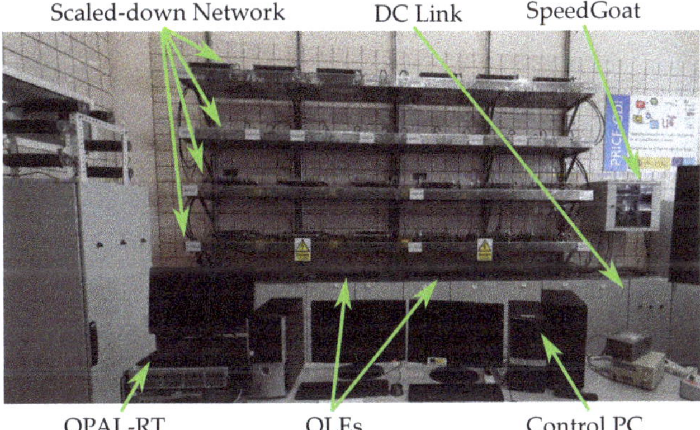


**Figure 4.** PSHIL layout of the Department of Electrical Engineering at the University of Sevilla.

- Real-time simulated network. OPAL-RT is available for simulating in real-time any transmission or distribution network.
- Physical electrical network. The PSHIL infrastructure has a scaled-down version of the MV distribution network proposed by the CIGRE Task Force C06.04.02 for the analysis of the distributed generation [58]. This physical system, whose one-line diagram is shown in Figure 5, is fully described in [24]. Basically, it is composed of two feeders with 14 nodes where different loads, generators, and DUTs can be connected. The lines between the different nodes are emulated using their corresponding lumped resistive and inductive parameters. This physical system is rated to 400 V and 100 kVA at the head of the feeders.
- Power supply. Three different possibilities are available: laboratory LV network, synchronous generators, and power amplifier. The laboratory LV network, 400 V and 100 kVA, is the common option for the grid-connected analysis of the network under study. For those tests where a precise control of the supply voltage is required, a Regatron TC.ACS power amplifier shown in Figure 6a, 400 V and 50 kVA, is used. This power amplifier can also be connected to the OPAL-RT platform for interfacing a DUT. The synchronous generators shown in Figure 6b, 3 units rated to 400 V and 15 kVA, can be used for testing the network in islanded conditions. In this case, the synchronous generators are driven by induction motors fed with a variable speed drive.
- Loads. The PSHIL infrastructure is equipped with active loads able to change their operating point (active and reactive powers) according to an external reference provided by the management layer. The active loads are based on the Omnimode Load Emulator (OLE) concept [24]. Basically, an OLE is composed pf a controlled VSC that is connected to a node of the scaled-down distribution network in its AC side. Additionally, the DC side is connected to a common DC bus that is controlled by an additional VSC in charge of providing the required active power of each OLE connected to the DC bus. Note that the OLEs can be used for emulating loads or generators, and the balanced VSC, which is connected to the laboratory LV grid, has to provide just the net power injected to the physical scaled-down distribution network. Each individual OLE, shown in Figure 6c, is based on a two-level VSC that is controlled following a classical PI approach in $dq$ coordinates [24]. All of the OLEs are rated to 400V and 20 kVA, while the balanced VSC in charge of regulating the DC bus voltage is rated to 400 V and 100 kVA. The OLEs are connected to the buses N3, N5, N6, N7, N8, N9, N10, and N14.
- DUTs. Two DUTs are so far available to be tested in this PSHIL infrastructure. First, a DC link between the nodes N8 and N14 for controlling the active power transfer between the feeders and also the reactive power injections at these nodes as shown in Figure 5. The DC link is rated to 400 V and 20 kVA. Second, a static OLTC based on thyristor technology coupled to a $400 \pm 5\%/400$ V transformer installed at the connection point with the laboratory network, which

allows regulating the supply voltage on load. In addition, it is important to highlight that the design of this PSHIL infrastructure is quite flexible, being possible to incorporate also as DUT some of the already analyzed OLEs, or any other device. In fact, [25] analyzes the minimization of power losses (AUT) in the scaled-down distribution system using the static OLTC, the DC link, and the reactive power injections of some distributed generators emulated with the OLEs.

- PSHIL management. It has been implemented in a real-time platform provided by SpeedGoat. Particularly, this management system provides a bidirectional communication channel with the OLEs, DUTs, and AUTs by means of an UDP/IP communication protocol. The time latency with the local controllers of OLEs and DUTs is in the range of hundreds of microseconds, while the characteristic latency of AUTs depends on the algorithm but are in the range of minutes.

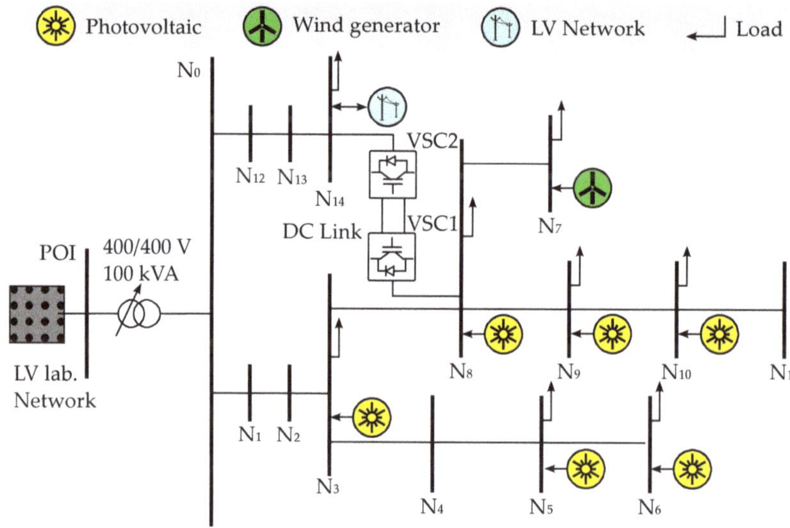


**Figure 5.** One-line diagram of the MV physical network within the PSHIL infrastructure at the University of Sevilla.

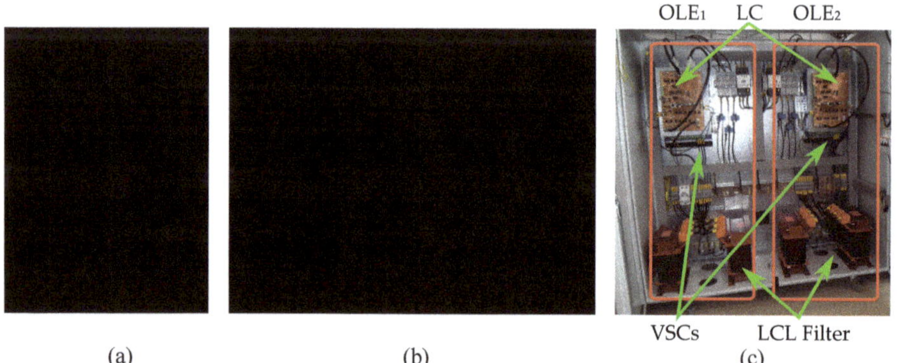


**Figure 6.** Elements of the PSHIL: (**a**) Regatron system, (**b**) synchronous generator, (**c**) OLEs Electrical cabinet.

## 5. PSHIL Example of Application

This section is devoted to describe a PSHIL application example using the University of Sevilla infrastructure presented in the above section. The objective is to evidence some of the main functionalities of a PSHIL testing environment to validate the integration of different DUTs and AUTs and their global impact on the power system under study. This application example revolves around the benefits that active distribution network management may bring to the operation of distribution systems with a high distributed energy resource (DER) penetration. This is usually studied in the MV and LV levels separately, but this section presents a simultaneous analysis of these two voltage levels. In order to do so, three different test cases following a step-by-step procedure will be presented to reach the pursued final goal.

## 5.1. PSHIL Infrastructure Used in the Application Example

The proposed application example comprises MV and LV distribution networks. The scaled-down distribution network supplied from the laboratory network previously described is used for this purpose. The LV network selected in this case is the residential feeder of the LV benchmark network proposed by the CIGRE Task Force C06.04.02 for the analysis of distributed generation in LV systems, shown in Figure 7 [58]. This LV network is simulated in real time using the OPAL-RT platform. It has been assumed that the LV network is connected to bus N14 of the MV scaled-down network as shown in Figure 5. The interface between these networks is done by the OLE connected at this bus following the scheme shown in Figure 8. To do so, the active and reactive powers of the LV network computed in the real-time simulation ($P^\star_{M0}$, $Q^\star_{M0}$) are sent to the corresponding OLE as references using the OPAL-RT analog outputs. Conversely, the voltage measured in bus N14 of the MV scaled-down network is sent to the real-time simulation to be considered as the bus M0 voltage using the OPAL-RT analog inputs. Note that the OLE connected to bus N14 also has to emulate an industrial load with power references ($P^\star_{14ind}$, $Q^\star_{14ind}$) sent from the management layer. Therefore, the OLE reference powers ($P^\star_{N14}$, $Q^\star_{N14}$) are the sum of these two setpoints.

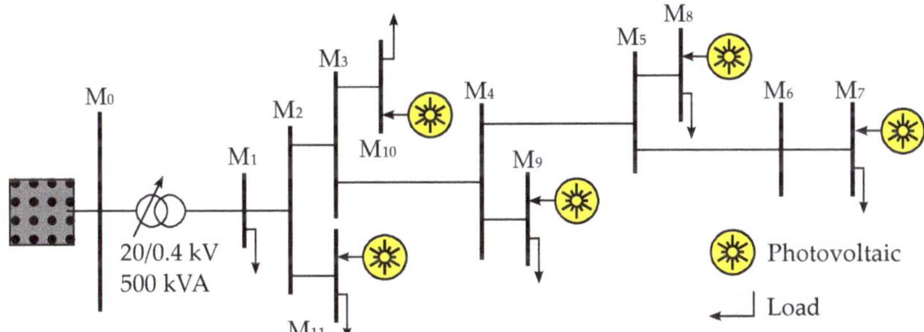


**Figure 7.** PSHIL application example. One-line diagram of the LV real-time simulated network.

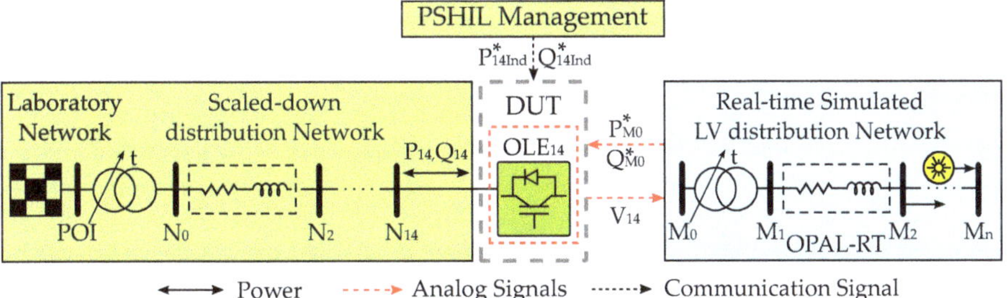


**Figure 8.** PSHIL application example. Integration of the LV real-time simulated network and the physical MV network.

Regarding the MV scaled-down physical network, the OLEs are connected to the following buses: N3, N5, N6, N7, N8, N9, and N10. Each bus has a combination of generators (photovoltaic and wind turbine) and loads (industrial and domestic types) with a peak power defined in Table 2. These peak power values correspond to those of the scaled-down system being required to apply the corresponding scale factor to obtain the values referred to the original system (20 MVA/100 kVA). A constant power factor of 0.87 for industrial loads and 0.98 for domestic loads has been assumed. The 24-h active power profiles of loads and generators are detailed in Figure 9 in per-unit values. The DUTs to be tested connected to this physical system are the distributed generation emulated by OLEs, the OLTC, and the DC link. These DUTs follow the optimal setpoints computed by an OPF that minimizes the power losses (AUT) [25].

**Table 2.** Peak power of loads and generators connected in each node of the MV scaled-down distribution network.

| Node | Generation (kW) | | Load (kW) | |
|------|-----|--------------|------------|-----------|
|      | PV  | Wind Turbine | Industrial | Household |
| N3   | 4.34 | 0.00 | 2.82 | 2.52 |
| N5   | 7.04 | 0.00 | 0.00 | 6.61 |
| N6   | 6.52 | 0.00 | 0.00 | 5.01 |
| N7   | 0.00 | 6.07 | 0.91 | 0.00 |
| N8   | 6.52 | 0.00 | 0.00 | 5.36 |
| N9   | 7.04 | 0.00 | 6.75 | 0.00 |
| N10  | 6.52 | 0.00 | 0.80 | 4.35 |
| N14  | 0.00 | 0.00 | 9.05 | 0.00 |

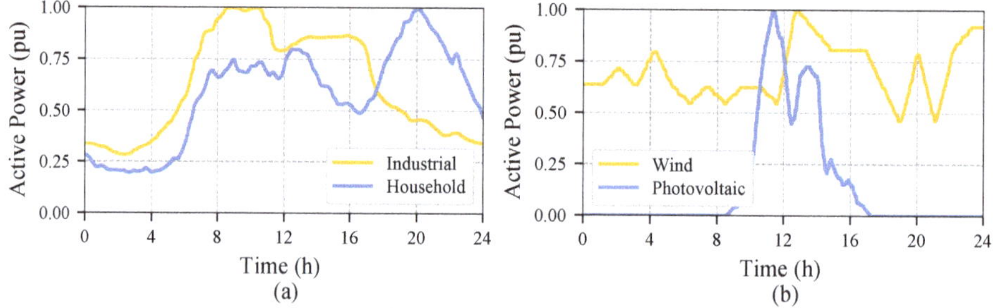


**Figure 9.** PSHIL application example. Daily profiles in per-unit (**a**) Load profiles, (**b**) Generation profiles.

Similarly, the loads and generators connected to the LV real-time simulated network get the same power profiles than the ones used in the physical system and represented in Figure 9. The peak power of the load and generators connected to each node is summarized in Table 3. Note that these values correspond to the original system rather than a scaled-down one because this network is simulated. Therefore, the corresponding scale factor must be applied to compute the reference powers ($P_{14}^{\star}$, $Q_{14}^{\star}$). The AUT to be tested considers the reduced communication infrastructure associated with LV distribution systems, which prevents the use of secondary controllers like the one applied in the MV network. In this case, it is proposed to test an AUT that assigns a voltage reference to each distributed generator of the network, computed on a planning stage based on historical data. The local controllers of the distributed generators apply a reactive power droop strategy to achieve the assigned voltage reference.

*5.2. PSHIL Test Cases*

The test cases used to demonstrate the benefits of an active management of MV and LV distribution systems have been organized in a step-by-step manner as follows:

- *PSHIL MV and LV network coupling* (PSHIL-1). This test case evaluates the performance of the DUT in charge of interfacing the physical network and the real-time simulated network. This corresponds to the OLE connected to bus N14 of the physical system.
- *PSHIL MV control assets* (PSHIL-2). The MV network DUTs (DERs, OLTC, and DC Link) are added to the previous case with setpoints computed by an AUT based on an OPF to minimize power losses. In addition to the assessment of each individual DUT, this case is focused on the impact of MV control assets to both MV and LV networks.
- *PSHIL MV/LV control assets* (PSHIL-3). This case adds to the previous case the voltage control on the LV network using a local reactive power droop strategy. Therefore, it should be possible to evaluate how the LV network control assets react with the MV and LV network changes.

**Table 3.** Peak power of loads and generators connected in each node of the LV original distribution network.

| Node | PV Generation (kW) | Household Load (kW) |
|------|--------------------|---------------------|
| M7   | 15                 | 15                  |
| M8   | 52                 | 52                  |
| M9   | 55                 | 55                  |
| M10  | 35                 | 35                  |
| M11  | 47                 | 47                  |

Table 4 collects the DUTs/AUTs and the functionalities tested in each case. This table clearly shows the step-by-step procedure followed to reach the final goal of simultaneous testing of an active management of MV and LV distribution systems with a high DER penetration.

**Table 4.** DUT/AUT and functionalities of the test cases.

| DUT/AUT | Case PSHIL-1 | Case PSHIL-2 | Case PSHIL-3 |
|---------|:------------:|:------------:|:------------:|
| MV/LV network interconnection | • | • | • |
| HV/MV OLTC | | • | • |
| MV DER reactive power | | • | • |
| DC link | | • | • |
| LV PV reactive power | | | • |
| **Functionalities** | | | |
| Current control (DER, DC link) | | • | • |
| Reactive power control (DER, DC link) | | • | • |
| Active Power control (DC link) | | • | • |
| OLTC tap Control | | • | • |
| OPF for MV power losses minimization | | • | • |
| Local droop for LV voltage control | | | • |

*5.3. PSHIL Experimental Results*

5.3.1. PSHIL-1 Results

Figure 10a shows the 24-h active and reactive power profiles of the OLE connected to bus N14 for the test case PSHIL-1. This OLE is in charge of emulating an industrial load, detailed in Table 2, and also the integration of the LV real-time simulated network into the physical MV scaled-down system. The continuous lines represent the actual OLE active and reactive powers, while the dashed lines correspond to the setpoints (sum of the industrial load and LV real-time simulated network). It can be observed that there is hardly any difference between the power references and the actual injected values. In addition, Figure 10b compares the voltages measured at bus N14 and the ones used in the real-time simulation for bus M0. Both voltages are practically coincident, similarly to what happens with the powers. Therefore, these results confirm an adequate integration of the LV real-time simulated network into the physical MV scaled-down network.

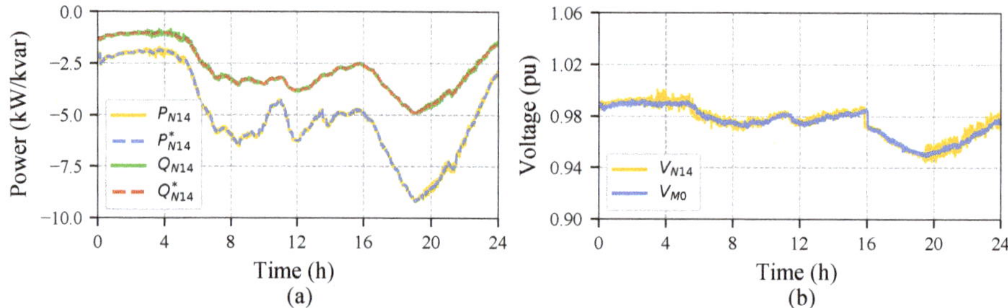

**Figure 10.** PSHIL-1: N14 industrial load and integration of the LV real-time simulated network within the MV scaled-down physical system. (**a**) 24-h evolution of active and reactive power injected by the OLE connected to bus N14. (**b**) 24-h evolution of voltages in bus N14 (physical network) and M0 (real-time simulated network).

Figure 11a shows the daily voltages of buses N3, N6, and N14 where undervoltage situations (voltages below 0.95 pu) during some periods are noticed. These undervoltages are more severe in those buses located farther away from the primary substation and mainly concentrated during the hours without DER injections. Conversely, these are responsible for the voltage peaks around 11:00 in nodes N3 and N14 because of the DER-related active power profile shown in Figure 9. It is interesting to note that this effect is not reproduced in bus N14 because the DC link is not connected in this test case and the net active power of the industrial load and the real-time simulated LV network is always negative (consumption), as shown in Figure 10a.

Figure 11b illustrates the nodal voltages of buses M1, M4, and M7 within the real-time simulated LV system. Similarly to the MV system, undervoltages arise around 20:00 and peak voltages around 11:00 when PV generation is peaking. Note that this effect is more prominent in bus M7, where PV generation is connected, becoming even higher than voltages in buses M4 and M1, which indicates a reverse power flow in the lines connecting these nodes. Bus M1 voltage is barely modified during this period of PV peaking because of its proximity to the MV/LV transformer. It can be observed, anyway, that the general voltage trend in the LV network is imposed by bus N14 of the MV network, especially in those periods without LV DER injections. These conclusions arise from the simultaneous analysis of the MV and LV networks facilitated by the proposed PSHIL approach.

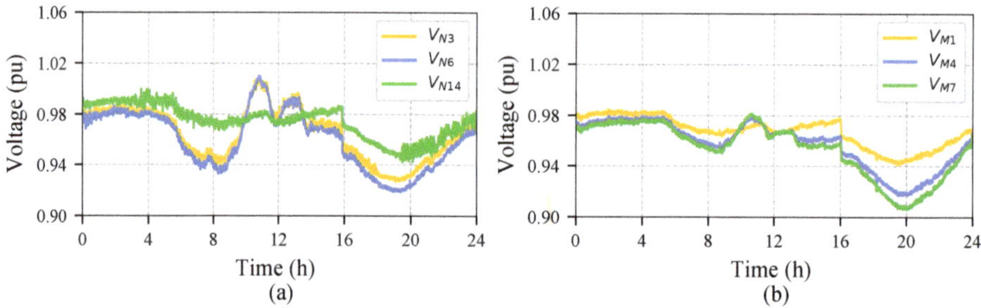

**Figure 11.** PSHIL-1. Impact on the MV and LV distribution networks. (**a**) 24-h evolution of the MV network voltages at buses N3, N6, and N14. (**b**) 24-h evolution of the LV network voltages at buses M1, M4, and M7.

### 5.3.2. PSHIL-2 Results

The AUT in the test case PSHIL-2 is an OPF for minimization of power losses in charge of defining the setpoints of the MV DUTs (OLTC, DERs, and DC link), whose daily evolution is detailed in Figure 12. The following comments can be stated for each of these DUTs:

- OLTC. Figure 12a shows the 24-h OLTC tap position and the voltage of bus N3. The tap position is maintained most of the time in the lowest position in order to increase the MV voltages to reduce the power losses as much as possible. However, the tap is changed to its central position during the period when the DER injects the maximum power to fulfill the maximum voltage constraint imposed by the OPF (1.05 pu).
- DERs. The reactive power injections of DERs connected to buses N3 and N10 are depicted in Figure 12b. These show a proper tracking of the setpoints computed by the AUT. Note that the reactive power injection in bus N10 is higher that in bus N3 because of the electrical distance. This bus is farther away from the initial node of the MV scaled-down network and, therefore, experiences a larger voltage drop caused by the loads. This requires a higher reactive power injection to maintain the voltages as high as possible.
- DC link. The 24-h active and reactive power flows through the DC link are depicted in Figure 12c and 12d, respectively. The setpoints computed by the AUT are tracked without errors by the DC link. Both DC link sides inject reactive power to their corresponding connection nodes to increase the voltages locally. In addition, the DC link transfers active power from N8 to N14 around noon when the PV generation injects the maximum power. The opposite situation happens in the rest of the day, where the DC link acts to equalize the load of each MV scaled-down distribution feeder to reduce the active power losses as much as possible.

Therefore, it can be claimed that the adequate control actions are computed by the AUT and executed correspondingly by the corresponding DUTs.

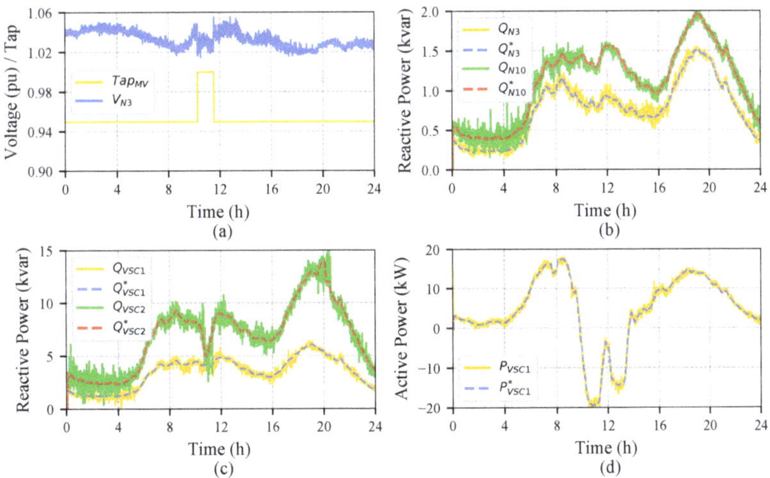

**Figure 12.** PSHIL-2. DUT performance using 24-h evolution of: (**a**) OLTC tap position and voltage of bus N3. (**b**) DER reactive power injections in buses N3 and N6. (**c**) DC-link reactive power injections. (**d**) DC-link active power transfer from N8 to N14.

Regarding the impact of the AUT on the tested power system, it is interesting to note the evolution of the MV and LV voltages shown in Figure 13. This figure represents the voltages of the same nodes represented in Figure 11 for the test PSHIL-1. Regarding the MV nodes, it can be clearly noticed that most of the time the voltages are above 1 pu and very close to 1.05 pu, which is the maximum voltage considered in the OPF. In addition, a flatter voltage profile can be observed compared to those of Figure 11a in spite of the highly variable DER generation. On the other hand, the evolution of the LV buses detailed in Figure 13b shows a similar evolution than the voltage of bus N14. This is reflected in a voltage increase of all LV buses thanks to the AUT and DUTs of the MV network, which avoid undervoltage situations (below 0.95 pu). This effect is especially remarked around hour 11, where the OLTC action drastically reduces the voltages of the LV buses. Finally, it is interesting to note the AUT effect on the power losses of the MV and LV networks, which have been reduced with respect to PSHIL-1 from 0.584% to 0.465% and 7.086% to 6.382%, respectively.

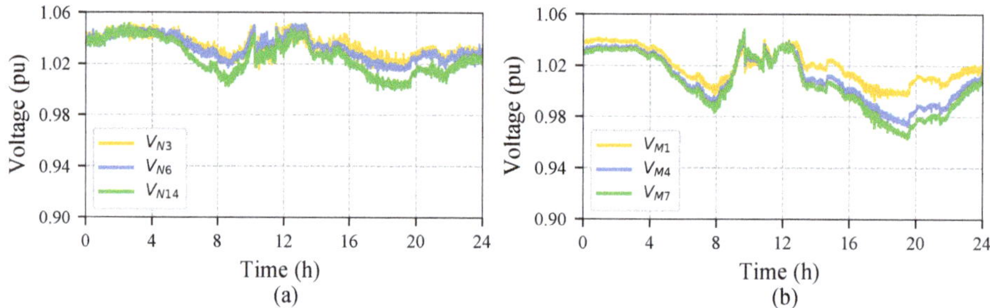

**Figure 13.** PSHIL-2. Impact on the MV and LV distribution networks. (**a**) 24-h evolution of the MV network voltages at buses N3, N6, and N14. (**b**) 24-h evolution of the LV network voltages at buses M1, M4, and M7.

### 5.3.3. PSHIL-3 Results

In this case, the same voltage reference was imposed for all of the LV PV generators, for simplification purposes. Figure 14 shows the voltage and the active and reactive powers injected in by the PV connected in node M7. The bus voltage is compared with the reference voltage (1.05 pu) to compute the reactive power injection using a linear droop strategy. Note that the reactive power injection shown in Figure 14b is always positive because the nodal voltage is most of the time below the reference voltage, as shown in Figure 14a. This is especially noticeable at night when the voltages are low due to the network loading and the PV uses most of its capacity to inject reactive power according to the rated power detailed in Table 3.

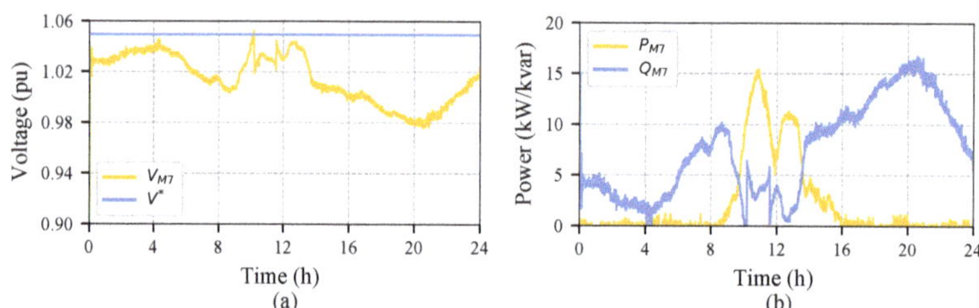

**Figure 14.** PSHIL-3. (**a**) 24-h evolution of voltage in bus M7. (**b**) 24-h evolution of PV active and reactive power injections in bus M7.

Regarding the impact of these control actions on the distribution system, Figure 15 shows the daily voltage profiles on the MV and LV networks. The MV voltages, shown in Figure 14a, are almost the same as those presented in PSHIL-2 because only one LV distribution network has been simulated and, therefore, its effect on the MV side is barely noticed. The analysis of the LV voltages, however, reveal some differences due to the voltage control algorithm implemented in this network. During the midday hours, the PV power injection causes the voltage in bus M7 to be higher than the voltage in bus M1, revealing an inverse power flow between these two LV nodes. During the evening, with the peak load and the absence of PV active power injection, the voltage decreases. However, the influence of the reactive power injection using the spare PV converter capacity is noticeable comparing the voltage of buses M4 and M7. Note that in this evening period, the voltage at bus M7 is very close to that of bus M4, even though its electrical distance to the head node is larger. This effect is also noticeable when comparing these nodal voltages in the test PSHIL-2 (Figure 13b), where no LV control actions are applied. Finally and regarding the power losses, the LV local voltage control achieves a reduction of power losses from 6.382% reported in PSHIL-2 to 4.590%.

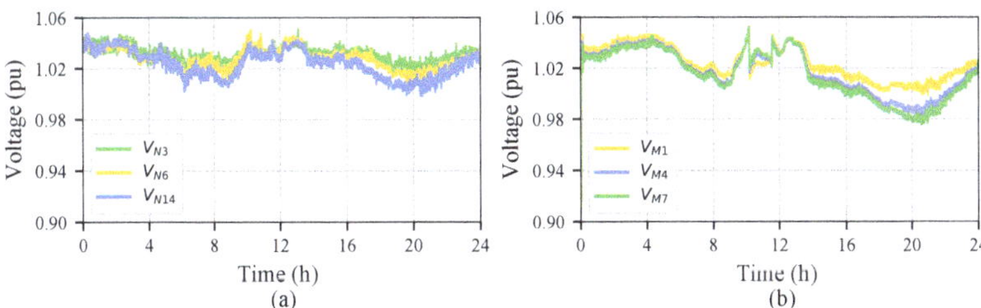

**Figure 15.** PSHIL-3. Impact on the MV and LV distribution networks. (**a**) 24-h evolution of the MV system voltage in buses N3, N6, and N14. (**b**) 24-h evolution of the LV system voltage in buses M1, M4, and M7.

## 6. Conclusions

This paper has presented a new HIL approach for simultaneously testing several DUTs and AUTs interacting with each other in a power system. In this way, the proposed PSHIL approach can be considered a natural evolution of the PHIL concept, which is mainly focused on the individual performance of a DUT connected to a network node. The PSHIL architecture has been depicted including a description of its main actual and simulated components classified according to their functionalities.

The PHSIL infrastructure available at the electrical engineering laboratory of the University of Sevilla has been used as an example of application to evidence the PSHIL testing functionalities, analyzing the active management of a MV and LV distribution system with high penetration of renewable generation. Different cases have been tested, each of which evidences some of the possible functionalities of the PSHIL approach.

The first test (PSHIL-1) shows how to integrate an LV real-time simulated network into a physical scaled-down MV system. The second one (PSHIL-2), using the first test as a starting point, elaborates on the use of an OPF (AUT) for power losses minimization in the physical MV scaled-down network combined with several DUTs (OLTC, DER reactive power, and a DC link) in charge of following the OPF-computed optimal setpoints. Finally, the third test case (PSHIL-3) introduces a voltage control algorithm in the LV network based on local conditions in buses where DERs are connected.

In conclusion, the staged testing procedure has demonstrated the PSHIL capacity of simultaneously analyzing an MV and LV distribution system in a hardware environment. Moreover, the results shown in the previous section evidence that the individual performance of each DUT can be successfully studied, as well as their impact on the power system. The PSHIL concept allows an evolution from a device- to a system-oriented testing approach, providing a powerful framework for the study of smart grid technologies and allowing an assessment of the potential benefits that new technologies may bring to the power system.

**Author Contributions:** Conceptualization, M.B.-V. and J.M.M.-O.; Methodology, M.B.-V.; Software, ; F.d.P.G.-L. and A.M.-M.; Validation, M.B.-V. and F.d.P.G.-L.; Formal Analysis, J.M.M.-O. and A.M.-M.; Investigation, M.B.-V.; Resources, J.M.M.-O.; Data Curation, F.d.P.G.-L.; Writing—Original Draft Preparation, M.B.-V., J.M.M.-O. and

A.M.-M.; Writing—Review & Editing, M.B.-V., J.M.M.-O. and A.M.-M.; Visualization, F.d.P.G.-L.; Supervision, M.B.-V. and J.M.M.-O.; Project Administration, J.M.M.-O.; Funding Acquisition, J.M.M.-O. All authors have read and agreed to the published version of the manuscript.

**Funding:** This research was funded by the Spanish Ministry of Economy and Competitiveness and the European Union Horizon 2020 under grant numbers ENE2017-84813-R and 764090, respectively. The APC was funded by the Ministry of Economy and Competitiveness under grant number ENE2017-84813-R.

**Conflicts of Interest:** The authors declare no conflict of interest.

## Abbreviations

The following abbreviations are used in this manuscript:

| | |
|---|---|
| AC | Alternating Current |
| AUT | Algorithm Under Test |
| CHIL | Controller-Hardware-in-the-Loop |
| DC | Direct Current |
| DER | Distributed Energy Resource |
| DUT | Device Under Test |
| FACTS | Flexible AC Transmission System |
| HIL | Hardware-in-the-Loop |
| HVDC | High-Voltage DC |
| LC | Local Control |
| LV | Low Voltage |
| MV | Medium Voltage |
| OLE | Omnimode Load Emulator |
| OLTC | On-Load Tap-Changer |
| OPF | Optimal Power Flow |
| PHIL | Power-Hardware-in-the-Loop |
| PSHIL | Power-System-Hardware-in-the-Loop |
| POI | Point of Interconnection |
| PV | Photovoltaic |
| RTCS | Real-Time Control System |
| VSC | Voltage Source Converter |

## References


1. International Energy Agency. Global EV Outlook 2018: Towards cross-modal electrification. In Proceedings of the Clean Energy Ministerial, Copenhagen, Denmark, 22–24 May 2018.
2. Kyoto Protocol to the United Nations Framework Convention on Climate Change. 1998. Available online: https://unfccc.int/resource/docs/convkp/kpeng.pdf (accessed on 16 March 1998).
3. Steinberg, D.; Bielen, D.; Eichman, J.; Eurek, K.; Logan, J.; Mai, T.; McMillan, C.; Parker, A.; Vimmerstedt, L.; Wilson, E. *Electrification and Decarbonization: Exploring U.S. Energy Use and Greenhouse Gas Emissions in Scenarios with Widespread Electrification and Power Sector Decarbonization*; Technical Report; NREL: Washington, DC, USA, 2017. doi:10.2172/1372620. [CrossRef]
4. De Quevedo, P.M.; Muñoz-Delgado, G.; Contreras, J. Impact of Electric Vehicles on the Expansion Planning of Distribution Systems Considering Renewable Energy, Storage, and Charging Stations. *IEEE Trans. Smart Grid* **2019**, *10*, 794–804, doi:10.1109/TSG.2017.2752303. [CrossRef]
5. Yu, X.; Jiang, Z.; Zhang, Y. Control of Parallel Inverter-Interfaced Distributed Energy Resources. In Proceedings of the 2008 IEEE Energy 2030 Conference, Atlanta, GA, USA, 17–18 November 2008; pp. 1–8. doi:10.1109/ENERGY.2008.4781030. [CrossRef]
6. Li, H.; Li, F.; Xu, Y.; Rizy, D.T.; Kueck, J.D. Adaptive Voltage Control With Distributed Energy Resources: Algorithm, Theoretical Analysis, Simulation, and Field Test Verification. *IEEE Trans. Power Syst.* **2010**, *25*, 1638–1647, doi:10.1109/TPWRS.2010.2041015. [CrossRef]
7. Li, H.; Li, F.; Xu, Y.; Rizy, D.T.; Kueck, J.D. Interaction of multiple distributed energy resources in voltage regulation. In Proceedings of the 2008 IEEE Power and Energy Society General Meeting—Conversion

and Delivery of Electrical Energy in the 21st Century, Pittsburgh, PA, USA, 20–24 July 2008; pp. 1–6. doi:10.1109/PES.2008.4596142. [CrossRef]

8. Yang, T.; Wu, D.; Stoorvogel, A.A.; Stoustrup, J. Distributed coordination of energy storage with distributed generators. In Proceedings of the 2016 IEEE Power and Energy Society General Meeting (PESGM), Boston, MA, USA, 17–21 July 2016; pp. 1–5. doi:10.1109/PESGM.2016.7741778. [CrossRef]

9. Xu, S.; Sun, H.; Zhang, Z.; Guo, Q.; Zhao, B.; Bi, J.; Zhang, B. MAS-Based Decentralized Coordinated Control Strategy in a Micro-Grid with Multiple Microsources. *Energies* **2020**, *13*, 2141, doi:10.3390/en13092141. [CrossRef]

10. Salcedo, R.; Corbett, E.; Smith, C.; Limpaecher, E.; Rekha, R.; Nowocin, J.; Lauss, G.; Fonkwe, E.; Almeida, M.; Gartner, P.; et al. Banshee distribution network benchmark and prototyping platform for hardware-in-the-loop integration of microgrid and device controllers. *J. Eng.* **2019**, *2019*, 5365–5373, doi:10.1049/joe.2018.5174. [CrossRef]

11. Kotsampopoulos, P.; Lagos, D.; Hatziargyriou, N.; Faruque, M.O.; Lauss, G.; Nzimako, O.; Forsyth, P.; Steurer, M.; Ponci, F.; Monti, A.; et al. A Benchmark System for Hardware-in-the-Loop Testing of Distributed Energy Resources. *IEEE Power Energy Technol. Syst. J.* **2018**, *5*, 94–103, doi:10.1109/JPETS.2018.2861559. [CrossRef]

12. Newaz, A.; Ospina, J.; Faruque, M.O. Controller Hardware-in-the-Loop Validation of a Graph Search Based Energy Management Strategy for Grid-Connected Distributed Energy Resources. *IEEE Trans. Energy Convers.* **2020**, *35*, 520–528, doi:10.1109/TEC.2019.2952575. [CrossRef]

13. Pellegrino, L.; Sandroni, C.; Bionda, E.; Pala, D.; Lagos, D.T.; Hatziargyriou, N.; Akroud, N. Remote Laboratory Testing Demonstration. *Energies* **2020**, *13*, 2283, doi:10.3390/en13092283. [CrossRef]

14. Kotsampopoulos, P.; Georgilakis, P.; Lagos, D.T.; Kleftakis, V.; Hatziargyriou, N. FACTS Providing Grid Services: Applications and Testing. *Energies* **2019**, *12*, 255 doi:10.3390/en12132554. [CrossRef]

15. Rizy, D.T.; Xu, Y.; Li, H.; Li, F.; Irminger, P. Volt/Var control using inverter-based distributed energy resources. In Proceedings of the 2011 IEEE Power and Energy Society General Meeting, Detroit, MI, USA, 24–28 July 2011; pp. 1–8. doi:10.1109/PES.2011.6039020. [CrossRef]

16. Huerta, F.; Gruber, J.K.; Prodanovic, M.; Matatagui, P. Power-hardware-in-the-loop test beds: evaluation tools for grid integration of distributed energy resources. *IEEE Ind. Appl. Mag.* **2016**, *22*, 18–26, doi:10.1109/MIAS.2015.2459091. [CrossRef]

17. Huerta, F.; Gruber, J.K.; Prodanovic, M.; Matatagui, P.; Gafurov, T. A laboratory environment for real-time testing of energy management scenarios: Smart Energy Integration Lab (SEIL). In Proceedings of the 2014 International Conference on Renewable Energy Research and Application (ICRERA), Glasgow, UK, 27–30 September 2014; pp. 514–519. doi:10.1109/ICRERA.2014.7016438. [CrossRef]

18. Darbali-Zamora, R.; Quiroz, J.E.; Hernández-Alvidrez, J.; Johnson, J.; Ortiz-Rivera, E.I. Validation of a Real-Time Power Hardware-in-the-Loop Distribution Circuit Simulation with Renewable Energy Sources. In Proceedings of the 2018 IEEE 7th World Conference on Photovoltaic Energy Conversion (WCPEC) (A Joint Conference of 45th IEEE PVSC, 28th PVSEC 34th EU PVSEC), Waikoloa, HI, USA, 10–15 June 2018; pp. 1380–1385. doi:10.1109/PVSC.2018.8547473. [CrossRef]

19. Kikusato, H.; Ustun, T.S.; Suzuki, M.; Sugahara, S.; Hashimoto, J.; Otani, K.; Shirakawa, K.; Yabuki, R.; Watanabe, K.; Shimizu, T. Microgrid Controller Testing Using Power Hardware-in-the-Loop. *Energies* **2020**, *13*, 44, doi:10.3390/en13082044. [CrossRef]

20. Kotsampopoulos, P.C.; Kleftakis, V.A.; Hatziargyriou, N.D. Laboratory Education of Modern Power Systems Using PHIL Simulation. *IEEE Trans. Power Syst.* **2017**, *32*, 3992–4001, doi:10.1109/TPWRS.2016.2633201. [CrossRef]

21. Oureilidis, K.; Malamaki, K.N.; Gallos, K.; Tsitsimelis, A.; Dikaiakos, C.; Gkavanoudis, S.; Cvetkovic, M.; Mauricio, J.M.; Maza Ortega, J.M.; Ramos, J.L.M.; et al. Ancillary Services Market Design in Distribution Networks: Review and Identification of Barriers. *Energies* **2020**, *13*, 917, doi:10.3390/en13040917. [CrossRef]

22. Maza-Ortega, J.M.; Mauricio, J.M.; Barragán-Villarejo, M.; Demoulias, C.; Gómez-Expósito, A. Ancillary Services in Hybrid AC/DC Low Voltage Distribution Networks. *Energies* **2019**, *12*, 3591 doi:10.3390/en12193591. [CrossRef]

23. Yu, X.; Tolbert, L.M. Ancillary services provided from DER with power electronics interface. In Proceedings of the 2006 IEEE Power Engineering Society General Meeting, Montreal, QC, USA, 18–22 June 2006; p. 8. doi:10.1109/PES.2006.1709460. [CrossRef]

24. Maza-Ortega, J.M.; Barragán-Villarejo, M.; García-López, F.d.P.; Jiménez, J.; Mauricio, J.M.; Alvarado-Barrios, L.; Gómez-Expósito, A. A Multi-Platform Lab for Teaching and Research in Active Distribution Networks. *IEEE Trans. Power Syst.* **2017**, *32*, 4861–4870, doi:10.1109/TPWRS.2017.2681182. [CrossRef]

25. García-López, F.d.P.; Barragán-Villarejo, M.; Marano-Marcolini, A.; Maza-Ortega, J.M.; Martínez-Ramos, J.L. Experimental Assessment of a Centralised Controller for High-RES Active Distribution Networks. *Energies* **2018**, *11*, 3364, doi:10.3390/en11123364. [CrossRef]

26. De Paula García-López, F.; Barragán-Villarejo, M.; Maza-Ortega, J.M. Grid-friendly integration of electric vehicle fast charging station based on multiterminal DC link. *Int. J. Electr. Power Energy Syst.* **2020**, *114*, 105341, doi:10.1016/j.ijepes.2019.05.078. [CrossRef]

27. Carpita, M.; Affolter, J.; Bozorg, M.; Houmard, D.; Wasterlain, S. ReIne, a flexible laboratory for emulating and testing the Distribution grid. In Proceedings of the 2019 21st European Conference on Power Electronics and Applications (EPE '19 ECCE Europe), Genova, Italy, 2–6 September 2019; pp. P.1–P.6. doi:10.23919/EPE.2019.8915190. [CrossRef]

28. Orue, I.; Larrieta, J.; Gilbert, I.P. Development and requirements of a new High Power Laboratory. In Proceedings of the IEEE PES T&D 2010, Minneapolis, MN, USA, 25–29 July 2010; pp. 1–6. doi:10.1109/TDC.2010.5484623. [CrossRef]

29. Torres, M.; Lopes, L. Inverter-Based Diesel Generator Emulator for the Study of Frequency Variations in a Laboratory-Scale Autonomous Power System. *Energy Power Eng.* **2013**, *5*, 274–283, doi:10.4236/epe.2013.53027. [CrossRef]

30. Liu, Y.; Lan, P.; Lin, H. Grid-connected PV inverter test system for solar photovoltaic power system certification. In Proceedings of the 2014 IEEE PES General Meeting | Conference Exposition, Washington, DC, USA, 27–31 July 2014; pp. 1–5. doi:10.1109/PESGM.2014.6939471. [CrossRef]

31. Chinchilla, M.; Arnaltes, S.; Rodriguez-Amenedo, J.L. Laboratory set-up for wind turbine emulation. In Proceedings of the 2004 IEEE International Conference on Industrial Technology, IEEE ICIT '04, Hammamet, Tunisia, 8–10 December 2004; Volume 1, pp. 553–557. doi:10.1109/ICIT.2004.1490352. [CrossRef]

32. Sanchez, S.; D'Arco, S.; Holdyk, A.; Tedeschi, E. An approach for small scale power hardware in the loop emulation of HVDC cables. In Proceedings of the 2018 Thirteenth International Conference on Ecological Vehicles and Renewable Energies (EVER), Monte Carlo, Monaco, 10–12 April 2018; pp. 1–8. doi:10.1109/EVER.2018.8362343. [CrossRef]

33. Aktas, A.; Erhan, K.; Ozdemir, S.; Ozdemir, E. Experimental investigation of a new smart energy management algorithm for a hybrid energy storage system in smart grid applications. *Electr. Power Syst. Res.* **2017**, *144*, 185–196, doi:10.1016/j.epsr.2016.11.022. [CrossRef]

34. Zecchino, A.; Hu, J.; Coppo, M.; Marinelli, M. Experimental testing and model validation of a decoupled-phase on-load tap-changer transformer in an active network. *IET Gener. Transm. Distrib.* **2016**, *10*, 3834–3843, doi:10.1049/iet-gtd.2016.0352. [CrossRef]

35. Papaspiliotopoulos, V.A.; Korres, G.N.; Kleftakis, V.A.; Hatziargyriou, N.D. Hardware-In-the-Loop Design and Optimal Setting of Adaptive Protection Schemes for Distribution Systems With Distributed Generation. *IEEE Trans. Power Deliv.* **2017**, *32*, 393–400, doi:10.1109/TPWRD.2015.2509784. [CrossRef]

36. Madureira, A.G.; Lopes, J.A.P. Coordinated voltage support in distribution networks with distributed generation and microgrids. *IET Renew. Power Gener.* **2009**, *3*, 439–454, doi:10.1049/iet-rpg.2008.0064. [CrossRef]

37. Fang, J.; Tang, Y.; Li, H.; Blaabjerg, F. The Role of Power Electronics in Future Low Inertia Power Systems. In Proceedings of the 2018 IEEE International Power Electronics and Application Conference and Exposition (PEAC), Shenzhen, China, 4–7 November 2018; pp. 1–6. doi:10.1109/PEAC.2018.8590632. [CrossRef]

38. Fang, J.; Li, H.; Tang, Y.; Blaabjerg, F. On the Inertia of Future More-Electronics Power Systems. *IEEE J. Emerg. Sel. Top. Power Electron.* **2019**, *7*, 2130–2146, doi:10.1109/JESTPE.2018.2877766. [CrossRef]

39. Dimitra Tragianni, S.; Oureilidis, K.O.; Demoulias, C.S. Supercapacitor sizing based on comparative study of PV power smoothing methods. In Proceedings of the 2017 52nd International Universities Power Engineering Conference (UPEC), Heraklion, Greece, 28–31 August 2017; pp. 1–6. doi:10.1109/UPEC.2017.8232029. [CrossRef]

40. Lashab, A.; Sera, D.; Guerrero, J.M. A Dual-Discrete Model Predictive Control-Based MPPT for PV Systems. *IEEE Trans. Power Electron.* **2019**, *34*, 9686–9697, doi:10.1109/TPEL.2019.2892809. [CrossRef]

41. Mei, Q.; Shan, M.; Liu, L.; Guerrero, J.M. A Novel Improved Variable Step-Size Incremental-Resistance MPPT Method for PV Systems. *IEEE Trans. Ind. Electron.* **2011**, *58*, 2427–2434, doi:10.1109/TIE.2010.2064275. [CrossRef]

42. Nieves-Portana, M.; Barragán-Villarejo, M.; Maza-Ortega, J.; Mauricio-Ferramola, J. Reduction of Zero Sequence Components in Three-Phase Transformerless Multiterminal DC link based on Voltage Source Converters. *Renew. Energy Power Qual. J.* **2013**, 1206–1211, doi:10.24084/repqj11.576. [CrossRef]

43. Harnefors, L.; Wang, X.; Yepes, A.G.; Blaabjerg, F. Passivity-Based Stability Assessment of Grid-Connected VSCs—An Overview. *IEEE J. Emerg. Sel. Top. Power Electron.* **2016**, *4*, 116–125, doi:10.1109/JESTPE.2015.2490549. [CrossRef]

44. Song, I.; Jung, W.; Kim, J.; Yun, S.; Choi, J.; Ahn, S. Operation Schemes of Smart Distribution Networks With Distributed Energy Resources for Loss Reduction and Service Restoration. *IEEE Trans. Smart Grid* **2013**, *4*, 367–374, doi:10.1109/TSG.2012.2233770. [CrossRef]

45. Fahmy, B.N.; Soliman, M.H.; Talaat, H.E. Active Voltage Control in Distribution Networks including Distributed Generations using Hardware-In-The-Loop Technique. In Proceedings of the 2019 21st International Middle East Power Systems Conference, MEPCON 2019, Cairo, Egypt, 17–19 December 2019; pp. 656–661. doi:10.1109/MEPCON47431.2019.9008063. [CrossRef]

46. Ko, M.S.; Lim, S.H.; Kim, J.K.; Hur, K. Distribution voltage regulation using combined local and central control based on real-time data. In Proceedings of the 2019 IEEE Milan PowerTech, PowerTech, Milan, Italy, 23–27 June 2019; pp. 1–6. doi:10.1109/PTC.2019.8810470. [CrossRef]

47. Fandi, G.; Ahmad, I.; Igbinovia, F.O.; Muller, Z.; Tlusty, J.; Krepl, V. Voltage Regulation and Power Loss Minimization in Radial Distribution Systems via Reactive Power Injection and Distributed Generation Unit Placement. *Energies* **2018**, *11*, 1399, doi:10.3390/en11061399. [CrossRef]

48. Kulmala, A.; Angioni, A.; Repo, S.; Giustina, D.D.; Barbato, A.; Ponci, F. Experiences of Laboratory and Field Demonstrations of Distribution Network Congestion Management. In Proceedings of the IECON 2018—44th Annual Conference of the IEEE Industrial Electronics Society, Washington, DC, USA, 21–23 October 2018; pp. 3543–3549. doi:10.1109/IECON.2018.8591412. [CrossRef]

49. Karagiannopoulos, S.; Gallmann, J.; Vayá, M.G.; Aristidou, P.; Hug, G. Active Distribution Grids Offering Ancillary Services in Islanded and Grid-Connected Mode. *IEEE Trans. Smart Grid* **2020**, *11*, 623–633, doi:10.1109/TSG.2019.2927299. [CrossRef]

50. Lauss, G.; Strunz, K. Multirate Partitioning Interface for Enhanced Stability of Power Hardware-in-the-Loop Real-Time Simulation. *IEEE Trans. Ind. Electron.* **2019**, *66*, 595–605, doi:10.1109/TIE.2018.2826482. [CrossRef]

51. Li, Y.; Nejabatkhah, F. Overview of control, integration and energy management of microgrids. *J. Mod. Power Syst. Clean Energy* **2014**, *2*, 212–222, doi:10.1007/s40565-014-0063-1. [CrossRef]

52. Nehrir, M.H.; Wang, C.; Strunz, K.; Aki, H.; Ramakumar, R.; Bing, J.; Miao, Z.; Salameh, Z. A Review of Hybrid Renewable/Alternative Energy Systems for Electric Power Generation: Configurations, Control, and Applications. *IEEE Trans. Sustain. Energy* **2011**, *2*, 392–403, doi:10.1109/TSTE.2011.2157540. [CrossRef]

53. Katiraei, F.; Iravani, R.; Hatziargyriou, N.; Dimeas, A. Microgrids management. *IEEE Power Energy Mag.* **2008**, *6*, 54–65, doi:10.1109/MPE.2008.918702. [CrossRef]

54. Colet-Subirachs, A.; Ruiz-Alvarez, A.; Gomis-Bellmunt, O.; Alvarez-Cuevas-Figuerola, F.; Sudria-Andreu, A. Centralized and Distributed Active and Reactive Power Control of a Utility Connected Microgrid Using IEC61850. *IEEE Syst. J.* **2012**, *6*, 58–67, doi:10.1109/JSYST.2011.2162924. [CrossRef]

55. Zhao, P.; Suryanarayanan, S.; Simoes, M.G. An Energy Management System for Building Structures Using a Multi-Agent Decision-Making Control Methodology. *IEEE Trans. Ind. Appl.* **2013**, *49*, 322–330, doi:10.1109/TIA.2012.2229682. [CrossRef]

56. Tehrani, K. A smart cyber physical multi-source energy system for an electric vehicle prototype. *J. Syst. Archit.* **2020**, *111*, 101804, doi:10.1016/j.sysarc.2020.101804. [CrossRef]

57. Guerrero, J.M.; de Vicuna, L.G.; Matas, J.; Castilla, M.; Miret, J. A wireless controller to enhance dynamic performance of parallel inverters in distributed generation systems. *IEEE Trans. Power Electron.* **2004**, *19*, 1205–1213, doi:10.1109/TPEL.2004.833451. [CrossRef]

58. CIGRE. *Benchmark Systems for Network Integration of Renewable and Distributed Energy Resources*; Electra: Paris, France, 2019.

*Article*

# Distributed Power Hardware-in-the-Loop Testing Using a Grid-Forming Converter as Power Interface

**Steffen Vogel** [1,*,†], **Ha Thi Nguyen** [2,†], **Marija Stevic** [1,†], **Tue Vissing Jensen** [2,†], **Kai Heussen** [2,†], **Vetrivel Subramaniam Rajkumar** [3] and **Antonello Monti** [1]

[1]   RWTH Aachen University, 52062 Aachen, Germany; mstevic@eonerc.rwth-aachen.de (M.S.);
      amonti@eonerc.rwth-aachen.de (A.M.)
[2]   Technical University of Denmark, 2800 Kgs. Lyngby, Denmark; thangu@elektro.dtu.dk (H.T.N.);
      tvjens@elektro.dtu.dk (T.V.J.); kh@elektro.dtu.dk (K.H.)
[3]   Delft University of Technology, 2628 CD Delft, The Netherlands; v.subramaniamrajkumar@tudelft.nl
*    Correspondence: svogel2@eonerc.rwth-aachen.de; Tel.: +49-241-8059-577
†    These authors contributed equally to this work.

Received: 20 June 2020; Accepted: 19 July 2020; Published: 22 July 2020

**Abstract:**   This paper presents an approach to extend the capabilities of smart grid laboratories through the concept of Power Hardware-in-the-Loop (PHiL) testing by re-purposing existing grid-forming converters. A simple and cost-effective power interface, paired with a remotely located Digital Real-time Simulator (DRTS), facilitates Geographically Distributed Power Hardware Loop (GD-PHiL) in a quasi-static operating regime. In this study, a DRTS simulator was interfaced via the public internet with a grid-forming ship-to-shore converter located in a smart-grid testing laboratory, approximately 40 km away from the simulator. A case study based on the IEEE 13-bus distribution network, an on-load-tap-changer (OLTC) controller and a controllable load in the laboratory demonstrated the feasibility of such a setup. A simple compensation method applicable to this multi-rate setup is proposed and evaluated. Experimental results indicate that this compensation method significantly enhances the voltage response, whereas the conservation of energy at the coupling point still poses a challenge. Findings also show that, due to inherent limitations of the converter's Modbus interface, a separate measurement setup is preferable. This can help achieve higher measurement fidelity, while simultaneously increasing the loop rate of the PHiL setup.

**Keywords:** geographically distributed real-time simulation; remote power hardware-in-the-Loop; grid-forming converter; hardware-in-the-loop; simulation fidelity; energy-based metric; energy residual; quasi-stationary

---

## 1. Introduction

Sustainable energy needs for the future are driving the increasing adoption of Renewable Energy Sources that have altered the makeup of the traditional power grid. The increasing complexity and scale of the power system requires tools that can carry out large scale simulations to study its interactions and interoperability with newer hardware and novel control schemes and develop advanced assessment methods. This entails a radical shift from static, offline load flow simulations towards dynamic, online real-time simulations with higher-fidelity. With the help of specialised and powerful computational tools, such as OPAL-RT and Real-time Digital Simulators (RTDS), real-time simulation and Hardware in the Loop (HiL) testing has emerged as an attractive option to study complex power system configurations in detail [1–4]. However, the capability of existing single real-time simulator-based infrastructures is still limited. Simulation of large-scale power grids with detailed distributed energy resource models and control systems is a major challenge. Furthermore, HiL-based testing has a major limitation in scalability, as only limited number of components or devices can be tested or investigated in a single scenario. Thus, scaling real-time power system simulations with HiL testing for large models is a challenging task [5]. Most importantly, the availability of adequate hardware interface of power HiL (PHiL) testing in the form of fast linear amplifiers is limited due to a high cost. These challenges can be addressed by Geographically Distributed Real-time Simulation (GD-RTS) [5,6], Geographically Distributed Power Hardware in the Loop (GD-PHiL) testing and reducing dynamic fidelity requirements on hardware in test platforms [7]. While a conventional PHiL setup consists of a digital real-time simulator (DRTS), power interface/amplifier and a hardware under Test (HuT), GD-PHiL extends it by a real-time virtual gateway, which handles data exchange between the two sites, as shown in Figure 1.

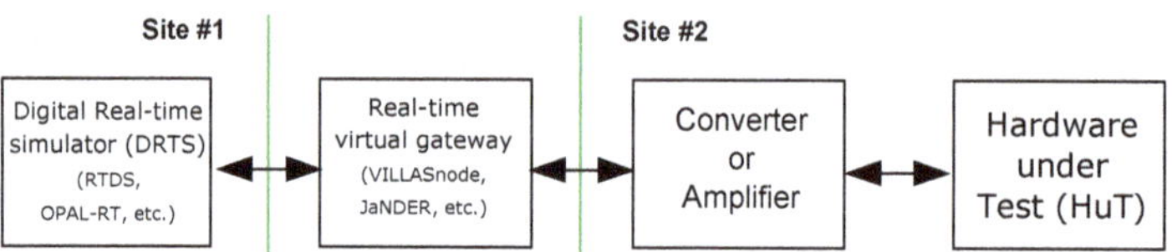


**Figure 1.** The general GD-PHiL layout.

These advanced concepts enable the ability to interface and test larger systems through distributed experiments, giving rise to *'Power System in the Loop'* based testing [8,9], which is also the objective of this work.

In this paper, we demonstrate the application of commercial grid-forming converter hardware for PHiL experiments, using the example of an ABB ship-to-shore converter, which overcomes the unavailability of a dedicated power amplifier. In contrast to purpose-built amplifiers for this application, a grid-forming converter cannot be controlled with instantaneous voltage signals but is rather controlled by a set of Root Mean Square (RMS) voltages and frequency. This limitation may pose challenges in the context of ensuring fidelity of PHiL setup with such amplifiers and requires evaluation. In addition, these types of converters usually lack digital or analog interfaces for fast and direct update of control parameters or the acquisition of measurements. However, laboratories may have grid-forming converters in their facilities, which could be utilised for PHiL experiments without the hefty costs of purchasing a new dedicated PHiL amplifier. Due to their comparatively slow interfaces and built-in protection measures, the PHiL integration cannot be dynamic, but instead is considered *'quasi-static'*. On the other hand, most of these commercial converters are relatively safe to operate and exhibit a reasonable robust control behaviour, which gives the opportunity to integrate complex multi-device systems with multiple power components and software controls in a *Power System-in-the-Loop* (PSiL) setup, such as to emulate a distribution feeder with multiple distributed energy resources.

GD-RTS and GD-PHiL environments enable the coupling of different software and hardware subsystems through discrete-time communication, time-varying delays and packet loss in the communication channel between two or more subsystems can result in artificial energy [10]. This is either stored or generated at the interface and may give rise to *artificial energy artifacts* in the system dynamics. Consequently, due to energy generation, these co-simulation systems can become unstable. Thus, it is advisable to observe this *artificial energy* generated at the interface and use it as metric, as well as possibly in the future online compensation. In [11], a non-iterative energy-conservation-based co-simulation algorithm is investigated using residual energies for adaptive step size control and evaluating the stability of the co-simulation, by which the accuracy and efficiency of the co-simulation is remarkably enhanced as compared to constant step sizes. A monitoring framework is proposed in [12] to keep track of the excess energy and then a dissipation method is implemented to eliminate that energy. Thus, artificial energy monitoring and compensation can help co-simulations achieve stability in spite of artificial interface energy. However, thus far, applications of dissipative co-simulation interfaces are limited to automotive and mechanical systems. Sources of fidelity degradation in real-time co-simulations using non-iterative solvers and interface algorithms differ from the factors that deteriorate fidelity in GD-PHiL, where communication delays, sampling rates of measurement and control signals and dynamic response of PHiL amplifiers are the main issues. However, the objective is in both applications to address sources of fidelity degradation by achieving energy conservation at the interface. To this end, a metric based on artificial energy at the interface is considered in this work for fidelity evaluation.

The key contributions of this paper are as follows:

1.  Experimental demonstration of a Quasi-static PHiL (QsPHiL) power interface using a grid-forming converter, in a geographically distributed multi-rate PSiL setup.
2.  Identification of limiting factors for the use of converters as interface hardware in QsPHiL experiments.
3.  A compensation method that takes advantage of the different timescales to accelerate convergence of the iterative quasi-static interface, to make up for inherent delays in the QsPHiL setup. This was experimentally verified by a reduction in voltage overshoot and convergence time.
4.  First application of a novel Energy-Based Metric (EBM) in GD-PHiL.

The rest of this paper is organised as follows. Section 2 discusses the important aspects of remote coupling and QsPHiL experiments. Metrics for fidelity evaluation are covered in Section 3, while Section 4 delves into the implementation details and applied interface algorithm along with compensation method. Further, Section 5 presents and discusses the key findings of this work, while, in Section 6, key conclusions are drawn.

## 2. Remote Coupling and Quasi-Static PHiL

The conventional PHiL setup leverages a simulator that performs electromagnetic transient simulation in real time and a power amplifier controlled by instantaneous values of voltage waveform provided by the simulator at sampling rates similar to the real-time simulator, such as 10 kHz to 20 kHz, which allows investigations of transient phenomena and hardware response to, e.g., anomalous or disturbed system states, such as protection response or fault-ride-through behaviour. By contrast, testing needs in smart energy systems also include long-term behaviours relevant, e.g., for the study of load behaviour connected in complex distribution networks, voltage support and flexibility through coupling to other energy domains [7]. Fidelity requirements here apply to sampling rates of 10 Hz and below. Thus, simplified forms of the conventional PHiL can be considered, with reduced model granularity, lower sampling rates for simulator and interfaces, and simplified dynamics. These simplifications allow for increased scalability, reduced demands on equipment and model accuracy, thus generally reduced testing cost. To accurately describe the method and impact of results, this section introduces the QsPHiL concept, terminology and related literature on simplified and mixed-fidelity PHiL setups, and then focuses on power interfacing issues.

### 2.1. QsPHiL and Other Reduced-Fidelity PHiL Configurations

The QsPHiL approach introduced and analysed here refers to a power interface (converter, communications and interface algorithm) that aims at fidelity for stationary RMS frequency and RMS voltage levels, corresponding to a time scale of 0.1 Hz and slower, and thus reducing requirements at the power interface [7]. *Quasi-static* here refers to ability to emulate the global energy balance and stationary power flows using stationary RMS frequency and voltage parameters. Alternate reduced-fidelity approaches achieve *quasi-dynamic* fidelity, meaning that the power interface and simulation are accurate and responsive such that RMS voltage and frequency dynamics can be accurately represented, i.e., above 1 Hz, sufficient to model frequency dynamics for inertia and primary frequency control [9]. Mathematically, *quasi-dynamic* setups reduce the represented dynamic states as compared to fully dynamic setups; *quasi-static* setups eliminate dynamic states and are only accurate as stationary representation. In a quasi-static interface setup, convergence of interface parameters is an iterative process where the test casts the number of iterations and time between stationary (equilibrium) states.

A quasi-static fidelity can thus also be considered when power-flow simulation models are applied in a PHiL setup. Quasi-static and quasi-dynamic PHiL setups therefore have many precursors in the literature (see, e.g., [13–15]). To classify a PHiL, in terms of dynamic fidelity, we consider: (a) the intra-emulator fidelity (at what resolutions are the emulated dynamics accurate?); and (b) the inter-emulator fidelity (at what rate is the mutual representation of the other emulator through the interface accurate?).

Reduced-Fidelity PHiL in the Literature

The authors of [13] referred to quasi-static time series (QSTS) based models in their setup, operating with a 2 s time step in the simulated systems, which is also employed in [14,15]. A quasi-static power system setup can be applied in a quasi-dynamic multi-domain setups, where fidelity for dynamic states, e.g., in the heat domain, is preserved [16]. In [17], the reduced fidelity setup operates with a model using PowerFactory "quasi-dynamic" simulations and thus their setup is referred to as "quasi-dynamic" PHiL. Note that PowerFactory definition of "quasi-dynamic" is only quasi-dynamic for frequency, but quasi-static for voltage regulation at distribution grids, such that definitions may overlap.

While conventional PHiL setups pose high challenges to be implemented in GD-PHiL manner as instantaneous values of waveforms must be exchanged between simulator and power amplifier, QsPHiL setups are suitable to be considered for GD-PHiL application [7]. A remote control hardware in the loop test, which remotely couples a distribution system simulated in RTDS located in Greece with an actual On-Load-Tap-Changer (OLTC) controller in Spain, is implemented in [18]. This test uses a virtual platform called *JaNDER* for data exchanging in real-time between two laboratories with the latency lower than 300 ms. A trans-pacific *quasi-static* closed-loop system test is conducted for a GD-PHiL setup, as presented in [13]. The work uses a power distribution network simulator with physical PV/battery inverter in the loop at the National Renewable Energy Laboratory in Golden, CO, USA and a physical PV inverter at power at the Commonwealth Scientific and Industrial Research Organisation's Energy Centre in Newcastle, NSW, Australia.

A common challenge in quasi-static co-simulation and PHiL setups where either dynamic simulators (as here) or a highly dynamic power interface is involved, as, e.g., in [14], is that a difference of sampling rates needs to be accommodated by a suitable interface model and algorithm.

*2.2. Mixed-Fidelity PHiL and PSiL*

The present setup is characterised as:

- Subsystem *A*: RTDS @ 10 kHz
- Subsystem *B*: SYSLAB @ 1 Hz to 10 Hz (measurement limitations)
- Inter-emulator $A \rightarrow B$: approx. 1 Hz: converter set point updates
- Inter-emulator $B \rightarrow A$: 10 Hz: measurements

Considering typical power system dynamics, accurate representation of both frequency and voltage dynamic parameters cannot be represented, even though both parameters can be set at the interface: the response of the Power Interface hardware is in the order of 2.5 Hz for voltage and several seconds, 0.3 Hz, for frequency, returning measurements of converter current and phase (RMS P/Q values) at a rate of 16 Hz. Several iterations of $A \rightarrow B$ and $B \rightarrow A$ are required for convergence of parameters at the interface, leading to an expected interface-based fidelity at about 0.5–1 Hz for voltage events and 0.1 Hz for frequency events.

*2.3. Interface Fidelity of GD-PHiL Setups*

In energy systems simulation, it is essential that energy is preserved across simulator interfaces. However, for most configurations of interface variables, this quality cannot be achieved in practice within a single step: e.g., when an interface is defined by transfer of P/Q information, package loss equates to loss of energy or simply due to the communication delay or inaccurate synchronisation of two simulations. It is possible, however, to log and recover the energy loss incurred through lost packages as in this example, using a corrective mechanism. Therefore, the multi-step interpretation of the exchange quality is important. With the dynamic of exchange in consideration, the convergence of interface variables and stability of invariant qualities within tight error bounds and a deterministic convergence horizon are key qualities to be achieved. An invariant, such as the conservation of energy, can thus be re-formulated as a residual value. This residual can be monitored for possible divergence. In a second step, this monitoring approach can be employed to correct the exchange variables, enabling a stabilisation of the coupling. For further study, this approach should be employed to first log, and then possibly improve the low-rate and multi-rate coupling between Co-simulation and hardware in the loop (QsPHiL) and co-simulation interfacing DRTS .

In addition to the measures for synchronisation, it is also critical to consider the qualities of the simulation interfaces and exchange variables. When sub-systems (simulations) are coupled at a physical layer, algebraic loops arise (cyclic dependencies). As the state of the exchange variables is computed independently in each simulator, in principle, only an iterative and explicit coupling would allow for accurate simulation. In all real-time HiL and SiL, one necessarily degrades the overall system state coherency to allow for the simulation to take place without such iterations. Under this notion of synchronisation, there is an inherent trade-off between the degree of coherency of the overall system state, i.e., to which extent it approximates a monolithic simulation, and the idle time allowed by the choice of simulator computational cycle time $T$. To wit, the greater one chooses $T$, the lower the degree of state coherency.

As a consequence, a concern is to preserve essential properties of the global system state as much as possible. To manage this trade-off, it is possible to identify qualities of global exchange variables to be maintained in spite of said coupling losses.

In [19], an integrated test setup for a remote coupling test is examined. The test couples a full-scale hybrid electric vehicle power system simulated in real time at the U.S. Army TARDEC simulation laboratory with a hybrid power train located at the another laboratory in San Jose, CA, USA. This application uses bi-directional real-time communications over the open Internet using Army assets of two different laboratories 2450 miles apart. In this study, leaked energy is calculated for both sites to evaluate how closely the model matches the behaviour of the real hardware and then proposes robust control techniques that compensate for asynchronous Internet communication delays for the closed loop operation.

However, the aforementioned work is missing a generic compensation method to account for different timescales and inherent delays in any remote QsPHiL setup. Hence, this is one of the key research contributions of this work.

## 3. Fidelity Evaluation and Monitoring Approach

GD-RTS and GD-PHiL are naturally limited by delays incurred by geographical distances. Thus, phenomena of interest on timescales shorter than these delays are generally not presentable in a GD-RTS implementation. Such experiments include fast transients in transmission networks, ripple due to power electronic switching or effects due to harmonics. Experiments in GD-RTS and GD-PHiL instead target facets of the power system described by the RMS-values of the system, including voltage and frequency response, energy exchange tracking or controller operation.

Due to this focus of GD-RTS and GD-PHiL compared to conventional PHiL, evaluation of the fidelity of a GD-RTS implementation can only to a limited extent apply methods of validation commonly used for PHiL systems, e.g., comparison of voltage and current waveforms during transient response with the objective to achieve high fidelity for a wide range of frequencies, including harmonics [20]. Instead, validation metrics in GD-RTS and GD-PHiL should focus on consistency of RMS values in the emulated grid across boundaries of emulators, and on longer-term drift of associated values, e.g., integrated complex energy exchange.

Another challenge in GD-RTS and GD-PHiL is the evaluation of simulation fidelity itself, as a key performance indicator. In the context of GD-RTS, fidelity can be defined as the degree of similarity of a system response in the GD-RTS environment to the response of a monolithic simulation model. A common approach for fidelity evaluation is to utilise a monolithic simulation to obtain reference results. However, in a real-world GD-RTS or GD-PHiL application scenarios, the monolithic model would typically be unavailable.

The lack of reference results means comparison, both between runs of the same implementation and across implementations, of model outcomes is the sole indicator of fidelity available. Metrics of comparison for these outcomes should be based on the target metric of the experiment, and they should be informed by any artefact of the interface that may impact this metric.

Our purpose here is then threefold:

1. Establish a general set of metrics under which GD-PHiL interfaces can be evaluated even when a reference signal is unavailable.
2. Provide sets of observables and parameters that allow characterisation testing of a GD-PHiL interface.
3. Develop a GD-PHiL interface that directly targets these metrics to demonstrate this characterisation.

We proceed in this section to discuss Points (1) and (2) in general terms, with Point (3) treated in Section 4.

### 3.1. GD-PHiL Interface Models as Two-Port Networks

A co-simulation interface can be modelled and analysed as a 2-port network, as illustrated in Figure 2. Here, a voltage measurement at the primary side is transmitted through an interfacing actuator ($IA_1$) to the secondary side across a communication channel (*Comm.*) and subsequently translated to be actuated by a voltage source on the secondary interfacing actuator $IA_2$. Similarly, a current measurement is transmitted from the secondary side and actuated on the primary side via a current source. Note that, as far as the interface is concerned, there is no difference if the voltages and currents are due to a simulation in an DRTS or represent physical flows due to a hardware component. The treatment in this section does not distinguish if either side is physical or simulated, only that they represent a power flow.

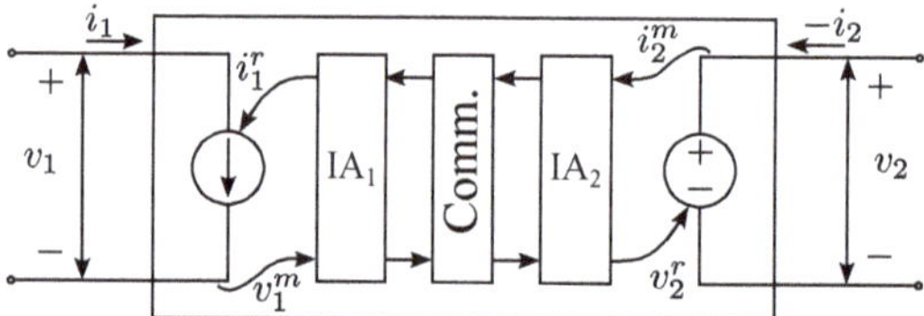


**Figure 2.** Generic GD-HiL/GD-RTS interface model represented as a two-port network.

For an ideal interface, voltage and current on primary and secondary sides match at all times, and thus in particular instantaneous powers match at all times:

$$v_1(t)i_1(t) = v_2(t)i_2(t). \tag{1}$$

When dealing with non-ideal interfaces, Equation (1) will be modified to account for, e.g., communication delay and jitter, intermittency of operation, multi-rate operation of primary and secondary loops and calibration offsets. As an example, suppose there is a consistent delay $\tau$ between $v_1$ occurring on the primary side and $v_2$ being actuated on the secondary side, with the same delay $\tau$ between $i_2$ and $i_1$. Neglecting other sources of error and assuming no compensation, the power balance now writes

$$v_1(t-\tau)i_1(t+\tau) = v_2(t)i_2(t), \tag{2}$$

which clearly does not respect energy balance for the two-port system in general.

### 3.2. Fidelity Evaluation Metrics

Given the ideal behaviour in Equation (1), a metric to examine an interface's fidelity can be derived as:

$$E_{obs}[n] = T_s \sum_{k=0}^{n} (i_1[k]v_1[k] - i_2[k]v_2[k]), \tag{3}$$

which compares the difference in instantaneous power between Ports 1 and 2 over a time horizon, with measurements $i_1[k], \ldots$ sampled isochronously at rate $1/T_s$.

In experiments where a system event causes the system to go to a steady state, one can use the new steady state of the system as a reference value from which fidelity metrics can be derived. For such an experiment where an event affects the signal $v_1(t)$ at time $t = 0$, leading to a new steady state value of $v_1$ at $\hat{v}_1$, the under- and overshoot of the signal are defined as

$$v_1^{\text{overshoot}} \equiv \max_{t \geq t'} (v_1(t)) \tag{4}$$

$$v_1^{\text{undershoot}} \equiv \min_{t \geq t'} (v_1(t)) \tag{5}$$

where $t'$ is the first crossing time where $v_1(t') = \hat{v}_1$.

Further, the stabilisation time under tolerance $d$ is defined as

$$t_{stab}(d) = \min \{t_s > 0 \mid \forall t > t_s : |v_1(t) - \hat{v}_1| < d\}, \tag{6}$$

and the RMS deviation is defined as

$$RMSD_{v_1} = \sqrt{\int_0^{\infty} (v_1(t) - \hat{v}_1)^2 \, dt} \tag{7}$$

In applying these formulations to signals which respond to the event through the interface, i.e, $v_2$, $i_2$ and $i_1$, one takes $t = 0$ as the first time step for which the event translates to these variables.

Suppose more than one event is included in the experiment at times $0 < \tau_2 < \tau_3 < \ldots$, with all variables reaching steady states for an event before the next event occurs. Then, the steady-state is defined for each event separately, and all formulations above span over the intervals $[0, \tau_2), \tau_2, \tau_3)$, etc. instead of the interval $[0, \infty)$. Note that the times of each event occurring should be taken according to when each variable responds.

### 3.3. Hypotheses

To conduct tests of a GD-RTDS interface, the testing regime must take into account both the action of the system under test and the action of the test bed on which that system is implemented. Considerations thus naturally divide along two axis: On the first axis, one distinguishes *parameters*, which are decision on free experimental variables made prior to the experiment, and *observables*, which are recorded during or calculated after the experiment. On the second axis, one distinguishes *intrinsic* aspects, which concern the signals transported by the interface, and *extrinsic* aspects, which concern the interface's handling of those signals.

The parameters which can be altered for a characterisation test can be divided into *intrinsic* parameters, which affect the signals transported by the interface, and *extrinsic* parameters, which affect the interface's handling of those signals.

Altering the intrinsic parameters reveal how well the interface is able to represent the electrical system connected at either side, i.e., to which extent the interface is able to maintain transparency of electrical operation. Examples of such parameters include: The size of voltage step on primary side; and the size of a current/impedance step on secondary side. In general, altering intrinsic parameters only affects intrinsic observables, as defined below.

In contrast to intrinsic parameters, extrinsic parameters seek to trace how the interface performs under performance degradation. By artificially degrading the performance of the interface in a quantifiable way, it may be possible to extrapolate observables to those that would be obtained using an ideal interface. Examples of such parameters include: Adding artificial delay on the primary or secondary side by inserting sleep commands in the loop; or reducing the rate of sending and actuating by only applying every $n = 1, 2, \ldots$ set points. Altering the extrinsic parameters of the system will affect both the intrinsic and extrinsic observables of the system, as defined below.

For both types of parameters, one can define analogous intrinsic and extrinsic observables for the interface; intrinsic observables relate to the signals transported, which are discussed in detail in Section 3.1, while extrinsic observables relate to the timing of the messages carrying these signals.

Pertinent extrinsic observables include:

1. the time between each message being sent from the primary side (primary side sending rate);
2. the delay between a voltage occurring on the primary side and a message being sent reflecting this voltage (primary side sending delay);
3. the delay between a message being sent from the primary side and the message being received at the secondary side (primary side communication delay);
4. the delay between a message arriving at the secondary side and the content of that message being reflected on the physical system (secondary side actuation delay); and
5. the equivalent timings for messages originating at the secondary side.

Taken together, these parameters and observables form a complete set of dimensions for characterisation. For a given interface between given electrical systems at the primary and secondary side, and for a given set of intrinsic and extrinsic variables, the ability of the interface to form a transparent interconnection is given by the intrinsic and extrinsic variables.

With these general considerations in mind, the following section relates the development of a compensation interface aiming to improve each intrinsic metric compared to an uncompensated interface.

## 4. Implementation in Distributed Environment

To further study the QsPHiL concept, a case-study was implemented and demonstrated in a real lab setup. This section describes system architecture and interfacing details while Section 5.1 introduces the test grid.

In this case-study, a RTDS real-time simulator at DTU Lyngby and a grid-forming Back-to-Back power converter (B2B) at SYSLAB were interconnected to a simple dump load. This setup poses two challenges to its implementation:

1. Latencies and jitter due to geographical distance between simulator and converter
2. Limited controllability and observability of the converter due to the Modbus interface of the converter

The objective is to achieve identical or at least comparable results in the distributed and monolithic environment without exchanging large packets with signal vectors. Here, the distributed environment was represented by the case-study using the SYSLAB laboratory. The monolithic environment was an equivalent model of the lab setup, which was purely simulated in the DRTS at Lyngby as a reference case.

This section describes the architecture and techniques used to achieve this goal. These include the exchange of GPS synchronised timestamps and sequence numbers, asynchronous programming and a fine tuning of the converters Modbus communication.

### 4.1. Architecture and System Setup

Figure 3 shows the layout of the GD-PHiL setup of this study. The distribution system was simulated in RTDS located in the Lyngby campus of DTU while the hardware part belonged to the SYSLAB laboratory situated in DTU Risø campus, which is approximately 40 km away. The locations are interconnected by the public Internet.

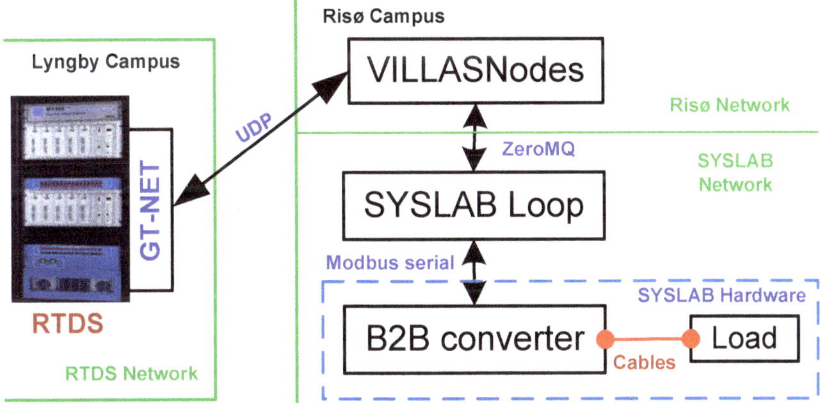


**Figure 3.** The PHiL setup layout.

A *VILLASnode* instance located in DTU Risø campus operates as a gateway for real-time data exchange between the RTDS simulator and the SYSLAB hardware. *VILLASnode* is a gateway for co-simulation interface data [21]. It converts various protocols and (de-)multiplexes signals to and from multiple sources. The gateway supports over 18 different protocols and interfaces, which include general purpose broker-based messaging protocols such as Advanced Message Queuing Protocol (AMQP) and Message Queuing Telemetry Transport (MQTT), as well as industry standard protocols such as IEC 61850, Web protocols like WebSockets and HTTP/REST and Interfaces to Simulators and custom Field Programmable Gate Arrays (FPGA). It collects statistics about the exchanged data such as packet loss, one-way delay and jitter and provides a network emulation to simulate the conditions of a geographically distributed co-simulation. *VILLASnode* is a C/C++ application executed on a Real-time optimised Linux machine.

Data exchange between RTDS and the *VILLASnode* gateway is performed via a GTNET card and using the UDP protocol. Communication between the *VILLASnode* gateway and the *SYSLAB loop* uses the ZeroMQ publish/subscribe protocol due to firewall restrictions which forbid direct UDP communication between SYSLAB and RTDS. Lastly, the *SYSLAB loop* communicates with the B2B converter via serial Modbus. The *SYSLAB loop* was implemented in Python and relies heavily on asynchronous programming (`asyncio`) to realise non-blocking forwarding of measurements and set points [22].

SYSLAB is an experimental facility at DTU Risø campus, designed as a test bed for advanced control and communication concepts for intelligent distributed power systems. Figure 4 shows a section of the 400 V, three-phase grid which, with a total of 16 bus bars and 119 automated coupling points serves as the electrical backbone of the facility and allows a large variety of different grid topologies. All equipment on the grid is automated and remote-controllable. Each unit is supervised locally by a dedicated computing node which acts as a communication gateway. Nodes contain measuring and network equipment, data storage, backup power and an interfaced computer, which can be used for hardware and software platform test. The whole system can be run and monitored remotely and as such is a good test bed for the (Power) System-in-the-Loop testing objective of this paper.

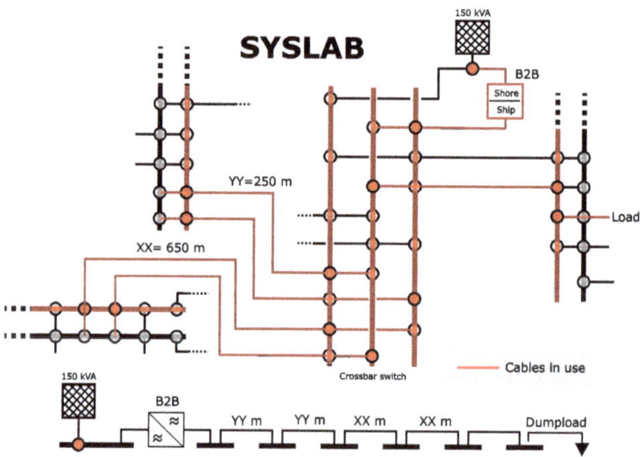


**Figure 4.** SYSLAB topology.

The HuT in this case-study is a dump load (DL) connected via a combination of long low-voltage underground cables to the PHiL interface. While the PHiL interface is our Objection under Investigation (OuI), it is not considered to be part of the HuT. The DL consists of eight different resistors which can be combined in any way possible by several remote controlled relays to achieve the power targeted set-point. The resistor sizes in kW are $0.42, 0.75, 1.45, 3.02, 6.0, 11.75, 23.4, 31.3$. Depending on the targeted set point, the DL controller activates a subset of these resistors by enabling the respective relays.

As PHiL interface, a B2B converter is used. Built by ABB, the PCS100 *Static Frequency Converter* is designed for ship-to-shore operation and operates at 50 or 60 Hz on low voltage distribution grids and is rated for a nominal output power of 125 kVA. Figure 5 shows the simplified electrical layout of the B2B converter. Its internal inverter is controlled by the set points provided from the simulation model in the DRTS in the form of RMS voltage ($V_{rtds}$) and frequency ($f_{rtds}$). Measured power at the converter output terminals $P_{b2b}$ and $Q_{b2b}$ is monitored and fed back into the simulation where they are injected by a controlled current (see Section 4.4). The B2B converter is a three-phase device. However, it is only controllable in a balanced mode by a single set of RMS voltage and frequency set point. Voltage measurements are provided individually for each phase, whereas power measurements are aggregated for all phases. This restricts the test system to balanced three phase grids. The required modifications to satisfy this constraint are described in Section 5.1.

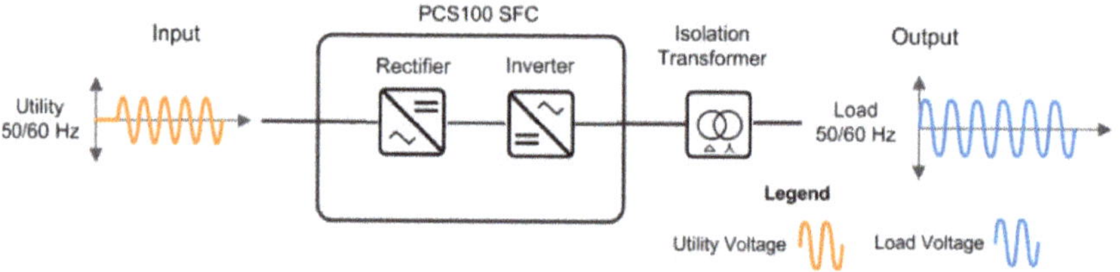


**Figure 5.** Simplified electrical layout of B2B Converter (ABB PCS100 Brochure).

For the purpose of Power System in the Loop (PSiL) testing, the paper follows the terminology of Heussen et al. [23]:

**System under Test (SuT)** is the IEEE benchmark system, the SYSLAB laboratory and OLTC within the simulation.

**Object under Investigation (OuI)** in the context of this paper is the power interface, namely the GD-PHiL setup, the B2B converter, Interface Algorithm and the Compensation.

**Hardware under Test (HuT)** is the hardware portion of the SuT, namely the cables and Dump Load, but not the B2B converter as it belongs to the OuI.

### 4.2. Synchronisation and Time Stamping

The sequence diagram in Figure 6 shows the communication of signals and timestamps between simulator and *SYSLAB loop*. Both sites of the setup provide their own time reference via either a dedicated GPS synchronised GTSYNC card for RTDS or an NTP server in the case of SYSLAB. While the NTP server exhibits an inferior accuracy in contrast to the GTSYNC card, it is still obtains its time reference via a direct attach GPS receiver (*Stratum 0*). The resulting time stamping precision is estimated in the range of a few hundred μs and still adequate for this setup as other sources of uncertainties in the system are around 2–3 magnitudes larger (jitter by B2B controller and ZeroMQ publish/subscribe). The exchange of timestamps alongside the signal value allows for the calculation of a round-trip time (RTT) and an estimated one-way delay (OWD).

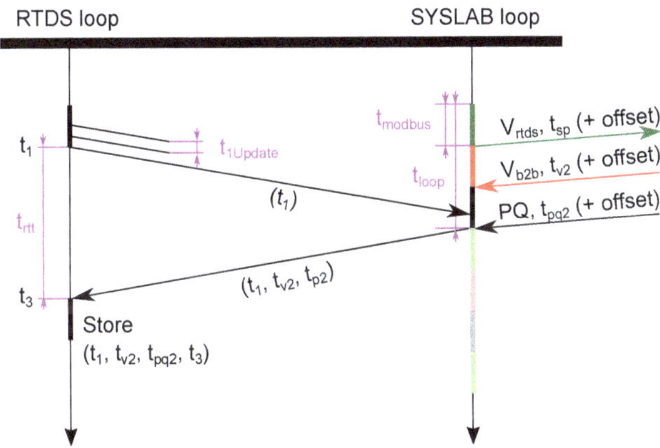


**Figure 6.** Time exchange protocol and associated loops.

Initial tests revealed that our experiment is dominated by the response time of the Modbus connection on the inverter $t_{modbus}$. The average response time of the converter is $t_{modbus} = t_{pq2} - t_{v2} \approx$ 20 ms for each measurement or set point update. The total round-trip time between simulator and converter is estimated as $t_{rtt} = t_3 - t_1 + t_{update}/2 \approx 5$ ms which results in an OWD of $t_{owd} \approx t_{rtt}/2 \approx 2.5$ ms that is around a magnitude lower than the Modbus request time. Therefore, the influence of the geographical distance in this experiment is minor in comparison to the peculiarities of the B2B converter due to its Modbus interface and internal control.

The time of a blocking Modbus request is the main limiting factor for a faster execution of the *SYSLAB loop*, as the execution of each request is strictly sequential. Ideally, the requests would be directly synchronised to signals received by the DRTS, which for a stable interconnection over the public internet should not exceed 1 kHz to 5 kHz [6]. To reach such high update rates, the number of Modbus requests must be reduced the minimum required for the HiL test and subsequent analysis (set points and measurements). A total of $n_{reg} = 3$ Modbus requests is used to exchange the following signal with satisfactory timing for the application.

The signals are listed in the order of their respective Modbus requests:

- RMS Voltage set point for the converter outputs
- RMS Voltage measurements for each phase of the converter $V_{b2b,abc}$ (RMS)
- Active/reactive power of the converter $P_{b2b}$, $Q_{b2b}$

These requests are continuously communicated in a tight loop with the B2B converter. This results in an an average loop cycle time $t_{loop} = n_{reg} \times t_{modbus} \approx 60ms$ or $f_{loop} = t_{loop}^{-1} \approx 16\,Hz$, which is at least a magnitude larger than the targeted update rate by the DRTS.

To simplify the experiment setup, all signals are collected by the simulator resulting in a single file per run. In post processing, timestamps for signals originating from the laboratory are corrected by $-t_{owd}$ in the case of $P_{b2b}$, $Q_{b2b}$ and $-(t_{owd} + t_{modbus})$ in the case of $V_{b2b,abc}$.

*4.3. Multi-Rate Interfacing*

When coupling discrete systems with differing rates, issues with lack of system state information arise. In the present case, a fast system (DRTS rate of 20 kHz) is coupled with a slow system (*SYSLAB loop* rate of $\approx 16\,Hz$). Due to the limited sampling rate of the *SYSLAB loop*, the simulation lacks information about the present current drawn by the *DL + Lines* in the SYSLAB environment which are needed for controlling the current injections in the model. Similarly, voltages from the DRTS ($V_{rtds}$) are sent at a rate faster than the converter can actuate due to the bottleneck of the Modbus interface.

As described in the previous subsection, the SYSLAB loop is running without synchronisation to the DRTS in order to maintain the maximum loop rate. To avoid congestion of the communication network between DRTS and *SYSLAB loop*, the DRTS should avoid sending update for every simulation time step as this would result in $1/\Delta t = 20\,kHz$ network packets per second. To avoid this, the communication rate from DRTS to *SYSLAB loop* was reduced to 100 Hz, which provides the *SYSLAB loop* with a new set point every 10 ms. The B2B converter will pick up this new set point on average with a 5 ms delay, as the time offset between arrival of a new set point and the next Modbus request can be expected to be uniformly distributed. In contrast, the measurements of the B2B converter are forwarded to the simulator after the respective Modbus read-register requests have been fulfilled. As shown in Figure 6, two Modbus requests are required for reading voltage and power measurements. The *SYSLAB loop* always performs both requests and sends the combined measurements to the DRTS. Consequently, the measurements are taken with a slight offset of $t_{pq2} - t_{v2}$ which are corrected later on.

One way to address this issue is to build a virtual model to represent what the slow system would do if it could respond quickly enough. For instance, given a voltage set point $u^f$ from the fast system, the interface would report a current value $i^f = f_s(u^f)$, using a function $f_s$ inferred from measured values of the slow system. Provided the characteristics of the slow system change slowly, $f_s$ will provide a good estimate of the actual current response of the slow system. If, however, parameters of the lab setup are changing, the virtual model must be capable of detecting these changes and adapt accordingly. An interface algorithm satisfying these requirements is introduced in Section 4.4.

*4.4. Interface Algorithm*

The PHiL setup in this work consists of a DRTS, which performs electromagnetic transient simulation, and a power converter, which amplifies the reference voltage signal received from the DRTS and electrically interfaces the HuT, in this case DL and lines (*DL + Lines*). While simulation time step of DRTS is 50 µs resulting in a 20 kHz, SYSLAB measurements at power converter are only available at a rate three magnitudes lower on average. Therefore, it is not feasible to apply Interface Algorithms (IA) based on instantaneous values of voltage and current waveforms.

Instead, the IA is based on RMS of voltage, frequency and power quantities. On DRTS side, the RMS of the voltage waveform is calculated at the fundamental frequency and forwarded to the power converter along with estimated system frequency. The power that is withdrawn or injected by *DL + Lines* is measured at the power converter terminals and forwarded to the DRTS. The IA in DRTS calculates current waveform based on the received power measurements and local voltage measurement. IA is illustrated in Figure 7.

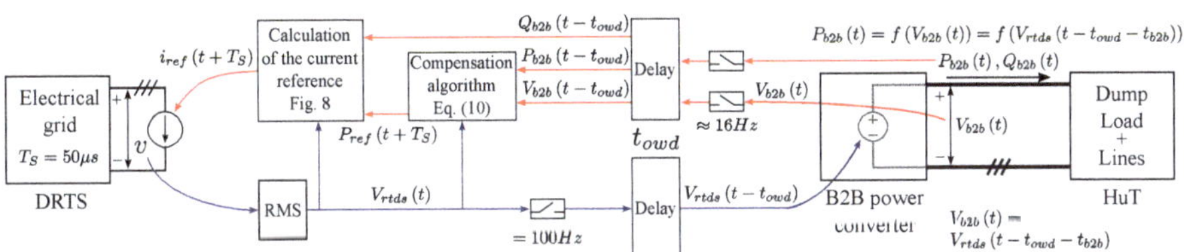

**Figure 7.** Overview of the interface algorithm.

The RMS voltage $V_{rtds}(t)$ at the DRTS terminal of the interface is sampled with a frequency of 100 Hz and available at B2B controller with time delay $t_{owd}$. Response time of B2B converter $t_{b2b}$ results in the delayed voltage $V_{b2b}(t) = V_{rtds}(t - t_{owd} - t_{b2b})$ at B2B terminal with respect to the voltage measured at DRTS terminal of the interface. Power measurements at the B2B terminal, $P_{b2b}(t)$ and $Q_{b2b}(t)$, represent HuT response to the delayed voltage, which results in $P_{b2b}(t) = f(V_{b2b}(t)) = f(V_{rtds}(t - t_{owd} - t_{b2b}))$. These power measurements and voltage measurement $V_{b2b}(t - t_{modbus})$ are sampled with $f_{loop} \approx 16$ Hz and communicated with the DRTS where they are available with additional $t_{owd}$ time delay. Therefore, power measurements of $DL + Lines$ that are available in DRTS are delayed for communication time delay $t_{owd}$ and for B2B time response $t_{b2b}$ since they represent $DL + Lines$ response to the voltage at B2B terminal that is delayed with respect to DRTS voltage. Calculation of the current reference value $i_{ref}(t + T_S)$ is conducted in direct-quadrature (dq) synchronous reference frame. The calculation is based on the power reference $P_{ref}(t + T_S)$ and requires a Phase Locked Loop unit to track the voltage phase at the interface to the simulated electrical grid. Figure 8 illustrates the calculation method. Note that $i_{ref}(t + T_S)$ represents a sample of the current waveform and it is updated at every simulation time step $T_S$.

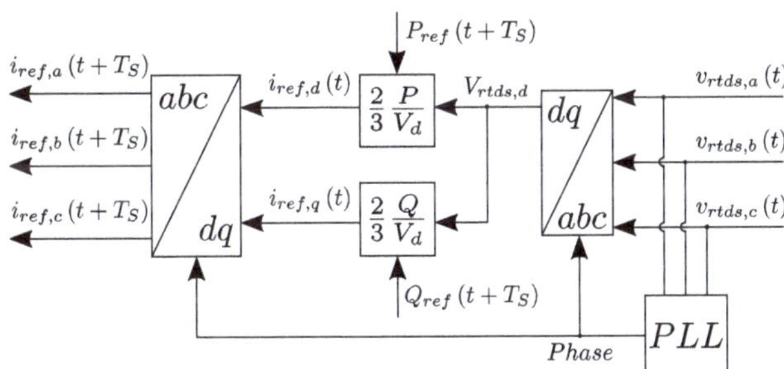

**Figure 8.** Calculation of the current reference.

### 4.4.1. Compensation Method

Limitations in terms of the Modbus communication rate and response dynamic of the converter represent the main factors of fidelity degradation. As it is not feasible to alter the response dynamic of the B2B converter, the compensation method utilised in this work aims at leveraging measurements available in the DRTS to compensate time delays. Power and voltage measurements of the B2B converter are utilised in the DRTS to estimate impedance of $DL + Lines$, which is used along with local voltage measurements in RTDS to correct power reference for current source in the DRTS.

The compensation method described above relies on following equations:

$$\Delta P(t + T_S) = (V_{rtds}^2(t) - V_{b2b}^2(t - t_{owd} - t_{modbus})) \cdot G_{DL+Lines} \tag{8}$$

$$G_{DL+Lines} = \frac{P_{b2b}(t - t_{owd})}{V_{b2b}^2(t - t_{owd} - t_{modbus})} \tag{9}$$

$$P_{ref}(t + T_S) = P_{b2b}(t - t_{owd}) + \Delta P(t + T_S) \tag{10}$$

In Equation (9), the conductance $G_{DL+Lines}$ of the HuT is estimated based on received measurements $P_{b2b}$ and $V_{b2b}$. The reference for the current source for the next time step in the DRTS is corrected by $\Delta P(t + T_S)$ and calculated based on Equation (10).

## 5. Results

### 5.1. Test System

The test system used in this study is a modified version of the IEEE 13-bus feeder, which was modelled in RSCAD and executed on a PB5-based RTDS simulator. It was interfaced with the HuT at the node 634 by the B2B converter connected to a controllable DL through a long electric cable as shown in Figure 9.

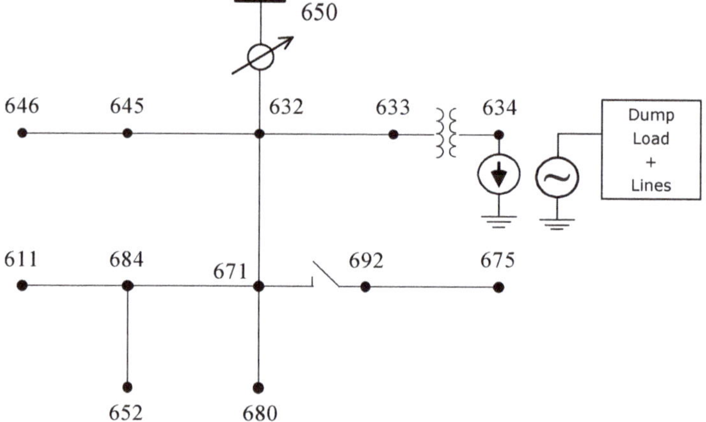


**Figure 9.** IEEE 13 node test feeder with PHiL test system.

To accommodate limitations of the laboratory setup, the original IEEE 13-bus feeder was modified as follows:

1. All loads of the system were balanced to account for the missing controllability of individual B2B converter phases.
2. The frequency of the main voltage source in the system was adjusted to 50 Hz to account for the frequency limits of the B2B converter.
3. A simple OLTC tap change control scheme was implemented based on [24].
4. The load at branch 634 was replaced by an ideal current injection controlled by the IA in Figure 7.
5. Current injections based of measurements from the real-world DL were scaled by a factor of 40.

The simulation model contains an OLTC between busses 650 and 632. This OLTC controls the voltage in the distribution network to compensate voltage drops caused by varying the DL in the lab setup. As a constant impedance load, the DL in turn reacts on voltage step caused the tap change event. Hence, the OLTC and DL are a suitable to demonstrate the interactions between simulation and lab setup by a sequence of events propagating between the two sides.

The simple OLTC control was implemented inside the test system with a voltage bandwidth $BW = 0.05$ pu, a step size $\Delta V = 0.025$ pu and a delay time $t_{delay} = 2$ s [24].

### 5.2. Test Scenario

A scenario of a load step increase from 0 kW to 20 kW of the DL (0 kW to 0.8 MW in RTDS simulation) was investigated in this test. The DL's maximum rated power was specified with 78 kW. Even when operated at its maximum rating, the load change of the DL will not produce a significant disturbance in the grid. Therefore, a scaling factor of 40 was used for the power of load in RTDS to overcome the hardware capacity limitation. Firstly, open loop tests were implemented to characterise connection timing and hardware components (DL and B2B converter). Then, the proposed compensation method was evaluated via a closed loop.

This test scenario triggers a series events of which only this step change of the DL set point originates from the real-world lab setup. The step change of the DL causes a voltage dip in the test system and leads to the activation of the OLTC in order to restore the voltage level in the distribution network. The OLTC controller delays the activation of the tap change for about 2 s after the initial transient produced by the DL. The tap change operation produces as a consequence another transient, which now originates from the simulation and propagates to the real-world lab setup.

This test scenario was chosen as it contains events both originating from the lab setup as well as from the simulation. Hence, the the compensation method described in Section 4.4 could be evaluated for both cases.

### 5.3. Test Procedure

The target test case for this study reviewed the suitability of a grid-forming converter for QsPHiL by studying the closed-loop interaction between a real-world DL and a simulated OLTC in the test model. At the same time, it targeted the evaluation of the introduced compensation method.

For this test case, many test runs were performed to demonstrate the repeatability of the test setup. In addition, parameters such as the measurement sampling rate were varied to study their impact.

Before the closed-loop test, two open-loop tests were performed to characterise the B2B converter and DL step responses. For the remaining closed-loop tests, the following procedure was used:

1.  Configure SYSLAB topology for connecting B2B converter with DL.
2.  Start *SYSLAB loop*.
3.  Initialise B2B converter.
4.  Initialise DL with initial set point of 0 kW.
5.  Start real-time simulation with disabled coupling (no current injection).
6.  Await steady-state of simulation.
7.  Check initial state of tap changer and other parameters.
8.  Activate coupling by enabling of current injections into the simulation model.
9.  Await steady-state of simulation.
10. Initiate transient by changing DL set point from 0 kW to 20 kW.

### 5.4. Comparison Metrics

To evaluate the the fidelity and accuracy of the PHiL setup, several metrics were selected to quantify the improvement by the compensation method.

$V_{overshoot}$  a voltage overshoot which can be observed after the DL step.
$V_{undershoot}$  a voltage undershoot which can be observed after the DL step.
$T_s$  a settling time until convergence of interface quantities (voltage/power).
$max(\Delta E_{bal})$  the energy balance before and after the event. It provides an indicator if the interface injects or absorbs energy.
$RMSD_V$  the root mean square deviations (RMSD) between bus voltage at the interface $V_{rtds}$ and the reference case of a monolithic simulation of the full setup (no PHiL).
$RMSD_P$  the root mean square deviations the active power measurement at the interface $P_{rtds}$ and the reference case.

For comparison, a monolithic simulation solely in the DRTS was performed. For this purpose, the laboratory was modelled by two constant impedance loads which were toggled between the on and off state of the DL and the respective steady state values seen by the converter in the real-world case. Initial open loop tests showed that these lines load the B2B with a significant amount of reactive power ($\approx 1$ kvar). The use of real-world steady state measurements instead of the known impedance of the DL for the monolithic simulation model was therefore motivated by the fact of other unknowns such as lines parameters connecting B2B and DL in the lab.

For the analysis, the monolithic simulation was used as the reference case for comparing the above mentioned metrics. Other metrics were compared only against other PHiL runs with different parameters such as the measurement sampling rate or (de)activated compensation.

### 5.5. Open-Loop Characterisation of Laboratory Hardware

To characterise the B2B converter, an open-loop test for a step-change in voltage from 400 V to 370 V with and without DL was examined. As can be seen clearly in Figure 10, the voltage response experiences various step changes before reaching the reference, which takes roughly $t_{b2b} \approx 400$ ms. This transient is represented by only 6–7 samples due to the limited sampling rate of the converter measurements.

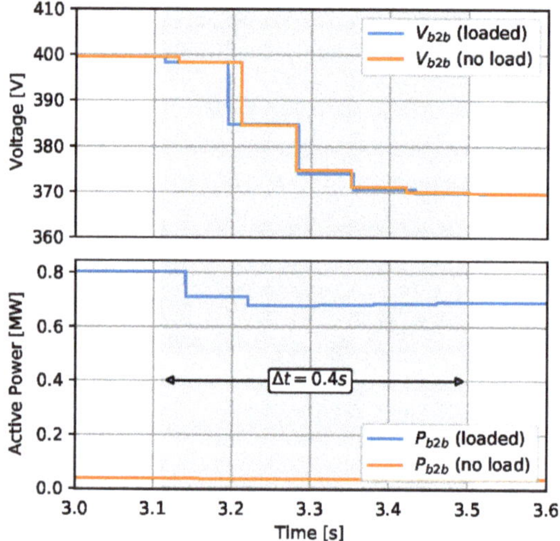


**Figure 10.** Open-loop B2B characterisation.

Similar results were reached for the inverse step as well as different step sizes. Overall, the voltage response of the B2B converter can be qualified as acceptable for the use in this study, due to the Quasi-stationary properties of the PHiL setup. Repeating similar tests with changes of the frequency set point show that the B2B converter reveals a slower response to frequency adjustments than voltage. Since the present scenario does not exhibit any frequency transients, the frequency of the B2B converter was fixed to $f = 50$ Hz. This simplification is permissible due the frequency invariance in the distribution grid caused by an ideal voltage source at the PCC. Besides, the lack of frequency set point updates saves an additional Modbus request and hence speeds up the *SYSLAB loop*.

Similarly, an open-loop test with a step change in the DL power from 0 kW to 20 kW at a constant voltage was investigated to characterise the DL. It takes approximately 100 ms for the DL relay response to hit the targeted set-point, which can impact the voltage response and compensation method in the closed loop. The results show that the B2B reacts immediately to this load change with only a negligible voltage dip. In addition, the constant impedance behaviour of the DL was verified during this open-loop test, as described in Section 4.

Especially for the fast DL step response, the sampling rate achievable over the B2B Modbus interface is hardly sufficient to reproduce the transient. These initial results necessitate the optimisation of the *SYSLAB loop's* Python code for a more efficient readout of measurements via the Modbus interface.

The internal design of the DL described in Section 4 suggests that the response of the DL should be an ideal step. However, Figure 11 shows two different variants of the DL step response which could be traced back to small timing differences in the switching behaviour of individual relays which activate the resistors. Out of 10 runs, five exhibit *one-step* response, while the others show the *two-step* response. This non-deterministic behaviour falsifies the closed-loop test runs. Therefore, subsequent test runs in which a *two-step* behaviour was detected were discarded from further analysis.

Another undesirable side-effect which was detected during open-loop characterisation tests is a an internal droop-control of the B2B converter. This droop control was subsequently disabled by reprogramming the converter control. The tests also showed internal limits for the voltage and frequency set points, which when exceeded trigger a safety shutdown of the converter. To avoid these shutdowns, signal limiters were added to the simulation model ($48\,\text{Hz} < f < 52\,\text{Hz}, 360\,\text{V} < V_{rms} < 440\,\text{V}$). As a result of the converter's frequency limit, the IEEE 13-bus distribution network was modified to a $50\,\text{Hz}$ system.

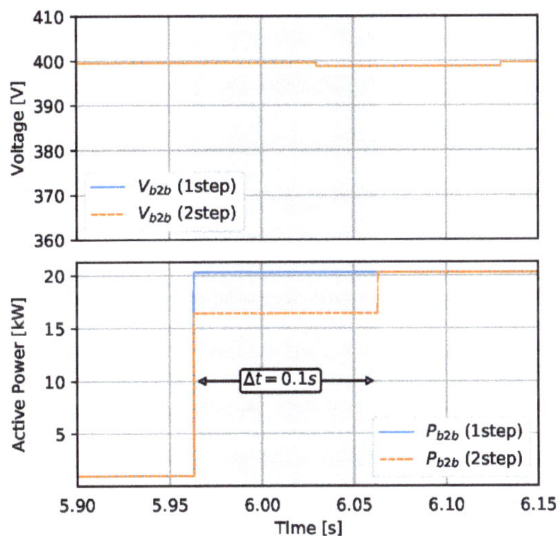


**Figure 11.** Open-loop DL characterisation.

### 5.6. Closed-Loop Characterisation of Compensation Method

Figures 12–14 show the results of a closed-loop test run with and without the compensation method, as described in Section 4.4.1. The DL step of $20\,\text{kW}$ ($0.8\,\text{MW}$ scaled) results in a voltage drop of about $25\,\text{V}$. This drop triggers the OLTC control, which in the following $4\,\text{s}$ of the DL step performs two tap changes, to restore the nominal voltage in the grid.

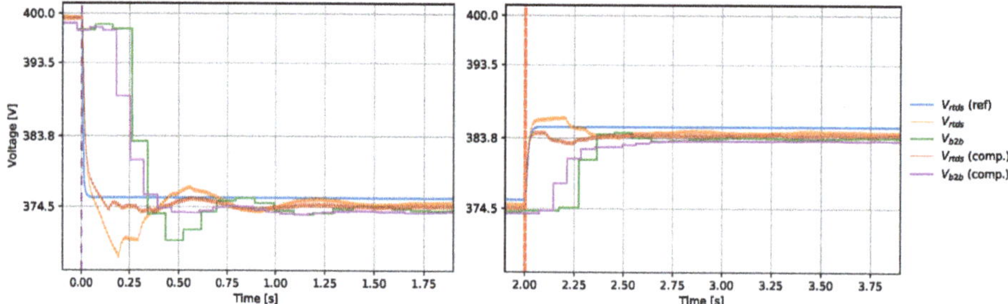

**Figure 12.** Inset graphs of voltage response during closed-loop test: DL step (**left**); and OLTC tap change (**right**).

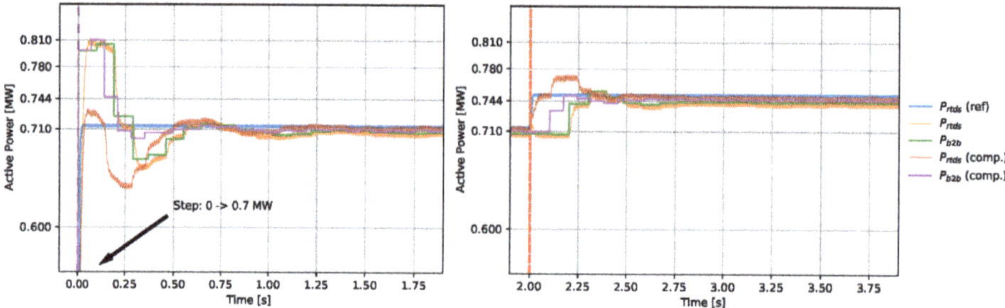



**Figure 13.** Inset graphs of power change during: DL step (**left**); and OLTC tap change (**right**).

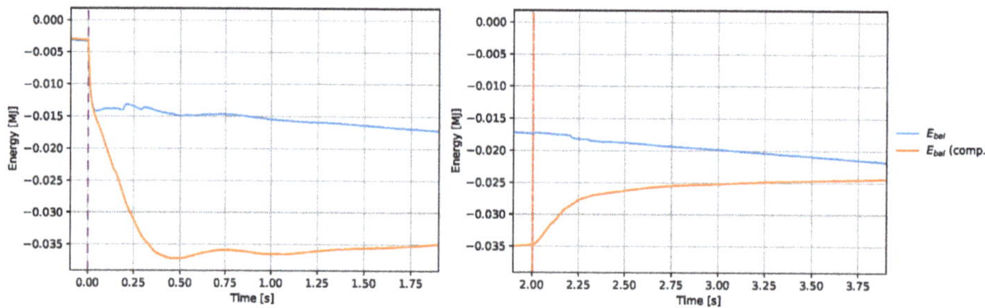

**Figure 14.** Inset graphs of interface energy during: DL step (**left**); and OLTC tap change (**right**).

A full version of the results can be found in Figure A1 in Appendix A.

The scenario generates two different types of transients. The first transients induced by the DL step ($t = 0\,\mathrm{s}$ - - -) propagates from the hardware setup to the simulation, whereas the two OLTC events ($t = 2\,\mathrm{s}, t = 4\,\mathrm{s}$ - - -) propagate in the opposite direction.

To discuss these two events, the results are represented as inset graphs and only OLTC tap change event following $t = 2\,\mathrm{s}$ is shown.

### 5.6.1. DL Step Change

The initiating load change of the DL causes a voltage dip in the feeder, as shown in Figure 12. The results show that the voltage response was improved by the compensation on the RTDS side (——— vs. ———). At the same time, the compensation caused no degradation of the response on the B2B side (——— vs. ———). Similarly, the overshoot in the power response on the RTDS side was improved by the compensation, as shown in Figure 13 (——— vs. ———). However, the respective undershoot in the power response on the RTDS side differs from the B2B side, and hence leads to a residual energy imbalance between the two ports. This energy imbalance is a natural consequence of improving power response on one side, but not the other.

In the case of the power response, the compensation failed to improve the initial DL step reaction, as the compensation distorts the current waveform. However, in the case of the OLTC tap changes, the compensation is capable of reducing the settling time (see Figure A1 $t = 2\,$s and $t = 4\,$s, compensated — vs. uncompensated —).

### 5.6.2. OLTC Tap Change Event

In contrast to the DL step event, the OLTC tap change originates from the RTDS simulation as the simulated OLTC control gets activated by the lasting under-voltage contingency on the feeder. The response of voltage and power was improved by the compensation in the same way as for the preceding DL step response. In addition, the power response on B2B side was also improved by compensation.

### 5.6.3. Energy-Based Metric

In Figures A1 and 13, the energy balance on PHiL interface is shown as $E_{bal}$. The energy balance accounts ingress and egress energy on both ports of the interface.

Notably, the DL step event results in a negative energy balance $\Delta E_{bal}$ across the interface while the subsequent OLTC events produce a positive contribution $\Delta E_{bal}$ and thereby absorb the generated energy of the DL step resulting in a more balanced outcome. However, this fact is only a coincidence caused by the present scenario. In the notation used here, a negative balance is caused by the interface injecting additional energy into the system, while a positive balance indicates absorption of energy (see Equation (11)).

$$E_{bal}(t) = \int_{-\infty}^{t} P_{rtds}\mathrm{d}t - \int_{-\infty}^{t} P_{b2b}\mathrm{d}t \tag{11}$$

The constant decline of $E_{bal}$ is an indicator for a small steady state error between simulation and laboratory signals. The step at $t = 0\,$s shows the absorption of energy during the transient.

### 5.7. Impact of Reduced Hardware Sampling Rate

To further investigate the influence of the measurement sampling rate, a series of tests was conducted with reduced rate. For the decimation factor $n_{ds}$, values $1, 2, 5, 7$ and $10$ were chosen, which reduces the sampling rate by a factor of $n_{ds}$.

Figure 15 shows the impact of the sampling rate reduction using the same scenario as in the previous sections. All runs were aligned at $t = 0$ to the drop of the simulated bus voltage $V_{rtds}$. This choice for alignment may be debatable, as we do not align the results to the DL step which triggers the event as in Section 5.6. However, with higher decimation ratio, the the sampling instant of the measurements differ more and more from the time of the DL step. Therefore, runs are not comparable due to the random alignment of DL step and sampling. With reduced sampling rates, we see a larger time lag before the first response in the simulation. For example, for $n_{ds} = 10$, the effective sampling rate is reduced to around $1.6\,$Hz, which in the worst case could delay the propagation of the DL step for up to $0.625\,$s.

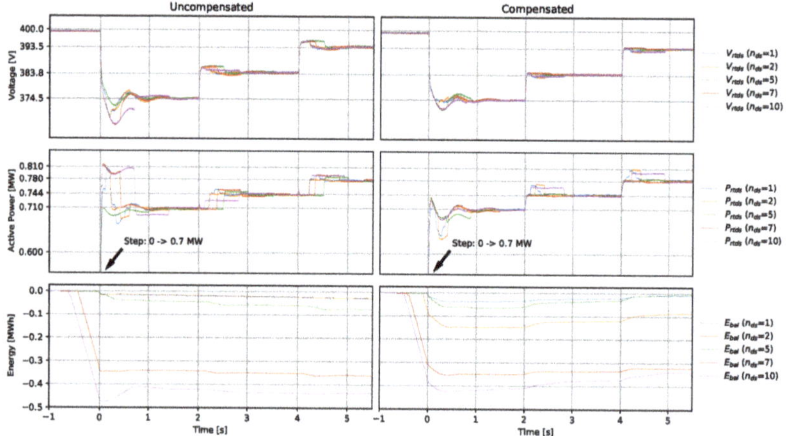

**Figure 15.** Voltage response of DL step at PHiL interface with varying sampling rates.

Table 1 lists the key metrics for the different decimation rates.

**Table 1.** Metrics under reduced sampling rates. $RMSD_V$ and $RMSD_p$ are given relative to the smallest value obtained.

| $n_{ds}$ | $V_{undershoot}$ [V] | | $V_{overshoot}$ [V] | | $RMSD_V$ [%] | | $RMSD_p$ [%] | | $max(\Delta E_{bal})$ [MWh] | |
|---|---|---|---|---|---|---|---|---|---|---|
| | Uncomp. | Comp. | Uncomp. | Comp. | Uncomp. | Comp. | Uncomp. | Comp. | Uncomp. | Comp. |
| 1 | 367.30 | 373.40 | 377.11 | 374.30 | 133 | 100 | 124 | 112 | 0.029 | 0.030 |
| 2 | 367.01 | 371.80 | 377.96 | 376.60 | 134 | 100 | 127 | 116 | 0.029 | 0.122 |
| 5 | 371.79 | 370.70 | 376.77 | 377.32 | 137 | 112 | 124 | 101 | 0.074 | 0.040 |
| 7 | 364.91 | 371.00 | 376.75 | 376.30 | 187 | 107 | 146 | 100 | 0.361 | 0.327 |
| 10 | 364.91 | 370.90 | 376.10 | 375.10 | 201 | 110 | 168 | 108 | 0.474 | 0.399 |

The results reveals two trends:

- The compensated runs show better metrics for all decimation ratios.
- For most runs, a higher decimation ratio yields worse results.

The convergence time $T_s$ is around 1.4 s to 1.5 s. No correlation of $n_{ds}$ or the compensation is visible.

## 6. Conclusions

The results demonstrate the feasibility of Geographically Distributed Power Hardware in the Loop experiments using a commercial grid-forming ship-to-shore converter. A first PHiL experiment was conducted in DTU's SYSLAB smart-grid laboratory with support of a RTDS real-time simulator located at a 40 km distant DTU campus.

A set of open-loop tests characterises the controllable dump load (DL) and converter for the application in the GD-PHiL setting.

The Modbus interface of the converter was optimised for a high throughput of measurement readouts and set point updates. However, the Modbus interface remains the main bottleneck for more frequent updates between simulator and the internal converter control. The maximum update rate of 16 Hz contributes significantly to the overall delay of the loop and directly limits the sampling rate for acquisition of interface quantities.

For similar low-cost PHiL setups, the authors suggest the enhancement of existing power converters with dedicated measurement equipment to achieve a higher sampling rate and reliable time stamping. A dedicated measurement setup would in turn allow for more frequent set point updates in the converter, as the Modbus interface is relived from the pressure of retrieving measurements. At the same time, set point updates of the converter could be synchronised to the real-time simulator.

The compensation method used in this study attained a fidelity improvement in the interface signals by reducing voltage over- and undershoot. However, after evaluation of the energy balance, it becomes evident that it failed to improve the conservation of energy across the interface. The residual energy imbalance indicates that a complementary compensation (e.g., based on an Integral controller) may be able to improve response.

Eventually, the chosen compensation method depends on the objective of the PHiL experiment. In the present study, the fidelity of the voltage signal is prioritised, while other studies may focus on more long-term phenomena in which an accurate reproduction of the exchanged energy is of capital importance (e.g., battery energy storage systems).

The studied power system in this paper is a common multi-bus low voltage distribution grid feeder. Modern distribution grids are however increasingly characterised by the presence of renewable energy generation (REG). Future work should therefore target new test-cases demonstrating the penetration of REG using the presented QsPHiL coupling methodology. New REG components offering new control functionalities requires improved management of distribution grids to ensure power quality and avoid stability issues. Thorough testing of such components must go beyond the common component level PHiL testing and could be realised by PSiL testing. An example for such a test case could be multi-level voltage control or efficient system integration of REG for electric vehicle (EV) charging.

The test case in this paper is characterised by a uni-directional power flow at the PHiL interface, in which the B2B converter only sources power to the Hardware-under-Test (HuT). Future test-cases might also result to a reversed power-flow at the PHiL interface. The employed power interface in the present laboratory setup could support this scenario due to its back-to-back topology.

Tests with a reduced sampling rate showed a reduced effect of the compensation method with lower sampling rates. Hence, it is inferred that higher sampling rates will yield an increasing improvement by the compensation.

**Author Contributions:** Conceptualisation, K.H., S.V. and M.S., methodology, T.V.J.; writing—review and editing, V.S.R., M.S., H.T.N., K.H. and S.V.; software, S.V.; validation, M.S.; data curation, H.T.N.; visualisation, H.T.N., T.V.J. and S.V.; writing—original draft preparation, S.V. and M.S.; project administration, K.H.; and supervision, A.M. All authors have read and agreed to the published version of the manuscript.

**Funding:** This work received funding from the European Regional Development Fund (EFRE-0500029) and the European Union's Horizon 2020 research and innovation programme under grant agreement No 654113.

**Acknowledgments:** The authors would like to thank the ERIGrid Transnational Access exchange program for enabling this work by providing access to DTU's Research Infrastructure and technical support.

**Conflicts of Interest:** The authors declare no conflict of interest.

## Abbreviations

The following abbreviations are used in this manuscript:

| | |
|---|---|
| AMQP | Advanced Message Queuing Protocol |
| B2B | Back-to-Back Converter |
| DL | Dump Load |
| DRTS | Digital Real-time Simulator |
| EBM | Energy-based Metric |
| FPGA | Field Programmable Gate Arrays |
| GD-PHiL | Geographically Distributed Power Hardware in the Loop |
| GD-RTS | Geographically Distributed Real-time Simulation |

| | |
|---|---|
| HuT | Hardware under Test |
| HiL | Hardware in the Loop |
| IA | Interface Algorithms |
| JaNDER | Joint Research Facility for Smart Energy Networks with Distributed Energy Resources |
| MQTT | Message Queuing Telemetry Transport |
| OLTC | On-Load-Tap-Changer |
| OuI | Object under Investigation |
| OWD | One-way delay |
| PHiL | Power Hardware in the Loop |
| PSiL | Power System in the Loop |
| Qs | Quasi Static |
| QSTS | Quasi-static time series |
| RMS | Root Mean Square |
| RMSD | Root Mean Square deviations |
| RTT | Round-trip time |
| SiL | System in the Loop |
| SuT | System under Test |
| VILLAS | Virtually Interconnected Laboratories for LArge systems Simulation/emulation |

## Appendix A. Complete Results

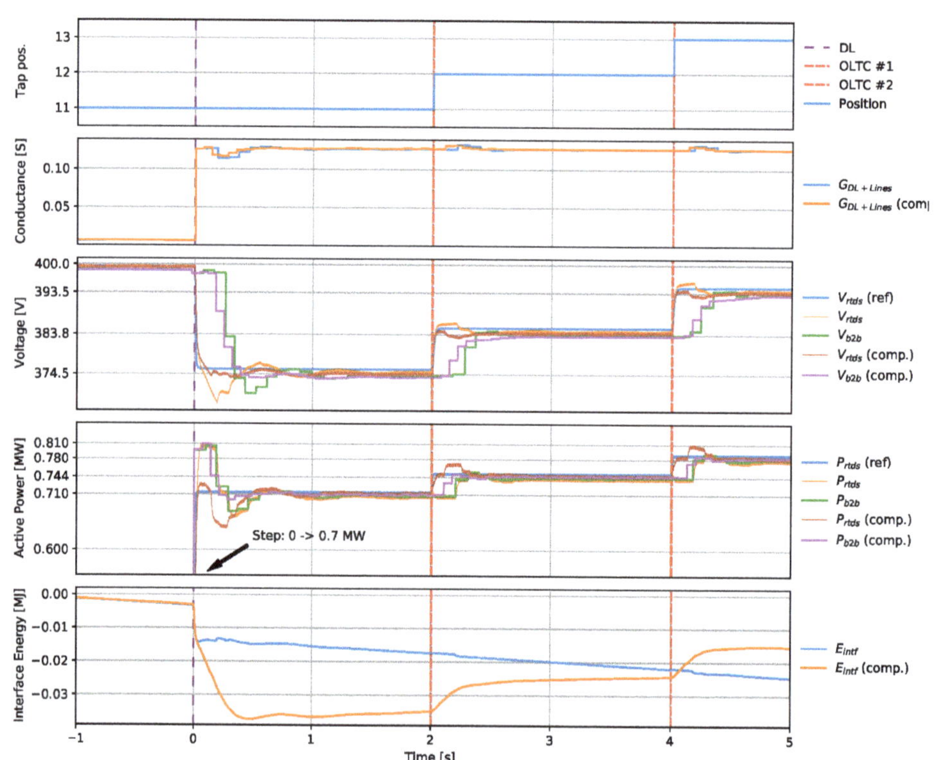



**Figure A1.** Closed-loop test results.

## References


1. Edrington, C.S.; Steurer, M.; Langston, J.; El-Mezyani, T.; Schoder, K. Role of Power Hardware in the Loop in Modeling and Simulation for Experimentation in Power and Energy Systems. *Proc. IEEE* **2015**, *103*, 2401–2409. [CrossRef]
2. Steurer, M.; Bogdan, F.; Ren, W.; Sloderbeck, M.; Woodruff, S. Controller and Power Hardware-In-Loop Methods for Accelerating Renewable Energy Integration. In Proceedings of the 2007 IEEE Power Engineering Society General Meeting, Tampa, FL, USA, 24–28 June 2007; pp. 1–4.
3. Lauss, G.F.; Faruque, M.O.; Schoder, K.; Dufour, C.; Viehweider, A.; Langston, J. Characteristics and Design of Power Hardware-in-the-Loop Simulations for Electrical Power Systems. *IEEE Trans. Ind. Electr.* **2016**, *63*, 406–417. [CrossRef]
4. Wang, Y.; Syed, M.H.; Guillo-Sansano, E.; Xu, Y.; Burt, G.M. Inverter-Based Voltage Control of Distribution Networks: A Three-Level Coordinated Method and Power Hardware-in-the-Loop Validation. *IEEE Trans. Sustain. Energy* **2019**, *99*. [CrossRef]
5. Monti, A.; Stevic, M.; Vogel, S.; De Doncker, R.W.; Bompard, E.; Estebsari, A.; Profumo, F.; Hovsapian, R.; Mohanpurkar, M.; Flickerand V.; et al. A Global Real-Time Superlab: Enabling High Penetration of Power Electronics in the Electric Grid. *IEEE Power Electr. Mag.* **2018**, *5*, 35–44. [CrossRef]
6. Vogel, S.; Rajkumar, V.S.; Nguyen, H.T.; Stevic, M.; Bhandia, R.; Heussen, K.; Palensky, P.; Monti, A. Improvements to the Co-simulation Interface for Geographically Distributed Real-time Simulation. In Proceedings of the IECON 2019—45th Annual Conference of the IEEE Industrial Electronics Society, Lisbon, Portugal, 14–17 October 2019; pp. 6655–6662. [CrossRef]
7. Strasser, T.; Jong, E.C.W.d.; Sosnina, M. (Eds.) *European Guide to Power System Testing: The ERIGrid Holistic Approach for Evaluating Complex Smart Grid Configurations*; Springer International Publishing: Cham, Switzerland, 2020. [CrossRef]
8. Brandl, R.; Montoya, J.; Strauss-Mincu, D.; Calin, M. Power System-in-the-Loop testing concept for holistic system investigations. In Proceedings of the 2018 IEEE International Conference on Industrial Electronics for Sustainable Energy Systems (IESES), Hamilton, New Zealand, 31 January–2 February 2018; pp. 560–565.
9. Montoya, J.; Brandl, R.; Vishwanath, K.; Johnson, J.; Darbali-Zamora, R.; Summers, A.; Hashimoto, J.; Kikusato, H.; Ustun, T.S.; Ninad, N.; et al. Advanced Laboratory Testing Methods Using Real-Time Simulation and Hardware-in-the-Loop Techniques: A Survey of Smart Grid International Research Facility Network Activities. *Energies* **2020**, *13*, 3267. [CrossRef]
10. Sicklinger, S.; Belsky, V.; Engelmann, B.; Elmqvist, H.; Olsson, H.; Wüchner, R.; Bletzinger, K.U. Interface Jacobian-based Co-Simulation. *Int. J. Num. Methods Eng.* **2014**, *98*, 418–444. [CrossRef]
11. Sadjina, S.; Kyllingstad, L.T.; Skjong, S.; Pedersen, E. Energy conservation and power bonds in co-simulations: non-iterative adaptive step size control and error estimation. *Eng. Comput.* **2017**, *33*, 607–620. [CrossRef]
12. González, F.; Arbatani, S.; Mohtat, A.; Kövecses, J. Energy-leak monitoring and correction to enhance stability in the co-simulation of mechanical systems. *Mech. Mach. Theory* **2019**, *131*, 172–188. [CrossRef]
13. Lundstrom, B.; Palmintier, B.; Rowe, D.; Ward, J.; Moore, T. Trans-oceanic remote power hardware-in-the-loop: multi-site hardware, integrated controller, and electric network co-simulation. *IET Generat. Transm. Distrib.* **2017**, *11*, 4688–4701. [CrossRef]
14. Palmintier, B.; Lundstrom, B.; Chakraborty, S.; Williams, T.; Schneider, K.; Chassin, D. A Power Hardware-in-the-Loop Platform With Remote Distribution Circuit Cosimulation. *IEEE Trans. Ind. Electr.* **2015**, *62*, 2236–2245. [CrossRef]
15. Williams, T.; Fuller, J.; Schneider, K.; Palmintier, B.; Lundstrom, B.; Chakraborty, S. Examining system-wide impacts of solar PV control systems with a power hardware-in-the-loop platform. In Proceedings of the 2014 IEEE 40th Photovoltaic Specialist Conference (PVSC), Denver, CO, USA, 8–13 June 2014; pp. 2082–2087.
16. Qin, X.; Shen, X.; Sun, H.; Guo, Q. A Quasi-Dynamic Model and Corresponding Calculation Method for Integrated Energy System with Electricity and Heat. *Energy Procedia* **2019**, *158*, 6413–6418. [CrossRef]
17. Ebe, F.; Idlbi, B.; Stakic, D.E.; Chen, S.; Kondzialka, C.; Matthias, C.; Heilscher, G.; Seitl, C.; Bründlinger, R.; Strasser, T.I. Comparison of Power Hardware-in-the-Loop Approaches for the Testing of Smart Grid Controls. *Energies* **2018**, *11*, 3381. [CrossRef]
18. Pellegrino, L.; Sandroni, C.; Bionda, E.; Pala, D.; Lagos, D.; Hatziargyriou, N.; Akroud, N. Remote Laboratory Testing Demonstration. *Energies* **2020**, *13*, 2283. [CrossRef]

19. Compere, M.; Goodell, J.; Simon, M.; Smith, W.; Brudnak, M. Robust Control Techniques Enabling Duty Cycle Experiments Utilizing a 6-DOF Crewstation Motion Base, a Full Scale Combat Hybrid Electric Power System, and Long Distance Internet Communications. In Proceedings of the Power Systems Conference, Atlanta, GA, USA, 29 October–1 November 2006; pp. 1–12.
20. Lentijo, S.; D'Arco, S.; Monti, A. Comparing the Dynamic Performances of Power Hardware-in-the-Loop Interfaces. *IEEE Trans. Ind. Electr.* **2010**, *57*, 1195–1207. [CrossRef]
21. Vogel, S.; Mirz, M.; Razik, L.; Monti, A. An open solution for next-generation real-time power system simulation. In Proceedings of the 2017 IEEE Conference on Energy Internet and Energy System Integration (EI2), Beijing, China, 26–28 November 2017. [CrossRef]
22. Hunt, J. Concurrency with AsyncIO. In *Advanced Guide to Python 3 Programming*; Springer International Publishing: Cham, Switzerland, 2019; pp. 407–417. [CrossRef]
23. Heussen, K.; Steinbrink, C.; Abdulhadi, I.F.; Nguyen, V.H.; Degefa, M.Z.; Merino, J.; Jensen, T.V.; Guo, H.; Gehrke, O.; Bondy, D.E.M.; et al. ERIGrid Holistic Test Description for Validating Cyber-Physical Energy Systems. *Energies* **2019**, *12*, 2722. [CrossRef]
24. Hartung, M.; Baerthlein, E.M.; Panosyan, A. Comparative Study of Tap Changer Control Algorithms for Distribution Networks with high Penetration of Renewables. In Proceedings of the International Conference on Electricity Distribution, CIRED Workshop, Rome, Italy, 11–12 June 2014; Paper 0376.

*Article*

# Facilitating the Transition to an Inverter Dominated Power System: Experimental Evaluation of a Non-Intrusive Add-On Predictive Controller

**Mazheruddin H. Syed** [1,*], **Efren Guillo-Sansano** [1], **Ali Mehrizi-Sani** [2] **and Graeme M. Burt** [1]

[1] Institute for Energy and Environment, University of Strathclyde, Glasgow G1 1RD, UK; efren.guillo-sansano@strath.ac.uk (E.G.S.); graeme.burt@strath.ac.uk (G.M.B.)

[2] Bradley Department of Electrical and Computer Engineering, Virginia Tech, Blacksburg, VA 24601, USA; mehrizi@vt.edu

* Correspondence: mazheruddin.syed@strath.ac.uk

Received: 2 July 2020; Accepted: 12 August 2020; Published: 16 August 2020

**Abstract:** The transition to an inverter-dominated power system is expected with the large-scale integration of distributed energy resources (DER). To improve the dynamic response of DERs already installed within such a system, a non-intrusive add-on controller referred to as SPAACE (set point automatic adjustment with correction enabled), has been proposed in the literature. Extensive simulation-based analysis and supporting mathematical foundations have helped establish its theoretical prevalence. This paper establishes the practical real-world relevance of SPAACE via a rigorous performance evaluation utilizing a high fidelity hardware-in-the-loop systems test bed. A comprehensive methodological approach to the evaluation with several practical measures has been undertaken and the performance of SPAACE subject to representative scenarios assessed. With the evaluation undertaken, the fundamental hypothesis of SPAACE for real-world applications has been proven, i.e., improvements in dynamic performance can be achieved without access to the internal controller. Furthermore, based on the quantitative analysis, observations, and recommendations are reported. These provide guidance for future potential users of the approach in their efforts to accelerate the transition to an inverter-dominated power system.

**Keywords:** control; inverters; inverter-dominated grids; power system transients; predictive control

---

## 1. Introduction

The integration of renewables within the electrical power system is a trend encouraged in many countries driven by environmental policies, fossil fuel restrictions, and energy security requirements. US Energy Information Administration projects that renewables (e.g., solar, wind, and hydro) will provide nearly half (49%) of the world's electricity demand by 2050 as end-use consumption experiences increasing electrification. For comparison, in 2018, only 28% of global electricity was generated from renewable energy sources, more than half of which was from hydropower.

In particular, both the United States and United Kingdom have increasingly installed more renewables over the past few years. In the United States, electricity generation from solar and wind resources has been steadily increasing since 2013. Wind and solar accounted for 6.8% and 2.8% of the total electricity generation in September 2019, a 31% and 15% increase from the same period in 2018 (while the total energy production increased by a mere 0.9%). Similarly in the United Kingdom, the share of generation from renewables grew to 38.9% in the third quarter of 2019, marginally exceeding generation from gas for the first time. On- and offshore wind dominated the renewables share at 19% with biomass second at 12%.

The increasing incorporation of power electronically interfaced renewables is accelerating the paradigm shift from a traditional synchronous generation dominated power system to an inverter-dominated power system. In a renewables-rich inverter-dominated grid, the operation and control under low and variable inertia conditions need renewed attention for the following reasons. First, lower inertia corresponds to a larger deviation of the system variables from their reference set points for any given system disturbance. Moreover, some disturbances may cause enough deviation to lead to instability. Second, the inverters themselves, due to their semiconductor switches, are more sensitive to such deviations which may cause physical damage. Third, the expected diurnal changes in the generation-load patterns can require frequent retuning of the controllers.

Several efforts have attempted to design robust set point tracking controllers to accommodate such emergent requirements. The established approaches in the literature, as summarized in Table 1, require either the system model or access to the internal parameters of the controller. These requirements present limitations due to vendor practices that do not always provide access to the internal parameters of the inverter and the unavailability of the system model with inverters. These approaches are often decentralized, implemented within an inverter, to ensure fast and high precision set point tracking-yielding a dynamically robust performance. The response of more than one inverter can be coordinated to support the response of a particular individual inverter or to achieve a global common objective. Such approaches, referred to as distributed control approaches, have been abundantly proposed in the literature [1–4]. However, the focus of this paper is on decentralized approaches to improve the dynamic response of inverters.

**Table 1.** Comparison of methods to improve the dynamic response of a system

| Approach | References | No Model | Noninternal |
|---|---|---|---|
| PI gain scaling | [5,6] | ✓ | X |
| Set point ramping | | ✓ | ✓ |
| Model predictive control | [7–11] | X | X |
| Adding a D-term to PI | [12–14] | X | X |
| Extremum seeking | [15–22] | ✓ | X |
| Posicast | [23–28] | X | ✓ |
| Sliding mode control | [29–32] | X | X |
| Model free control | [33–38] | ✓ | X |

It is also important to note that the control approaches summarized in Table 1 are mutually exclusive, i.e., only one chosen approach can be implemented for a given set point tracking objective. An alternative approach is to supplement the existing controllers with an add-on controller to enhance its performance under abnormal circumstances. Examples of such control approaches in the literature are adaptive gain adjustments of controllers using methods such as fuzzy logic [39–41] or more recently machine learning (also referred to as data-driven or artificial intelligence methods) [42,43]. Although such approaches are supplementary to the existing controllers, modification of the gain/parameters requires access to the internal control itself.

In [44], an add-on controller, referred to as SPAACE (set point automatic adjustment with correction enabled), that employs predictive set point modulation (SPM) to ensure robust performance subject to uncertain system conditions is proposed. SPAACE modifies the reference set point without the need for access to the low-level controller or knowledge of the system, in contrast to formerly described controls. Furthermore, SPAACE can be incorporated within any set point tracking application, independent of the primary control approach utilized (such as Proportional-Integral or model predictive control). Such an approach can be incorporated within existing devices within the field, improving their response by (i) reducing the overshoot, (ii) reducing the settling time, or (iii) both—thereby enhancing the utilization of existing infrastructure. The effectiveness of SPAACE to enhance dynamic performance of a system with (i) no access to its internal controller and (ii) minimal system knowledge is shown in [45]. SPAACE uses prediction of the response behavior, and a number of predictors are proposed, with their applicability for varied applications demonstrated in [46–49].

Theoretically, the SPAACE approach is well-established, with extensive mathematical formulations, stability analysis, and simulation-based performance evaluation published. However, the following research gaps still exist:

- The choice of predictor for a given application has not been discussed.
- The fundamental hypothesis of SPAACE, i.e., the feasibility of its incorporation without requiring access to an internal controller, has not been verified. One reported experimental evaluation of SPAACE was limited to the implementation of SPAACE within the development environment of an inverter controller itself [45].
- SPAACE is a generic approach with wide application potential, yet its adoption and deployment are limited due to a lack of established evidence of its practical feasibility for real-world applications.

This paper addresses these three identified gaps via the following contributions:

- The practical feasibility and robustness of SPAACE have been established utilizing a high fidelity hardware-in-the-loop systems level test bed. Conventional approaches to DER validation, commonly undertaken as equipment testing [50], are unable to de-risk the performance of the control when connected to an unknown and continuously changing system. The systems level test bed, with its reproduction of close to real-world environments, enabled the rigorous validation of SPAACE, endowing high confidence in its performance.
- For the DER controller application under consideration, a comprehensive evaluation with a set of practical variations is undertaken, i.e., analysis of choice of predictors, implementations (within and outwith the DER inherent controller), expected modes of operation and subject to a selected range of day to day operational scenarios.
- The methodological approach to ascertain computational and dynamic performance, and the subsequent detailed quantitative analysis, form the basis for the observations and recommendations reported. These are intended to serve as a reference guide for potential future users of the approach.

Therefore, this paper, with its key contributions and findings, provides archived evidence of the feasibility and applicability of SPAACE for DERs, building confidence for its adoption in the real world (by distribution system operators, virtual power plant operators or demand aggregators) and thereby facilitating the transition of the power system to an inverter-dominated grid.

## 2. Predictive Set Point Modulation

The predictive set point modulation approach, referred to as SPAACE, is a control method proposed to enhance the dynamic performance of a system with limited model information and no access to its internal controller parameters. Its incorporation enables the realization of a dynamically robust overall control architecture. In this section, the fundamentals of SPAACE are explained, including the theoretical foundation and the role of prediction strategies, and this is followed by the practicalities of parameter selection and real-world applications.

### 2.1. Fundamentals of Space

#### 2.1.1. Theoretical Foundation

Consider a single-input single-output (SISO) unit (dashed box in Figure 1a); denoting the reference set point of this controlled unit as $x_{sp}$ and its measured output as $x(t)$, the system representation with the incorporation of SPAACE as an add-on to the existing controllers of the unit is shown in Figure 1a. The primary controller is typically a PI-based controller designed for set point tracking; the secondary controller is a higher-level controller that decides the system reference set points. SPAACE monitors the response $x(t)$ and based on its trend and proximity to the constraints, applies a temporary change in the reference set point $x_{sp}(t)$, denoted by $x'_{sp}$, to achieve the desired response trajectory.

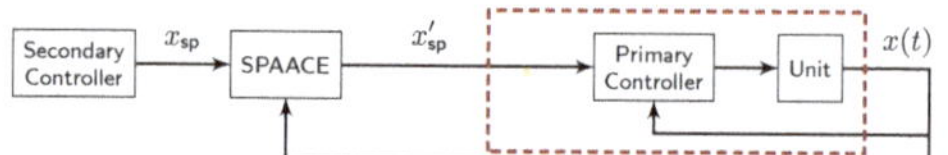

(**a**) Relationship between SPAACE and existing controllers.

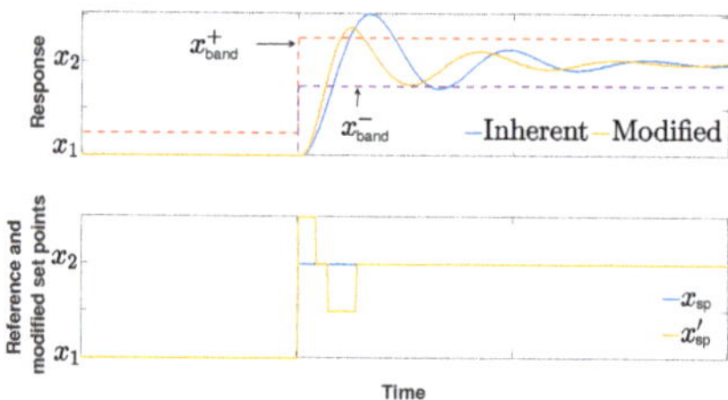

(**b**) Demonstration of SPAACE performance. Top: the response with and without SPAACE; bottom: the modified set point created by SPAACE.

**Figure 1.** Fundamentals of the SPAACE operation.

To explain this further, consider the example presented in Figure 1b. The original system response is referred to as the inherent response while the response with SPAACE is referred to as the modified response. For a step change in the reference set point $x_{sp}$ from $x_1$ to $x_2$ ($x_2 > x_1$), the inherent and modified responses of the system are presented. The inherent response exhibits undesired dynamic characteristics such as large overshoot and a long settling time. Conventionally, such characteristic behavior of the controller can be alleviated by controller redesign- requiring knowledge of system parameters and access to the internal controller parameters. In contrast to the conventional approach, SPAACE responds to such a scenario by treating the original controller as a black box. As shown in Figure 1b, SPAACE defines a tolerance band around the set point and upon digression of the response outside the defined tolerance band $\epsilon$, a modified set point $x'_{sp}$ is issued to remedy the situation as

$$x'_{sp} = \begin{cases} x_2, & x^+_{band} > x(t) > x^-_{band} \\ (1-m)x_2, & x(t) > x^+_{band} \\ (1+m)x_2, & x(t) < x^-_{band} \end{cases} \qquad (1)$$

where $m$ is a scaling factor. By shaping the response to be limited to the band around the set point, smaller overshoot and shorter settling times are obtained represented by the modified response. A negative step change ($x_1 > x_2$) is dealt with similarly.

SPAACE does not limit itself to issuing only one temporary set point modification; if after one set point modification the response still does not follow the desired trajectory, SPAACE issues subsequent set point changes. However, SPAACE includes logic to avoid (i) issuing too many set point modifications within a short period of time which may subject the system (especially if mechanical) to stress, and (ii) applying indefinite set point modifications if for any reason they do not improve the system response.

The theoretical foundations pertaining to the existence of an improved response for several classes of dynamic systems can be found in [46–48].

2.1.2. Prediction Strategies

The operation of the SPAACE as defined in Equation (1) is responsive. The performance of SPAACE can be further improved if a preemptive operation is enabled by utilizing a predicted value of $x(t)$, denoted by $\hat{x}(t)$ [49]. For a given response $x(t)$, the prediction method estimates the response at time $T_{pred}$ into the future, where $T_{pred}$ is the horizon of prediction. Any prediction method relies on $n$ past historical data points and the number $n$ depends on both the prediction method and the emphasis on historical data points: a smaller $n$ produces predictions that rely only on recent past data but may suffer from noise, whereas a larger $n$ produces smoother predictions but may slow down the response.

Each prediction method fits a parametrized function to the past measurements of the controlled variable $x(t)$ and uses this fit to estimate the future value of $x(t)$. Data fitting is performed using well-established methods such as least-square error (LSE). To ensure real-time implementation, the predictors typically employ cases of LSE that allow for significant simplifications, enabling their use for SPAACE with a smaller sampling time.

In this paper, we discuss and compare the results for three prediction methods as reported in the literature: linear, quadratic, and exponential. Other prediction methods have also been employed in SPAACE, e.g., lead compensator in [45] and Lagrangian polynomials in [47].

(a) Linear prediction: Linear prediction, arguably the simplest predictor, uses the following fitting function:

$$\hat{x}(t_0 + T_{\mathrm{pred}}) = x(t_0) + r(t_0)T_{\mathrm{pred}}, \tag{2}$$

where $r(t_0)$ is the average rate of change calculated over the historical data based on LSE. With only one historical data point, linear prediction essentially results in the current value of $x(t)$ being equal to the average of the historical data point and the predicted term:

$$\hat{x}(t_0 + T_{\mathrm{pred}}) = 2x(t_0) - x(t_0 - T_{\mathrm{pred}}) \tag{3}$$

(b) Quadratic prediction: In quadratic prediction, two parameters need to be calculated based on LSE:

$$\hat{x}(t_0 + T_{\mathrm{pred}}) = x(t_0) + r(t_0)T_{\mathrm{pred}} + \frac{1}{2}q(t_0)T_{\mathrm{pred}}^2 \tag{4}$$

where $r(t_0)$ is as defined previously and $q(t_0)$ is a measure of the second derivative of historical data—both calculated based on the LSE formulation.

(c) Exponential prediction: Exponential prediction assumes that in the short-term, the response of the controlled system can be approximated as an exponential rise or decay. Consequently, it fits an exponential function to the historical data as

$$\hat{x}(t + T_{\mathrm{pred}}) = ae^{b(t_0 + T_{\mathrm{pred}})} \tag{5}$$

where $a$ and $b$ are calculated based on LSE.

## 2.2. Implementation Considerations

### 2.2.1. Parameter Selection

The fundamental premise of SPAACE is to ensure that parameter selection does not represent a complex task. This is achieved through the adaptive nature of SPAACE as it continually monitors the response and its reference tracking performance. However, the discussion below offers some insights on best practices for choosing SPAACE parameters:

(a) Scaling factor: Selecting a smaller $m$ creates modified set points that are closer to the original set point and thus changes to the set point are smaller and smoother. In contrast, a larger $m$ introduces more severe set point changes that can help with achieving faster damping. In most cases, a modest value of $m = 0.2$ seems to produce reasonably fast results. A more detailed discussion of the effect of $m$ on the system performance is presented in [47,49]; an analytical discussion is presented in [46].

(b) Prediction strategy and prediction horizon: Several parameters can affect the accuracy of the prediction. In addition to the nature of the system response which suggests the type of fitting function (prediction strategy), prediction horizon $T_{\mathrm{pred}}$, and window of prediction history $T_{\mathrm{past}}$ are important design parameters. In general, $T_{\mathrm{pred}}$ should correspond to the inverse of the system dominant natural frequency $\omega_0$. That is, a faster system needs a smaller $T_{\mathrm{pred}}$ and vice versa. Heuristically, an acceptable rule of thumb is to choose $T_{\mathrm{pred}} = 10 \times 2\pi/\omega_0$.

(c)    Tolerance band: Typically the band is chosen to be symmetrical around the set point, but this is not a requirement for SPAACE operation, rather a design preference.

### 2.2.2. Real-World Implications

(a)    Feasibility: Most real-world systems are non-linear in nature, for which linear (taking advantage of established linearization methods) and non-linear control approaches have been widely utilized. As SPAACE is an add on controller, its incorporation is independent of the control approach (linear or non-linear) and the non-linearities within the unit under consideration. Therefore, in theory, SPAACE can be incorporated within a unit under consideration for any application where precise and high-speed set point tracking is imperative. By enabling the unit under consideration to be treated effectively as a black box, SPAACE can increase the potential utilization of the existing infrastructure by reducing transients, which in turn reduces the need to overdesign the system.

Furthermore, the stability of SPAACE was discussed in [46]. It was shown that the incorporation of SPAACE within an apparatus does not impact the stability of the wider system, i.e., if the system with the apparatus was stable without SPAACE, it will remain stable with the incorporation of SPAACE. Therefore, SPAACE can readily be adopted within power system applications, ranging from small isolated systems such as electric propulsion applications to larger systems such as HVDC segmented power systems.

This paper, in particular, discusses the incorporation of SPAACE in DERs coupled to the remainder of the power system via power electronic interfaces (inverters or converters). This choice is motivated by the current practices and challenges faced by the community: transition to a renewable rich inverter-dominated grid (utility systems), interest in incorporation within small-scale systems with variable and flexible architectures (remote microgrids and propulsion applications), to name a few. Furthermore, it provides a more relevant example for demonstration as DER manufacturer's practices historically present limitations in access to the internal parameters of the inverter and the unavailability of the system model with inverters.

(b)    Architecture and Communications: SPAACE is designed for incorporation within an individual apparatus with a view to improving its dynamic response. SPAACE does not require any additional information outwith the apparatus, and therefore no communication is required. From an architectural perspective, the implementation of SPAACE can be referred to as decentralized. Depending on whether SPAACE is incorporated within the controller of the DER or on an external controller, very short distance communications via analog connection may be present.

(c)    Responsibility: The responsibility to incorporate SPAACE would essentially lie with the owner of the DER. An end user, such as a residential customer with a battery energy storage system (BESS) interfaced via an inverter, might not have an incentive for the incorporation of SPAACE. However, the benefits the approach brings to the wider grid system might interest distribution system operators, virtual power plant operators, or demand aggregators to provide incentives to the end user for its incorporation. Alternatively, this can be set forth as a requirement by a demand aggregator for any end user BESS unit to participate within ancillary service markets.

(d)    Commercialization: The details of the implementation of SPAACE and its corresponding prediction strategies have been summarized in this work. SPAACE does not exist as a commercial product, however, the entities identified above (distribution system operators, virtual power plant operators, or demand aggregators) can develop their own solutions with confidence instilled in real-world applicability through contributions of this paper.

## 3. Experimental Setup

This paper validates several variants of the SPAACE algorithm in both an experimental setup and a real-time HIL environment to determine its practicality.

### 3.1. Test Rig

The performance evaluation of SPAACE is undertaken at the University of Strathclyde's Dynamic Power Systems Laboratory (DPSL). DPSL Microgrid is a 115 kVA, 400 V three-phase facility; its simplified one-line diagram is shown in Figure 2a. DPSL comprises equipment representing both conventional and non-synchronous generation, static and dynamic loads, arranged in such a way as to be able to run as three independent islands (or cells, as in Figure 2a) or brought together in any combination as a single system. State-of-the-art HIL techniques such as seamless initialization and synchronization of large test setups [51], experimental setup time-delay characterization [52] and measurements delay identification [53] and compensation [54] enable the realization of a high fidelity systems-level testing facility able to robustly de-risk novel control solutions for emerging power systems.

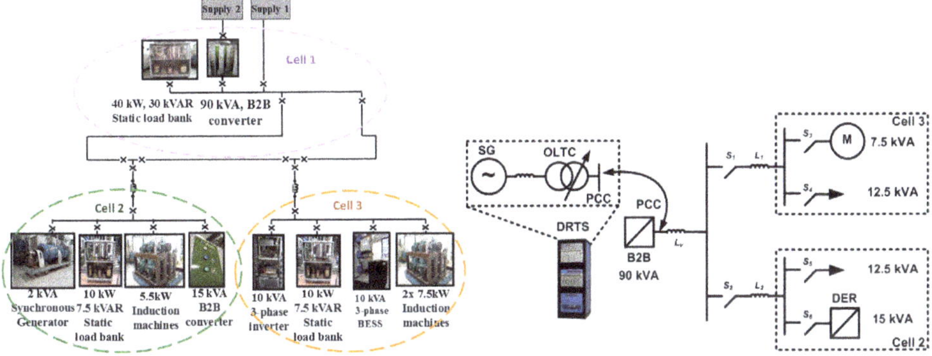


(**a**) Simplified one-line diagram of DPSL microgrid.   (**b**) DPSL hardware-in-the-loop configuration.

**Figure 2.** Test rig and hardware-in-the-loop configuration.

### 3.2. Test Configurations

The DPSL hardware-in-the-loop configuration utilized for the validation is presented in Figure 2b. A 15 kVA back to back (B2B) power module by Triphase is used as the DER unit, referred to as TP15 henceforth. The rest of the network constitutes a 90 kVA B2B Triphase power module serving as a grid emulator, two 12 kVA load banks, and a 7.5 kVA induction motor. A simple network, comprising a synchronous generator (SG) and an on load tap changer (OLTC) simulated within a digital real-time simulator (DRTS), provides the reference voltage for reproduction by the 90 kVA power interface. The chosen configuration is one of many configurations made possible by the flexible architecture of the DPSL Microgrid (Figure 2a). The two bus configuration is representative of a community microgrid with a lumped representation of residential customers, water pumps, and energy storage systems [55,56]. Larger system studies can be undertaken owing to the power hardware-in-the-loop capability of the DPSL, although this has not been fully exploited here. Examples of such setups for varied applications can be found in [57–60].

As SPAACE is an add-on controller, two different implementation approaches for its real-world applications are possible as described next.

### 3.2.1. Spaace within the Der Controller

If the controller of the DER is open for modifications by the user, the SPAACE algorithm can be implemented within the development environment of the DER controller. This is the most straightforward implementation of SPAACE. This implementation within the DPSL is shown in Figure 3a where the SPAACE algorithm is implemented within the controller (real-time target–RTT) of the TP15 power module.

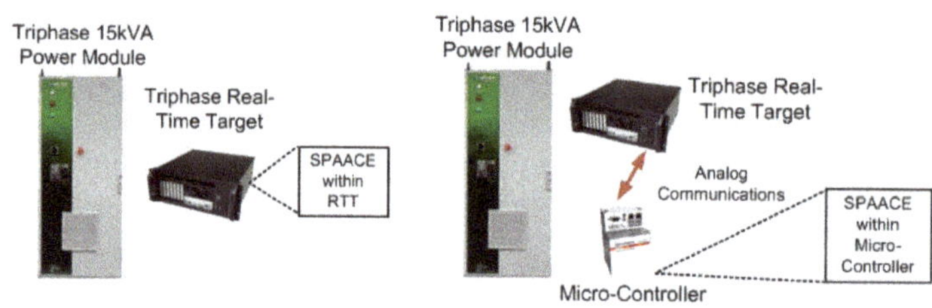


(**a**) Implementation within the DER.      (**b**) Implementation external to the DER.

**Figure 3.** Network and test configurations within the DPSL.

### 3.2.2. Spaace Outwith the Der Controller

As discussed previously, SPAACE does not necessarily need access to the lower-level controls of the DER, it can be implemented within an external RTT or microcontroller. This still requires the DER to accept analog or digital communications, which most vendors provide, specifically when the development environment is closed for users. This implementation within the DPSL is shown in Figure 3b. The network configuration remains the same as explained earlier except that SPAACE is implemented within an external controller communicating with TP15 using analog communication. A twisted pair copper wire is used for the analog communications, with a propagation speed of $\sim 2 \times 10^8$ m/s and the distance between the microcontroller and TP15 of approximately 50 m, yielding a delay of approximately 2.5 µs.

### 3.3. Controls for Grid-Connected and Islanded Operation

To demonstrate the applicability of SPAACE, two scenarios are considered for the operation of the DER: (i) grid-connected, and (ii) islanded. The incorporation of SPAACE in these two modes are discussed below:

### 3.3.1. Grid-Connected Operation

In the grid-connected mode, the DER is operated as a current-controlled voltage source. The real and reactive power set points are received by the DER controller. The incorporation of SPAACE within such DER control is shown in Figure 4a, where $P_{sp}$, $Q_{sp}$ are the real and reactive power set points; $P_{meas}$, $Q_{meas}$ are the measured real and reactive powers; and $P'_{sp}$, $Q'_{sp}$ are the SPAACE-modified real and reactive power set points. $V_{dq}$ is the direct and quadrature components of the voltage obtained by Park's transform. The inputs of the inner current control loop are the direct and quadrature components of the current $I_{dq}$; its output is sent to the modulation block for the generation of the gating signals. SPAACE modulates the real and reactive power reference set points to improve the dynamic response.

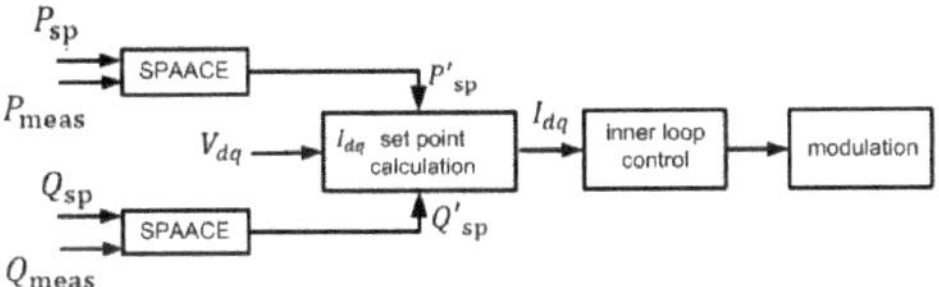

(**a**) Grid connected DER control with SPAACE.

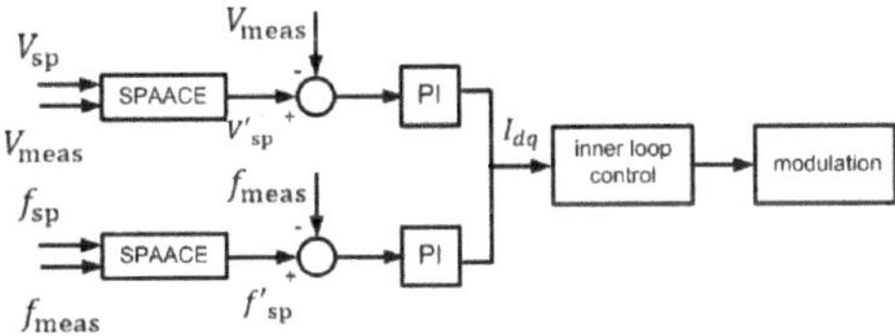

(**b**) Islanded DER control with SPAACE.

**Figure 4.** SPAACE incorporation within the DER control.

### 3.3.2. Islanded Operation

In the islanded mode of operation, a DER is responsible for regulating the voltage and frequency at its output terminal. A single DER responsible for maintaining the voltage and frequency of a small microgrid is considered in this paper. The incorporation of SPAACE within such DER control is shown in Figure 4b, where $V_{sp}$, $f_{sp}$ are the voltage and frequency set points; $V_{meas}$ and $f_{meas}$ are the measured voltage and frequency; $V'_{sp}$, $f'_{sp}$ are the SPAACE-modified set points of voltage and frequency.

## 4. Performance Evaluation

This section analyzes the performance of SPAACE. The methodology for the evaluation is presented followed by the computational and dynamic performance evaluation.

### 4.1. Methodology

The performance of the three variants of SPAACE discussed in this paper are quantitatively evaluated within both grid-connected and islanded modes of operation of the DER based on the following metrics:

### 4.1.1. Computational Complexity

Computational complexity refers to the processing resources required for the execution of an algorithm. Computational complexity plays an important role in experimental validations due to the limited resources of RTTs and microcontrollers. This analysis presents the complexity of the different variants of the SPAACE algorithm with respect to the RTT utilized for its run and therefore is application- and scenario-agnostic. The following two key indicators are defined:

- Execution time: This refers to the time taken for the execution of SPAACE represented by $T_{exec}$.
- Memory requirement: This refers to the memory requirement of the algorithm as represented by $M$.

### 4.1.2. Dynamic Response

The performance of SPAACE is assessed for a change in the DER reference set point and by subjecting DER to a set of predefined common external disturbances based on the following key indicators:

- Settling time: The settling time $T_{\text{settling}}$ is the time elapsed from the time of the disturbance to the time when the measured output signal $x(t)$ stays within the defined tolerance band $\epsilon$.

$$T_{\text{settling}} = \text{argmin}\{T_{\text{settling}} \in \mathcal{R} \mid \forall\, t > T_{\text{settling}} : \\ x^{+}_{\text{band}} < x(t) < x^{-}_{\text{band}}\} \tag{6}$$

- Overshoot: Defining the maximum excursion of $x(t)$ subject to an external disturbance or after a step change in the reference set point as $x_{\max}$, the overshoot is calculated as

$$x_{\text{os}} = \left| \frac{x_{\max} - x_{\text{sp}}}{x_{\text{sp}}} \right| \tag{7}$$

- Cumulative tracking error: Cumulative tracking error is defined as the sum of the tracking errors at every time step $T_s$ from the application of the step or the disturbance to the time when the measured output signal has settled, i.e., $T_{\text{settling}}$.

$$S_e = \sum_{k=0}^{N} \left( x_{\text{sp}}[k] - x[k] \right) T_s \tag{8}$$

where $N = T_{\text{settling}} / T_s$. A smaller $S_e$ corresponds to better set point tracking capability.

### 4.2. Computational Complexity Evaluation

The time and space complexity comparison of the three variants of SPAACE are presented in Figure 5 and Table 2 respectively. All implementations of SPAACE (linear, exponential, or quadratic) have a few common functions within and for the purpose of comparison only the mutually exclusive functions are considered. As can be observed from Figure 5, the linear predictor is computationally most efficient with an average $T_{\text{exec}} = 4.01\,\mu\text{s}$, followed by quadratic with an average $T_{\text{exec}} = 4.95\,\mu\text{s}$. The exponential predictor is computationally the most expensive with an average $T_{\text{exec}} = 185\,\mu\text{s}$.

The space complexity is calculated in terms of the number of bits of storage required at a given point in time to allow for the prediction strategy to operate. The linear implementation requires only 32 bits (1 float value), followed by quadratic that requires 64 bits (2 float values) while the exponential prediction strategy has the highest space complexity requiring 320 bits (10 float values). The space complexity of linear and quadratic prediction strategies does not increase with larger prediction horizons for real-time implementation. On the other hand, the space complexity of exponential strategy increases linearly with the increase in the prediction horizon.

**Table 2.** Memory requirement comparison.

|  | Linear | Exponential | Quadratic |
| --- | --- | --- | --- |
| $M$ | 32 | 320 | 64 |

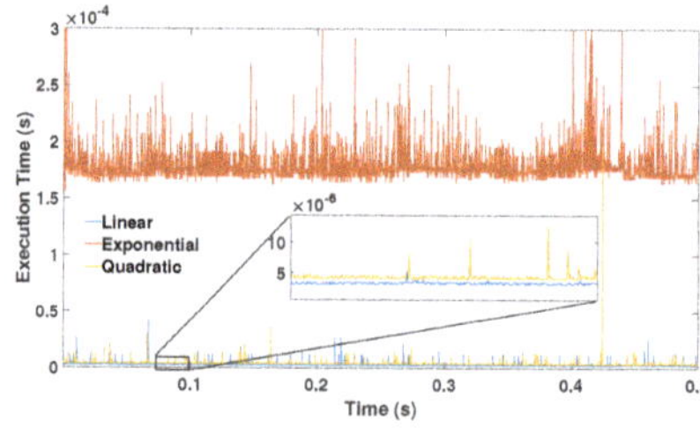


**Figure 5.** Execution time comparison.

### 4.3. Dynamic Response

#### 4.3.1. Grid-Connected Operation

The DER dynamic response improvement is characterized subject to seven different events as summarized in Table 3. These events are chosen to be indicative of common scenarios in an inverter-dominated power system. The response of the DER subject to these seven events is presented in Figure 6a,b, where the former presents results when SPAACE is implemented within the DER while the latter where the SPAACE is implemented outwith the DER. The subfigures (a), (b), and (c) in both figures represent linear, exponential, and quadratic prediction strategy respectively. The results are further discussed below.

**Table 3.** Cases under consideration for the grid-connected operation

| No. | Event | Change |
|---|---|---|
| 1 | Real power set point change | 2 kW to 6 kW |
| 2 | Voltage change | 1 pu to 1.1 pu |
| 3 | Voltage change | 1.1 pu to 1 pu |
| 4 | Virtual impedance change | 50 μH to 5 μH |
| 5 | Virtual impedance change | 5 μH to 50 μH |
| 6 | Real power set point change | 6 kW to 2 kW |
| 7 | Induction motor energization | 7.5 kVA soft-start |

(a) Set point change in real power: Events 1 and 6 emulate the day to day operations of a DER where, for example, the reference real power set point of a DER is changed based on its commitment to contribute to frequency control. The response of the DER subject to a real power set point change from 2 kW to 6 kW at $t = 0.1$ s and from 6 kW to 2 kW at $t = 5.1$ s is shown in columns 1 and 6 of Figure 6a,b with the performance indicators presented in Table 4. When SPAACE is incorporated, a minimum overshoot reduction of 16.52% for set point increase and 11.25% for set point decrease is observed. For step increase in set point, a minimum of 84.4% reduction in settling time is observed while for step decrease in set point the system response did not violate the tolerance band $\epsilon$ and therefore the settling time reported as 0 s.

(b) Grid voltage set point change: Events 2 and 3 represent a set of transient conditions that can be expected within a distribution grid with high penetration of solar photovoltaic installations. A sudden cloud cover across a distribution feeder can lead to a drop in voltage, and the clearing of the cloud cover can lead to an increase in voltage. In this case study, the voltage magnitude is changed using the 90 kVA B2B module. The response of the DER subject to a grid voltage set point change from 1 pu to 1.1 pu at $t = 1.1$ s and from 1.1 pu to 1 pu at $t = 2.1$ s is shown in columns 2 and 3 of Figure 6a,b with the performance indicators presented in Table 4. The inherent controller employed for the grid-connected mode, in this case, works as a poorly tuned controller, for which the increase in system voltage causes oscillations in the power output. With SPAACE, the response of the system is well damped with improvement in overshoot observed.

(c) Impact of system short-circuit ratio (SCR): Events 4 and 5 evaluate the dynamic performance of the DER subject to a change in system short circuit ratio (SCR). For 100% inverter-penetrated microgrids, changing the virtual impedance $Z_v$ of individual inverters is an effective solution to improve overall system stability [61]. Changing $Z_v$ of individual inverters in turn impacts the SCR of the system. The response of the DER to a change in virtual impedance is shown in columns 4 and 5 of Figure 6a,b, where $Z_v$ is reduced from 50 μH to 5 μH at $t = 3.1$ s and increased back to 50 μH at $t = 4.1$ s. The minimum improvement with SPAACE observed for overshoot reduction is 5.67%. With SPAACE, the response of the system does not violate the tolerance band $\epsilon$, thus having a 0 s settling time.

**Table 4.** Dynamic response performance indicators for the seven events under the grid-connected mode for operation.

| | Inherent | | | Linear | | | | | | Exponential | | | | | | Quadratic | | | | | |
| | Within DER | | | Within DER | | | Outwith DER | | | Within DER | | | Outwith DER | | | Within DER | | | Outwith DER | | |
| Event | $x_{os}$ | $T_s$ | $S_e$ | $x_{os}$ | $T_s$ | $S_e$ | $x_{os}$ | $T_s$ | $S_e$ | $x_{os}$ | $T_s$ | $S_e$ | $x_{os}$ | $T_s$ | $S_e$ | $x_{os}$ | $T_s$ | $S_e$ | $x_{os}$ | $T_s$ | $S_e$ |
| | (%) | (s) | $\times 10^6$ | (%) | (s) | $\times 10^6$ | (%) | (s) | $\times 10^6$ | (%) | (s) | $\times 10^6$ | (%) | (s) | $\times 10^6$ | (%) | (s) | $\times 10^6$ | (%) | (s) | $\times 10^6$ |
|---|---|---|---|---|---|---|---|---|---|---|---|---|---|---|---|---|---|---|---|---|---|
| 1 | 21.75 | 0.44 | 2.34 | 4.72 | 0.07 | 0.55 | 5.23 | 0.06 | 0.42 | 4.13 | 0.05 | 0.59 | 4.36 | 0.06 | 0.49 | 4 | 0.06 | 0.38 | 5.1 | 0.07 | 0.46 |
| 2 | 18.95 | - | 5.34 | 2.72 | - | 0.22 | 3.63 | - | 0.28 | 2.87 | - | 0.08 | 3.52 | - | 0.28 | 3.03 | - | 0.25 | 5.25 | - | 0.47 |
| 3 | 15.64 | 0.34 | 2.01 | 0 | 0 | 0.02 | 0 | 0 | 0.06 | 0 | 0 | 0.06 | 0 | 0 | 0.13 | 0 | 0 | 0.03 | 0 | 0 | 0.14 |
| 4 | 8.72 | 0.18 | 0.43 | 2.06 | 0 | 0.01 | 2.57 | 0 | 0.04 | 2.2 | 0 | 0.09 | 3.32 | 0 | 0.18 | 2.15 | 0 | 0.01 | 3.33 | 0 | 0.09 |
| 5 | 9 | 0.17 | 0.36 | 1.93 | 0 | 0.06 | 1.37 | 0 | 0.15 | 2.96 | 0 | 0.06 | 2.97 | 0 | 0.15 | 1.83 | 0 | 0.01 | 3.33 | 0 | 0.1 |
| 6 | 16.2 | 0.22 | 0.91 | 4.95 | 0 | 0.66 | 1.8 | 0 | 0.63 | 4.45 | 0 | 0.65 | 2.5 | 0 | 0.63 | 4.65 | 0 | 0.64 | 2.3 | 0 | 0.5 |
| 7 | 96.75 | 0.33 | 1.53 | 91.9 | 0.31 | 0.9 | 84.55 | 0.246 | 0.89 | 91.15 | 0.25 | 0.94 | 90.85 | 0.25 | 0.96 | 91.1 | 0.25 | 0.92 | 86.95 | 0.25 | 0.93 |

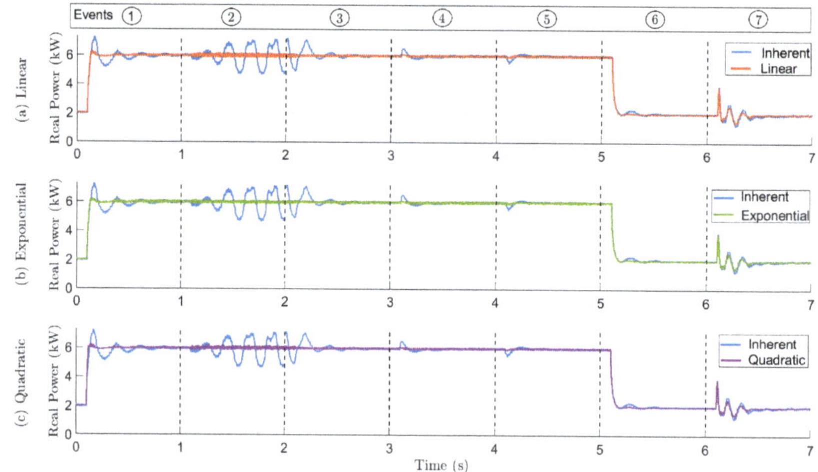


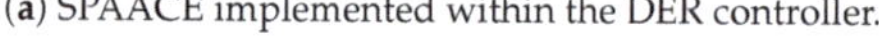

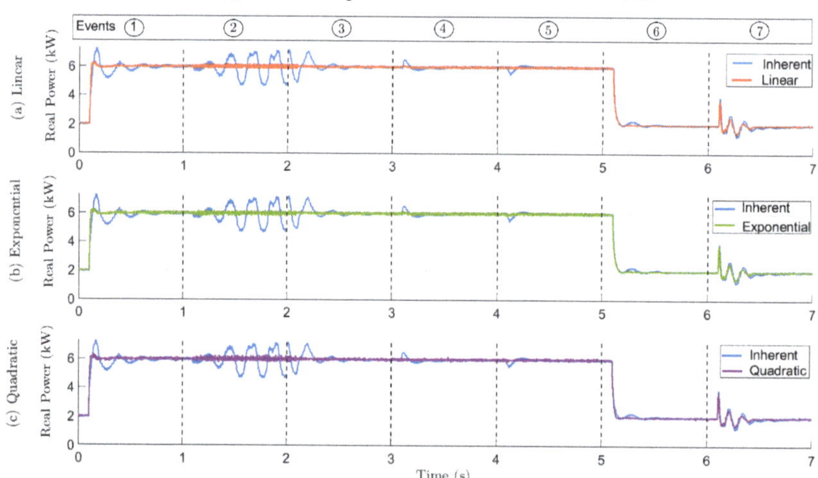


(**a**) SPAACE implemented within the DER controller.



(**b**) SPAACE implemented outwith DER controller.

**Figure 6.** Dynamic response evaluation of SPAACE under grid connected mode of operation with.

(d)    Induction motor energization: Event 7 evaluates the response of the DER subject to a switching transient on the network. For the system under study, the switching transient is emulated by the energization of a 7.5 kVA induction motor. The response of the DER is shown in column 7 of Figure 6a,b. From the performance indicators presented in Table 4, it is evident that SPAACE improves the system performance with minimum overshoot reduction of 4.85% and a 7.4% reduction in settling time.

(e)     Cumulative tracking error: Table 4 also presents a comparison of the cumulative tracking error ($S_e$) for the inherent controller and the three variants of SPAACE under both implementations (within and outwith the DER). All variants of SPAACE reduce $S_e$ for each of the seven events under consideration under both implementations. There is no distinctive performance difference between the three variants and the two implementations.

(f)     Summary: In this subsection, the dynamic performance of a weak controller operating in the grid-connected mode is evaluated. For the seven events under consideration, SPAACE considerably improves the dynamic performance with respect to each of the key indicators defined. This paper shows the feasibility of the implementation of SPAACE outwith the DER. In most cases, the performance of SPAACE when implemented outwith the DER is slightly inferior to its performance when it is implemented within the DER controller. This can be attributed to the communications delay to communicate the modified set point from the microcontroller to the DER internal controller. SPAACE's performance can be further optimized by considering the time delay during the design of the predictor.

### 4.3.2. Islanded Operation

The performance of SPAACE in the islanded mode of operation is evaluated for the DER subject to two events: (i) set point change in voltage from 1 pu to 0.9 pu at $t = 1.1$ s and (ii) increase in the real power of the load from 2 kW to 8 kW at $t = 2.1$ s. Figure 7 shows the DER response with SPAACE implemented outwith the controller, where subfigures (a), (b), and (c) represent linear, exponential, and quadratic prediction strategies, respectively. The performance indicators for islanded operation are summarised in Table 5.

**Table 5.** Dynamic response performance indicators for the two events under the islanded mode of operation.

| | Inherent | | | Linear | | | | | | Exponential | | | | | | Quadratic | | | | | |
| --- | --- | --- | --- | --- | --- | --- | --- | --- | --- | --- | --- | --- | --- | --- | --- | --- | --- | --- | --- | --- | --- |
| | Within DER | | | Within DER | | | Outwith DER | | | Within DER | | | Outwith DER | | | Within DER | | | Outwith DER | | |
| Event | $x_{os}$ | $T_s$ | $S_e$ | $x_{os}$ | $T_s$ | $S_e$ | $x_{os}$ | $T_s$ | $S_e$ | $x_{os}$ | $T_s$ | $S_e$ | $x_{os}$ | $T_s$ | $S_e$ | $x_{os}$ | $T_s$ | $S_e$ | $x_{os}$ | $T_s$ | $S_e$ |
| | (%) | (s) | $\times 10^4$ | (%) | (ms) | $\times 10^4$ | (%) | (ms) | $\times 10^4$ | (%) | (ms) | $\times 10^4$ | (%) | (ms) | $\times 10^4$ | (%) | (ms) | $\times 10^4$ | (%) | (ms) | $\times 10^4$ |
| 1 | 0 | 2.81 | 0.05 | 0 | 1.75 | 0.02 | 0 | 1.81 | 0.03 | 0 | 1.76 | 0.03 | 0 | 1.81 | 0.03 | 0 | 1.75 | 0.03 | 0 | 1.82 | 0.04 |
| 2 | 0 | 0 | 3.24 | 0 | 0 | 3.22 | 0 | 0 | 3.24 | 0 | 0 | 3.23 | 0 | 0 | 3.23 | 0 | 0 | 3.23 | 0 | 0 | 3.24 |

The DER controller utilized for islanded operation typically needs to be well-tuned and robust to be able to maintain the voltage under varied load conditions. As can be observed, although the inherent controller does not present any overshoot for a change in voltage set point, SPAACE still does improve the speed of its response. The increase in load causes a transient in voltage that is well mitigated by the inherent controller; SPAACE brings the system to the steady-state even faster. This behavior is corroborated by the cumulative tracking error $S_e$ presented in Table 5, where reduced $S_e$ for event 1 corresponds to the increase in response speed. No significant difference in performance under a step increase in load is observed.

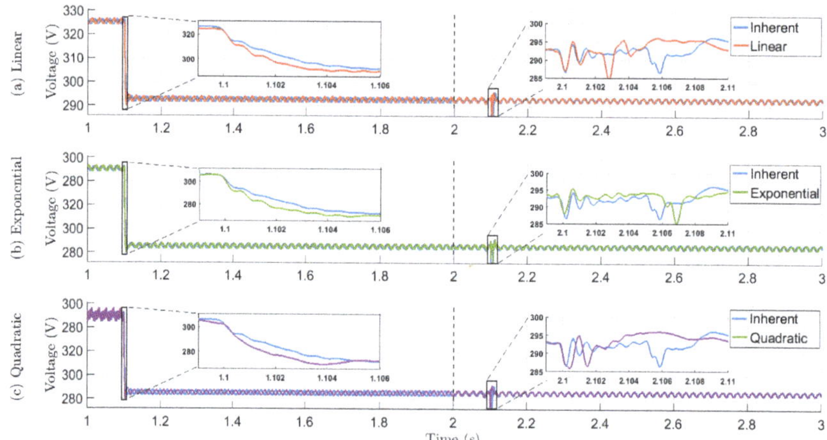


**Figure 7.** Dynamic response evaluation of SPAACE under the islanded mode of operation. SPAACE is implemented outwith the DER controller.

## 5. Conclusions

In this paper, a high-fidelity hardware-in-the-loop experimental test bed has been utilized to prove the practical real-world feasibility of an add-on controller, SPAACE, designed to improve the performance of existing controllers in an inverter-dominated power system. Following a methodological approach, first, an application- and scenario-agnostic computational performance evaluation of three prediction strategies for SPAACE is undertaken; followed by an application-oriented dynamic performance evaluation. With the defined computational and dynamic performance metrics, the following observations are highlighted:

- The linear predictor is computationally the most efficient with the least time and space complexity. In addition, the time and space complexity of a linear predictor does not increase with an increase in the prediction horizon.
- The choice of predictor only marginally impacts the dynamic performance and therefore, based on the computational performance evaluation, the use of a linear predictor is recommended. This is an application-specific recommendation, however, the trade-off between computational and dynamic performance should be assessed, specifically given the demonstrated close performance of linear and quadratic predictors.
- With a weak controller (in grid-connected mode), the incorporation of SPAACE leads to improvement in the dynamic properties (at least one of settling time, overshoot, and tracking error) of the response subject to the studied disturbances. With a strong controller (in islanded mode), the impact of SPAACE is limited to bringing the system to steady-state faster.
- The set of events chosen are non-exhaustive but sufficient for demonstration of the performance of the approach under a broad range of circumstances. However, given the versatility of its operation, the improvement of dynamics under unknown system conditions can be expected.

Furthermore, in this paper, SPAACE's implementation external to a primary controller has been demonstrated. A slight deterioration in the performance of SPAACE compared to its implementation within the primary controller is observed. This deterioration in performance is associated with the time delay of the implementation that highlights an important aspect of future work, i.e., to incorporate time delay within the design of the predictor.

The requirement for incorporation of SPAACE is the existing capability DER controller to accept external setpoints (implementation outwith DER) or access to the internal control (implementation within DER). However, this does not present as a limitation as without either one of the above requirments, the DER would not be able to participate within ancillary services provision.

**Author Contributions:** Conceptualization, M.H.S. and A.M.-S; funding acquisition, A.M.-S. and G.M.B.; methodology, M.H.S. and A.M.-S.; validation, M.H.S. and E.G.-S.; writing—original draft, M.H.S. and A.M.-S.; writing—review and editing, E.G.-S. and G.M.B. All authors have read and agreed to the published version of the manuscript

**Funding:** This work was supported in part by the European Commission under the H2020 Programme's ERIGRID and ERIGRID2.0 projects (grant no: 654113 and 870620) and in part by the U.S. National Science Foundation under grants EPCN-1509895 and CPS-1837700. Any opinions, findings, and conclusions or recommendations expressed in this material are those of the authors and do not necessarily reflect those of the funding bodies.

**Conflicts of Interest:** The authors declare no conflict of interest.

## Nomenclature

### Abbreviations

| | |
|---|---|
| SPAACE | Set point automatic adjustment with correction enabled. |
| SPM | Set point modulation. |
| DER | Distributed energy resource. |
| HIL | Hardware-in-the-loop. |
| SISO | Single input single output. |
| LSE | Least square error. |
| DPSL | Dynamic power systems laboratory. |
| B2B | Back to back. |
| RTT | Real time target. |
| SCR | Short circuit ratio. |
| HVDC | High voltage direct current. |
| BESS | Battery energy storage system. |
| DRTS | Digital real-time simulator. |

### Variables

| | |
|---|---|
| $x_{sp}$ | Reference set point. |
| $x$ | Measured output. |
| $x_1$ | Initial reference set point. |
| $x_2$ | Final reference set point. |
| $x'_{sp}$ | Modified set point. |
| $m$ | Scaling factor. |
| $T_{pred}$ | Horizon of prediction. |
| $n$ | An integer value. |
| $\hat{x}$ | Predicted value of $x$. |
| $r$ | Average rate of change of historical data. |
| $q$ | Second derivative of historical data. |
| $P_{sp}$ | Real power set point. |
| $Q_{sp}$ | Reactive power set point. |
| $P_{meas}$ | Measured real power. |
| $Q_{meas}$ | Measured reactive power. |
| $P'_{sp}$ | Modified real power set point. |
| $Q'_{sp}$ | Modified reactive power set point. |
| $V_{dq}$ | Direct and quadrature components of voltage. |
| $I_{dq}$ | Direct and quadrature components of current. |
| $V_{sp}$ | Voltage set point. |
| $f_{sp}$ | Frequency set point. |
| $V_{meas}$ | Measured voltage. |
| $f_{meas}$ | Measured frequency. |
| $V'_{sp}$ | Modified voltage set point. |

$f'_{\text{sp}}$      Modified frequency set point.
$S_{1-6}$      Switches within the utilized network.
$L_{1-2,v}$      Line impedances within the utilized network.
$T_{\text{exec}}$      Execution time.
$M$      Memory requirement.
$T_{\text{settling}}$      Settling time.
$x_{\text{max}}$      Absolute maximum value of $x$.
$T_s$      Time step.
$S_e$      Cumulative tracking error.
$x_{\text{os}}$      Overshoot.
$Z_v$      Virtual impedance.
$T_{\text{past}}$      Window of prediction history.
$x^+_{\text{band}}$      Upper threshold of tolerance band.
$x^-_{\text{band}}$      Lower threshold of tolerance band.
$\epsilon$      Tolerance band.
$\omega_0$      Dominant natural frequency.

## References


1. Fan, B.; Peng, J.; Yang, Q.; Liu, W. Distributed Periodic Event-Triggered Algorithm for Current Sharing and Voltage Regulation in DC Microgrids. *IEEE Trans. Smart Grid* **2020**, *11*, 577–589. [CrossRef]
2. Fan, B.; Guo, S.; Peng, J.; Yang, Q.; Liu, W.; Liu, L. A Consensus-Based Algorithm for Power Sharing and Voltage Regulation in DC Microgrids. *IEEE Trans. Ind. Inform.* **2020**, *16*, 3987–3996. [CrossRef]
3. Ding, D.; Han, Q.-L.; Wang, Z.; Ge, X. A Survey on Model-Based Distributed Control and Filtering for Industrial Cyber-Physical Systems. *IEEE Trans. Ind. Inform.* **2019**, *15*, 2483–2499. [CrossRef]
4. Molzahn, D.K.; Dorfler, F.; Sandberg, H.; Low, S.H.; Chakrabarti, S.; Baldick, R.; Lavaei, J. A Survey of Distributed Optimization and Control Algorithms for Electric Power Systems. *IEEE Trans. Smart Grid* **2017**, *8*, 2941–2962. [CrossRef]
5. Li, H.; Li, F.; Xu, Y.; Rizy, D.T.; Kueck, J.D. Adaptive Voltage Control With Distributed Energy Resources: Algorithm, Theoretical Analysis, Simulation, and Field Test Verification. *IEEE Trans. Power Syst.* **2010**, *25*, 1638–1647. [CrossRef]
6. Leithead, W. Survey of Gain-Scheduling Analysis Design. *Int. J. Control* **1999**, *73*, 1001–1025. [CrossRef]
7. Namara, P.M.; Negenborn, R.R.; Schutter, B.D.; Lightbody, G. Optimal Coordination of a Multiple HVDC Link System Using Centralized and Distributed Control. *IEEE Trans. Control Syst. Technol.* **2013**, *21*, 302–314. [CrossRef]
8. Moradzadeh, M.; Boel, R.; Vandevelde, L. Voltage Coordination in Multi-Area Power Systems via Distributed Model Predictive Control. *IEEE Trans. Power Syst.* **2013**, *28*, 513–521. [CrossRef]
9. Roshany-Yamchi, S.; Cychowski, M.; Negenborn, R.; Schutter, B.; Delaney, K.; Connell, J. Kalman Filter-Based Distributed Predictive Control of Large-Scale Multi-Rate Systems: Application to Power Networks. *IEEE Trans. Control. Syst. Technol.* **2013**, *21*, 27–39. [CrossRef]
10. Liu, X.; Kong, X.; Lee, K.Y. Distributed model predictive control for load frequency control with dynamic fuzzy valve position modelling for hydro–thermal power system. *IET Control Theory Appl.* **2016**, *10*, 1653–1664. [CrossRef]
11. Mehmood, F.; Khan, B.; Ali, S.M.; Rossiter, J.A. Distributed model predictive based secondary control for economic production and frequency regulation of MG. *IET Control Theory Appl.* **2019**, *13*, 2948–2958. [CrossRef]
12. Åström, K.J.; Hägglund, T. *PID Controllers: Theory, Design and Tuning*, 2nd ed.; Instrument society of America: Research Trinangle Park, NC, USA, 1995.
13. Åström, K.J. Auto-Tuning, Adaptation and Expert Control. In Proceedings of the American Control Conference, Boston, MA, USA, 19–21 June 1985; pp. 1514–1519.
14. Åström, K.J.; Murray, R.M. *Feedback Systems: An Introduction for Scientists and Engineers*, 2nd ed.; Princeton University Press: Princeton, NJ, USA, 2008.
15. Åström, K.J.; Hägglund, T. Automatic Tuning of PID Controllers. In *The Control Handbook*; Levin, W., Ed.; Number 53; CRC Press: Boca Raton, FL, USA, 1996.

16. Killingsworth, N.; Krstic, M. PID Tuning Using Extremum Seeking: Online, model-free performance optimization. *IEEE Control Syst. Mag.* **2006**, *26*, 70–79.

17. Hjalmarsson, H.; Gevers, M.; Gunnarsson, S.; Lequin, O. Iterative feedback tuning: Theory and applications. *IEEE Control. Syst.* **1998**, *18*, 26–41.

18. Lequin, O.; Bosmans, E.; Triest, T. Iterative feedback tuning of PID parameters: Comparison with classical tuning rules. *Contr. Eng. Pract.* **2003**, *11*, 1023–1033. [CrossRef]

19. Lequin, O.; Gevers, M.; Triest, T. Optimizing the settling time with iterative feedback tuning. In Proceedings of the 15th IFAC World Congress, Beijing, China, 5–9 July 1999; pp. 433–437.

20. Burns, D.J.; Laughman, C.R.; Guay, M. Proportional–Integral Extremum Seeking for Vapor Compression Systems. *IEEE Trans. Control Syst. Technol.* **2020**, *28*, 403–412. [CrossRef]

21. Hunnekens, B.; Dino, A.D.; van de Wouw, N.; van Dijk, N.; Nijmeijer, H. Extremum-Seeking Control for the Adaptive Design of Variable Gain Controllers. *IEEE Trans. Control Syst. Technol.* **2015**, *23*, 1041–1051. [CrossRef]

22. Lu, X.; Krstić, M.; Chai, T.; Fu, J. Hardware-in-the-Loop Multiobjective Extremum-Seeking Control of Mineral Grinding. *IEEE Trans. Control Syst. Technol.* **2020**, 1–11. [CrossRef]

23. Smith, O.J.M. Posicast Control of Damped Oscillatory Systems. *Proc. IRE* **1957**, *45*, 1249–1255. [CrossRef]

24. Hung, J.Y. Feedback Control with Posicast. *IEEE Trans. Ind. Electron.* **2003**, *50*, 94–99. [CrossRef]

25. Vaughan, J.; Smith, A.; Kang, S.J.; Singhose, W. Predictive Graphical User Interface Elements to Improve Crane Operator Performance. *IEEE Trans. Syst. Man Cybern. Part A Syst. Hum.* **2011**, *41*, 323–330. [CrossRef]

26. Cook, G. An Application of Half-Cycle Posicast. *IEEE Trans. Autom. Control* **1966**, *11*, 556–559. [CrossRef]

27. Hung, J.Y. Posicast Control Past and Present. *IEEE Multidiscip. Eng. Educ. Mag.* **2007**, *2*, 7–11.

28. Feng, Q.; Nelms, R.M.; Hung, J.Y. Posicast-Based Digital Control of the Buck Converter. *IEEE Trans. Ind. Electron.* **2006**, *53*, 759–767. [CrossRef]

29. Mi, Y.; Song, Y.; Fu, Y.; Wang, C. The Adaptive Sliding Mode Reactive Power Control Strategy for Isolated Wind-Diesel Power System Based on Sliding Mode Observer. *IEEE Trans. Sustain. Energy* **2019**. [CrossRef]

30. Zhang, Y.; Li, G. Non-causal Linear Optimal Control of Wave Energy Converters with Enhanced Robustness by Sliding Mode Control. *IEEE Trans. Sustain. Energy* **2019**. [CrossRef]

31. Onyeka, A.E.; Xing-Gang, Y.; Mao, Z.; Jiang, B.; Zhang, Q.; Yan, X. Robust decentralised load frequency control for interconnected time delay power systems using sliding mode techniques. *IET Control Theory Appl.* **2020**, *14*, 470–480. [CrossRef]

32. Sarkar, M.K.; Dev, A.; Asthana, P.; Narzary, D. Chattering free robust adaptive integral higher order sliding mode control for load frequency problems in multi-area power systems. *IET Control Theory Appl.* **2018**, *12*, 1216–1227. [CrossRef]

33. Zhang, H.; Zhou, J.; Sun, Q.; Guerrero, J.M.; Ma, D. Data-Driven Control for Interlinked AC/DC Microgrids Via Model-Free Adaptive Control and Dual-Droop Control. *IEEE Trans. Smart Grid* **2017**, *8*, 557–571. [CrossRef]

34. Safaei, A.; Mahyuddin, M.N. Adaptive Model-Free Control Based on an Ultra-Local Model With Model-Free Parameter Estimations for a Generic SISO System. *IEEE Access* **2018**, *6*, 4266–4275. [CrossRef]

35. Hou, Z.; Zhu, Y. Controller-Dynamic-Linearization-Based Model Free Adaptive Control for Discrete-Time Nonlinear Systems. *IEEE Trans. Ind. Inform.* **2013**, *9*, 2301–2309. [CrossRef]

36. Chen, J.; Yang, F.; Han, Q.-L. Model-Free Predictive $H_\infty$ Control for Grid-Connected Solar Power Generation Systems. *IEEE Trans. Control Syst. Technol.* **2014**, *22*, 2039–2047. [CrossRef]

37. Duan, L.; Hou, Z.; Yu, X.; Jin, S.; Lu, K. Data-Driven Model-Free Adaptive Attitude Control Approach for Launch Vehicle With Virtual Reference Feedback Parameters Tuning Method. *IEEE Access* **2019**, *7*, 54106–54116. [CrossRef]

38. Lu, C.; Zhao, Y.; Men, K.; Tu, L.; Han, Y. Wide-area power system stabiliser based on model-free adaptive control. *IET Control Theory Appl.* **2015**, *9*, 1996–2007. [CrossRef]

39. Wang, H.; Yang, J.; Chen, Z.; Ge, W.; Li, Y.; Ma, Y.; Dong, J.; Okoye, M.O.; Yang, L.; Onyeka, O.M. Analysis and Suppression for Frequency Oscillation in a Wind-Diesel System. *IEEE Access* **2019**, *7*, 22818–22828. [CrossRef]

40. Zolfaghari, M.; Hosseinian, S.H.; Fathi, S.H.; Abedi, M.; Gharehpetian, G.B.; Fathi, S.H. A New Power Management Scheme for Parallel-Connected PV Systems in Microgrids. *IEEE Trans. Sustain. Energy* **2018**, *9*, 1605–1617. [CrossRef]

41.  M'Zoughi, F.; Garrido, I.; Garrido, A.J.; De La Sen, M. Fuzzy Gain Scheduled-Sliding Mode Rotational Speed Control of an Oscillating Water Column. *IEEE Access* **2020**, *8*, 45853–45873. [CrossRef]

42.  Sun, H.; Zong, G.; Ahn, C.K. Quantized Decentralized Adaptive Neural Network PI Tracking Control for Uncertain Interconnected Nonlinear Systems With Dynamic Uncertainties. *IEEE Trans. Syst. Man Cybern. Syst.* **2019**, 1–14. [CrossRef]

43.  Su, X.; Liu, Z.; Zhang, Y.; Chen, C.L.P. Event-Triggered Adaptive Fuzzy Tracking Control for Uncertain Nonlinear Systems Preceded by Unknown Prandtl-Ishlinskii Hysteresis. *IEEE Trans. Cybern.* **2019**, 1–14. [CrossRef]

44.  Mehrizi-Sani, A.; Iravani, R. Online Set Point Adjustment for Trajectory Shaping in Microgrid Applications. *IEEE Trans. Power Syst.* **2012**, *27*, 216–223. [CrossRef]

45.  Yazdanian, M.; Mehrizi-Sani, A.; Seebacher, R.; Krischan, K.; Muetze, A. Smooth reference modulation to improve dynamic response in drive systems. *IEEE Trans. Power Electron.* **2018**, *33*, 6434–6443. [CrossRef]

46.  Mehrizi-Sani, A.; Iravani, R. Online Set Point Modulation to Enhance Microgrid Dynamic Response: Theoretical Foundation. *IEEE Trans. Power Syst.* **2012**, *27*, 2167–2174. [CrossRef]

47.  Ghaffarzadeh, H.; Stone, C.; Mehrizi-Sani, A. Predictive set point modulation to mitigate transients in lightly damped balanced and unbalanced systems. *IEEE Trans. Power Syst.* **2016**, *32*, 1041–1049. [CrossRef]

48.  Ghaffarzadeh, H.; Mehrizi-Sani, A. Predictive set point modulation to mitigate transients in power systems with a multiple-input-multiple-output control system. In Proceedings of the IEEE Innovative Smart Grid Technologies Conference (ISGT), Minneapolis, MN, USA, 6–9 September 2016.

49.  Mehrizi-Sani, A.; Iravani, R. Performance Evaluation of a Distributed Control Scheme for Overvoltage Mitigation. In Proceedings of the CIGRÉ International Symposium on The Electric Power System of The Future: Integrating supergrids and microgrids, Bologna, Italy, 13–15 September 2011.

50.  Sandia National Laboratories. *Test Protocols for Advances Inverter Interoperability Functions*; Sandia Corporation: Albuquerque, NM, USA, 2013; pp. 1–22.

51.  Guillo-Sansano, E.; Syed, M.H.; Roscoe, A.J.; Burt, G.M. Initialization and Synchronization of Power Hardware-In-The-Loop Simulations: A Great Britain Network Case Study. *Energies* **2018**, *11*, 1087. [CrossRef]

52.  Guillo-Sansano, E.; Syed, M.H.; Roscoe, A.J.; Burt, G.M.; Coffele, F. Characterization of Time Delay in Power Hardware in the Loop Setups. *IEEE Trans. Ind. Electron.* **2020**, *50*, 1–10. [CrossRef]

53.  Blair, S.M.; Syed, M.H.; Roscoe, A.J.; Burt, G.M.; Braun, J. Measurement and Analysis of PMU Reporting Latency for Smart Grid Protection and Control Applications. *IEEE Access* **2019**, *7*, 48689–48698. [CrossRef]

54.  Guillo-Sansano, E.; Roscoe, A.J.; Burt, G.M. Harmonic-by-harmonic time delay compensation method for PHIL simulation of low impedance power systems. In Proceedings of the 2015 International Symposium on Smart Electric Distribution Systems and Technologies (EDST), Vienna, Austria, 8–11 September 2015; pp. 560–565. [CrossRef]

55.  Li, Q.; Yu, S.; Al-Sumaiti, A.; Turitsyn, K. Modeling and Co-Optimization of a Micro Water-Energy Nexus for Smart Communities. In Proceedings of the 2018 IEEE PES Innovative Smart Grid Technologies Conference Europe (ISGT-Europe), Sarajevo, Bosnia-Herzegovina, 21–25 October 2018; pp. 1–5.

56.  Hao, H.; Somani, A.; Lian, J.; Carroll, T.E. Generalized aggregation and coordination of residential loads in a smart community. In Proceedings of the 2015 IEEE International Conference on Smart Grid Communications (SmartGridComm), Miami, FL, USA, 2–5 november 2015; pp. 67–72.

57.  Syed, M.H.; Guillo-Sansano, E.; Avras, A.; Downie, A.; Jennett, K.; Burt, G.M.; Coffele, F.; Rudd, A.; Bright, C. The Role of Experimental Test Beds for the Systems Testing of Future Marine Electrical Power Systems. In Proceedings of the 2019 IEEE Electric Ship Technologies Symposium (ESTS), Washington, DC, USA, 14–16 August 2019; pp. 141–148.

58.  Kontis, E.O.; Nousdilis, A.I.; Papagiannis, G.K.; Syed, M.H.; Guillo-Sansano, E.; Burt, G.; Papadopoulos, T.A. Power Hardware-in-the-Loop Setup for Developing, Analyzing and Testing Mode Identification Techniques and Dynamic Equivalent Models. In Proceedings of the 2019 IEEE Milan PowerTech, Milan, Italy, 23–27 June 2019; pp. 1–6.

59.  Syed, M.H.; Sansano, E.G.; Blair, S.M.; Burt, G.; Prostejovsky, A.M.; Rikos, E. Enhanced load frequency control: Incorporating locational information for temporal enhancement. *IET Gener. Transm. Distrib.* **2019**, *13*, 1865–1874. [CrossRef]

60. Wang, Y.; Syed, M.H.; Guillo-Sansano, E.; Xu, Y.; Burt, G.M. Inverter-Based Voltage Control of Distribution Networks: A Three-Level Coordinated Method and Power Hardware-in-the-Loop Validation. *IEEE Trans. Sustain. Energy* **2019**. [CrossRef]
61. Lu, X.; Sun, K.; Guerrero, J.M.; Vasquez, J.C.; Huang, L.; Wang, J. Stability Enhancement Based on Virtual Impedance for DC Microgrids With Constant Power Loads. *IEEE Trans. Smart Grid* **2015**, *6*, 2770–2783. [CrossRef]

*Article*

# Reduced-Scale Models of Variable Speed Hydro-Electric Plants for Power Hardware-in-the-Loop Real-Time Simulations

**Baoling Guo** [1,2,*,†], **Amgad Mohamed** [3], **Seddik Bacha** [2], **Mazen Alamir** [3], **Cédric Boudinet** [2] and **Julien Pouget** [1]

1   School of Engineering, (HES-SO) University of Applied Sciences and Arts of Western Switzerland, 1950 Sion, Valais, Switzerland; julien.pouget@hevs.ch
2   Institute of Engineering Grenoble, University Grenoble Alpes, CNRS, G2Elab, F-38000 Grenoble, France; seddik.bacha@g2elab.grenoble-inp.fr (S.B.); cedric.boudinet@g2elab.grenoble-inp.fr (C.B.)
3   GIPSA-Lab, Institute of Engineering Grenoble, University Grenoble Alpes, CNRS, F-38400 Saint Martin D'Hères, France; amgad.mohamed@gipsa-lab.grenoble-inp.fr (A.M.); mazen.alamir@gipsa-lab.grenoble-inp.fr (M.A.)
*   Correspondence: baoling.guo@hevs.ch
†   Current address: Rue du Rawil 47, 1950 Sion 2, Valais, Switzerland.

Received: 30 September 2020; Accepted: 29 October 2020; Published: 3 November 2020

**Abstract:** Variable Speed Hydro-Electric Plant (VS-HEP) equipped with power electronics has been increasingly introduced into the hydraulic context. This paper is targeting a VS-HEP Power Hardware-In-the-Loop (PHIL) real-time simulation system, which is dedicated to different hydraulic operation schemes tests and control laws validation. Then, a proper hydraulic model will be the key factor for building an efficient PHIL real-time simulation system. This work introduces a practical and generalised modelling hydraulic modelling approach, which is based on 'Hill Charts' measurements provided by industrial manufacturers. The hydraulic static model is analytically obtained by using mathematical optimization routines. In addition, the nonlinear dynamic model of the guide vane actuator is introduced in order to evaluate the effects of the induced dynamics on the electric control performances. Moreover, the reduced-scale models adapted to different laboratory conditions can be established by applying scaling laws. The suggested modelling approach enables the features of decent accuracy, light computational complexity, high flexibility and wide applications for their implementations on PHIL real-time simulations. Finally, a grid-connected energy conversion chain of bulb hydraulic turbine associated with a permanent magnet synchronous generator is chosen as an example for PHIL design and performance assessment.

**Keywords:** hydro-electric plant; variable speed operation; 'Hill Charts'; reduced-scale model; power hardware-in-the-loop; real-time simulation; testing and validation

---

## 1. Introduction

Hydropower, as a type of cost-competitive renewable source, has played a significant role in electricity mix [1,2]. In recent years, hydropower contributes to balancing the intermittent electricity resources [3,4]. However, the conventional hydro-electric plants do not perfectly meet these new demands because of its slow reaction dynamics. The power adjustment dynamics can be enhanced by introducing Power Electronics (PE) converters [1]. Moreover, the efficiency of hydraulic energy conversion can be improved through the variable speed operation when the hydraulic conditions change frequently [5,6]. In addition, the previous research indicates that variable speed operations can alleviate water hammer effects and optimize various hydraulic transients [2]. Therefore, a Variable Speed Hydro-Electric Plant (VS-HEP) equipped with PE has been increasingly introduced into the hydraulic context [1,2,6–8]. Limited to the PE capacity, the variable speed technique is mainly implemented into small hydro-electric plants [9–13]. The Run-of-River (RoR) scheme is economic and environmentally friendly as no dam is required, which have been suggested for small hydropower applications [9]. In an RoR scheme, a class of hydraulic turbines such as semi-Kaplan or Kaplan [5,10,13], propeller [11], and bulb turbine [12] has become the favourable prime mover because of their high efficiency under low water heads and fast flow-rates. In the work, to highlight, we will therefore focus on a class of small-scale hydraulic turbines.

VS-HEPs present many advantages in hydraulic applications but bring problems of the controls, disturbances rejections, and induced constraints as well [7,8]. Advanced control techniques continue to be needed to achieve higher robustness and enable faster control dynamics for VS-HEPs [14–17]. Reduced-scale PHIL benchmarks become necessary to test and validate control laws in hydraulic schemes because repetitive deterministic experiments cannot be conducted under a real disturbed environment [18–20]. The key question of designing a successful VS-HEP PHIL real-time simulation system is how to build a proper mathematical model for the concerned hydraulic turbine. However, unlike well-developed wind turbines [19], there is no general efficiency expression for hydraulic turbines. The hydraulic efficiency depends on both the turbine geometrical design and its practical operating conditions [21]. Various models have been proposed in previous works. The Kaplan turbine is described by a linear decreasing line of the mechanical torque versus the turbine speed in [10], while both the pitch adjustment and the guide vane opening are not considered. The hydraulic losses model of a bulb turbine is approximated by a 2D expression of the flow rate and the turbine speed in [12]. In another work [11], a 2D efficiency model of a propeller turbine is achieved by numerical approximations as well. The 2D models generally fix the guide vane angle or neglect its effects on the obtained hydraulic models. The real-time efficiency model is considered in [13]; however, precise measurements of water flow rate are still a critical challenge. A group of static data, measured from a real-life full-scale hydraulic electric plant, are then discretely stored in the look-up tables in [22,23]. If a control method works on the condition that the system must be continuous, such as the optimal control (Casadi, IPopt, etc.), a continuous model would be required [24]. Then, a class of look-up table based models is not able to be employed in such control schemes. A class of nonlinear or linear dynamic hydraulic models has been discussed in [25,26]. They are commonly used for hydraulic dynamics control test under offline simulations environments. However, the computational burden would increase by introducing large hydraulic dynamics. If the digital processor in use cannot satisfy the computational rapidity for real-time simulations, the modelling complexities of hydraulic turbine have to be sacrificed. Otherwise, the behaviours of implemented controllers could be deteriorated [19]. Hydraulic features are regressed to be singly linked to the rotational speed in [27], which is particularly suitable for a propeller turbine with a fixed guide vane angle. In a recent work of [28], the water flow rate is approximated by a third-degree polynomial, and the efficiency model is determined by the Multi-Layer Perception (MLP) of a neural network. However, the model determination process based MLP is complicated; meanwhile, a large scale of measurements are required to ensure the efficiency accuracy [28]. The Euler turbine equations are considered in [17,29]. The input is the turbine's main geometry at the best efficiency point of operation. However, the  performance of hydraulic turbine can be affected by the hydraulic conditions of its localisation; moreover, the system efficiency feature can also change due to the ageing [28]. The discussions above indicate that a practical and generalised modelling approach regarding various hydraulic turbines is still required for achieving efficient PHIL real-time simulations.

In this paper, a generalized hydraulic modelling approach is presented for achieving efficient PHIL real-time simulations. The static models are based on the commonly used 'Hill Charts' measurements provided by industrial manufacturers. The regression models of water flow rate and hydraulic efficiency are computed by using optimization routines. The optimization routine is defined with the objective of minimizing the difference between the measured data points and the determined regression models. In addition, we can update the models with recent 'Hill Charts' measurements to avoid the model deviations due to the ageing or localisations. Furthermore, this work introduces the nonlinear dynamic model of the guide vane actuator. The dynamic model helps to more accurately evaluate how the hardware parts react to the induced dynamics and what would be the impacts on the whole closed-loop control performances. The scaling rules of hydraulic turbine and the mechanical drive train are described as well. Then, reduced-scale models can be flexibly obtained according to various laboratory conditions. Some features are enabled for its implementations on PHIL real-time simulations: decent accuracy, high computation efficiency, significant flexibility, and wide applications. Furthermore, a bulb hydraulic turbine associated with a Permanent Magnet Synchronous Generator (PMSG) is chosen as an example for the performance assessment. Based on the flexible hybrid PHIL benchmark in the laboratory [19], a VS-HEP experimental benchmark is built, which is adapted to the obtained reduced-scale turbine model.

This rest of this paper is organized as follows: A 'Hill Charts' based modelling approach is presented in Section 2, where a bulb turbine with adjustable guide vane actuator is chosen as an example to assess the performance. In addition, scaling rules of hydraulic turbine and the mechanical drive train have been described in this section as well. In Section 3, a VS-HEP PHIL real-time simulation system is developed that is adapted to the obtained reduced-scale turbine model. Experiments have been conducted to assess the performance of the proposed reduced-scale models under the built PHIL simulation benchmark in Section 4. Remarks are concluded in Section 5.

## 2. 'Hill Charts' Based Hydraulic Modelling

In this section, the static features are firstly modelled based on the 'Hill Charts' data. The dynamic model of guide vane actuator is then considered and described. Moreover, the scaling rules regarding both the hydraulic turbine and the mechanical drive train are discussed as well. Finally, the modellings discussed in Section 1 and the proposed hydraulic model in this work for their implementations on PHIL simulations are generally compared and analysed. The following assumptions are made before the modelling:

(1) The RoR scheme is considered for for small hydropower applications in this work. A constant water head is generally required in RoR schemes [13,30]. A separate Proportional-Integral (PI) controller can be used to regulate the water levels [13]. The variations of water level are much smaller compared to the water head. Therefore, the influences of water head variations on the hydraulic static models can be neglected, which then helps to reduce the modelling complexities [28]. Such modelling rule is commonly employed by industrial manufacturers such as General Electric (GE), as it is sufficiently practical to reflect the real hydraulic behaviours.

(2) Most RoR hydro-electric plants are equipped with either Kaplan turbines or bulb turbines [31]. There are very few turbines with penstocks, which only happen when the water heads reach above 45 m up to 60 m [32]. Regarding bulb turbines, there are usually no real penstocks for a RoR regime, while a concrete intake can be built. The dynamics of penstocks are finally not taken into account in this work [33].

(3) A fixed-pitch bulb turbine is considered. The turbine is assumed to be single regulated in this paper. Only varying guide vane ratio $\gamma$ is considered, and the pitch angle ($\delta$) is fixed at 0.27.

### 2.1. Static Features Modelling

The general procedure of obtaining the reduced-scale hydraulic model is schematically described in Figure 1, which will be detailed in the following parts. This method differs from the scenario discussed in [34], as an actual full-scale 'Hill Charts' data of a bulb turbine is used in computing the mathematical model, without the need of removing the unstable region (S-region) [35]. In addition, being different from the modelling approach proposed in [28], the models of water flow rate and efficiency are regressed through solving optimization problems.

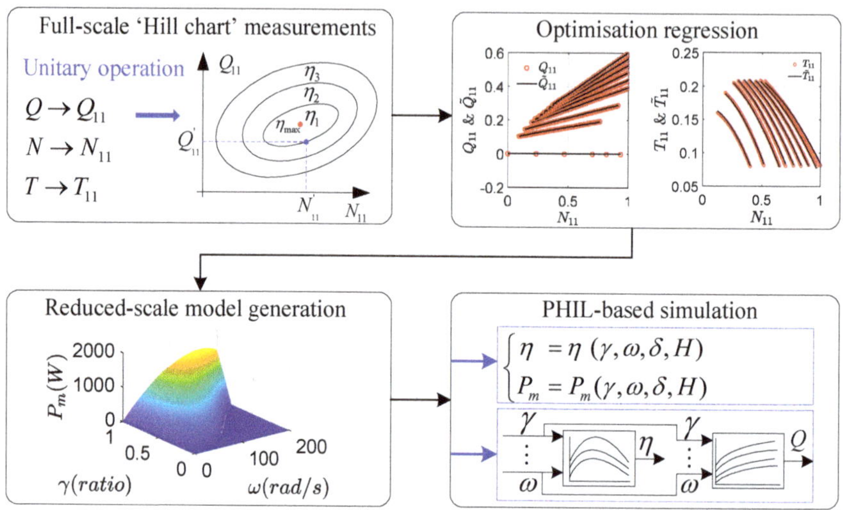



**Figure 1.** Procedure of a 'Hill Charts' based modelling.

### 2.1.1. Static Model Generation

Hydraulic turbines conventionally use the 'Hill Charts' (see Figure 2) to describe the steady-state relationships of the turbine speed $N$ (r/min), the guide vane opening $\gamma$ (ratio) if it exists, the water head $H$ (m), the water flow rate $Q$ (m$^3$/s), and the mechanical torque $T$ (Nm). The unitary variables $N_{11}$, $Q_{11}$, and $T_{11}$ are generally used to represent the 'Hill Charts' in the hydraulic context. The 'Hill Charts' measurements are usually provided by the industrial manufacturers.

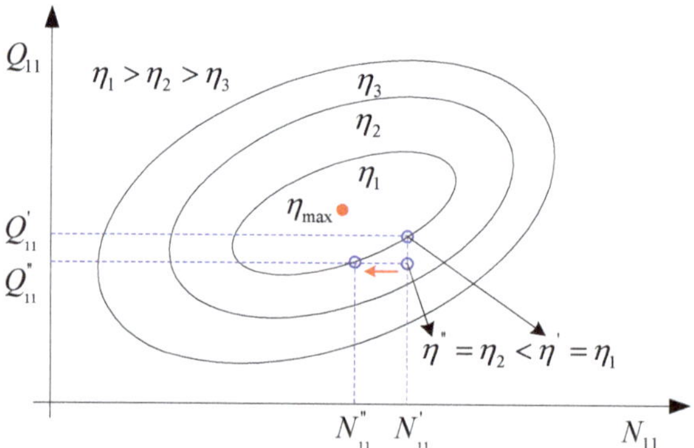



**Figure 2.** Diagram of 'Hill Charts' and the variable speed operation, $\eta$ represents the turbine efficiency.

The variables $N_{11}$, $Q_{11}$, and $T_{11}$ with turbine's diameter $D(m)$ for each guide vane opening $\gamma$ can be derived as follows [35]:

$$N_{11} = \frac{ND}{\sqrt{H}} \; ; \; Q_{11} = \frac{Q}{D^2\sqrt{H}} \; ; \; T_{11} = \frac{T}{D^3 H} \tag{1}$$

Based on the data points $(Q_{11}, T_{11}, N_{11}, \gamma)$ provided by the industrial manufacturer, the regression models $\tilde{Q}_{11}$ and $\tilde{T}_{11}$ are defined by

$$\tilde{Q}_{11}(B_Q, \gamma, N_{11}) = A_Q(\gamma) B_Q C_Q(N_{11})^T \tag{2}$$

$$\tilde{T}_{11}(B_T, \gamma, N_{11}) = A_T(\gamma) B_T C_T(N_{11})^T \tag{3}$$

where:

$$A_Q(\gamma) = \begin{bmatrix} 1 & \gamma \dots \gamma^{m_Q} \end{bmatrix} \in \mathbb{R}^{1 \times (m_Q+1)}, \quad A_T(\gamma) = \begin{bmatrix} 1 & \gamma \dots \gamma^{m_T} \end{bmatrix} \in \mathbb{R}^{m_T+1}$$

$$C_Q(N_{11}) = \begin{bmatrix} 1 & N_{11} \dots N_{11}^{n_Q} \end{bmatrix} \in \mathbb{R}^{1 \times (n_Q+1)}, \quad C_T(N_{11}) = \begin{bmatrix} 1 & N_{11} \dots N_{11}^{n_T} \end{bmatrix} \in \mathbb{R}^{n_T+1}$$

where the parameters $m_Q$, $n_Q$, $m_T$, and $n_T$ need to be chosen. The optimization routine *lsqlin* defined in Matlab is chosen to solve the related optimization problems (4) and (5) to get the regression parameters $B_Q \in \mathbb{R}^{(m_Q+1) \times (n_Q+1)}$ and $B_T \in \mathbb{R}^{(m_T+1) \times (n_T+1)}$. By solving the least square optimization problems (4) and (5), we minimize the mismatch between the chosen regression model and the provided data points, while making sure that, whenever $\gamma = 0$, we get $Q = 0$ regardless of the rotational speed, in addition to ensuring that $T = 0$ when $\gamma = 0$ and $N = 0$:

$$\min_{B_Q} \quad \tfrac{1}{2} \left[ Q_{11} - \tilde{Q}_{11}(B_Q, \gamma, N_{11}) \right]^2$$
$$\text{s.t.} \quad \tilde{Q}_{11}(\gamma = 0) = 0 \tag{4}$$

$$\min_{B_T} \quad \tfrac{1}{2} \left[ T_{11} - \tilde{T}_{11}(B_T, \gamma, N_{11}) \right]^2$$
$$\text{s.t.} \quad \tilde{T}_{11}(\gamma = 0, N_{11} = 0) = 0 \tag{5}$$

The choice of the regression parameters $m_Q$, $n_Q$, $m_T$, and $n_T$ is made such that the relative regression errors can be minimized; meanwhile, the high sensitivity due to over-fitting must be avoided. If higher order polynomials are considered, then we may get undesirable oscillatory behaviours, since it does not represent the nature of the measured data used for regressions and causes higher regression errors. The procedure is as follows: firstly, the parameters $m_Q$, $n_Q$, $m_T$, and $n_T$ can be gradually increased; the regression errors are then computed; the optimal values of $m_Q$, $n_Q$, $m_T$, and $n_T$ are chosen just before the data points start to be over-fitted, which in turn can cause oscillatory models. Figures 3 and 4 show the comparative curves the original unitary industrial data points and the generated regression models with the defined parameters $m_Q = 3$, $n_Q = 2$, $m_T = 2$, and $n_T = 5$.

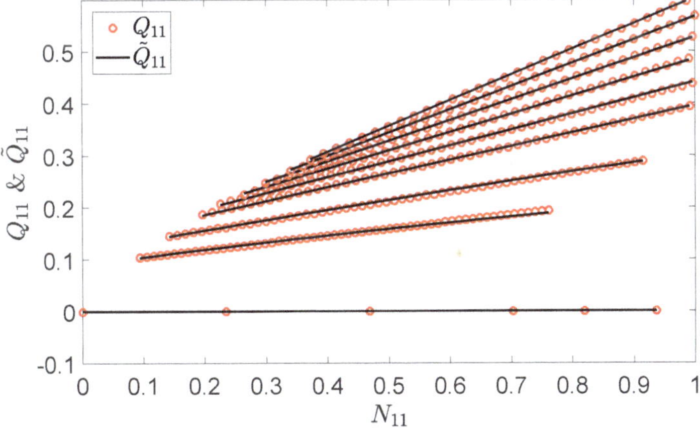


**Figure 3.** $\tilde{Q}_{11}$ vs. $N_{11}$ for different values of $\gamma$.

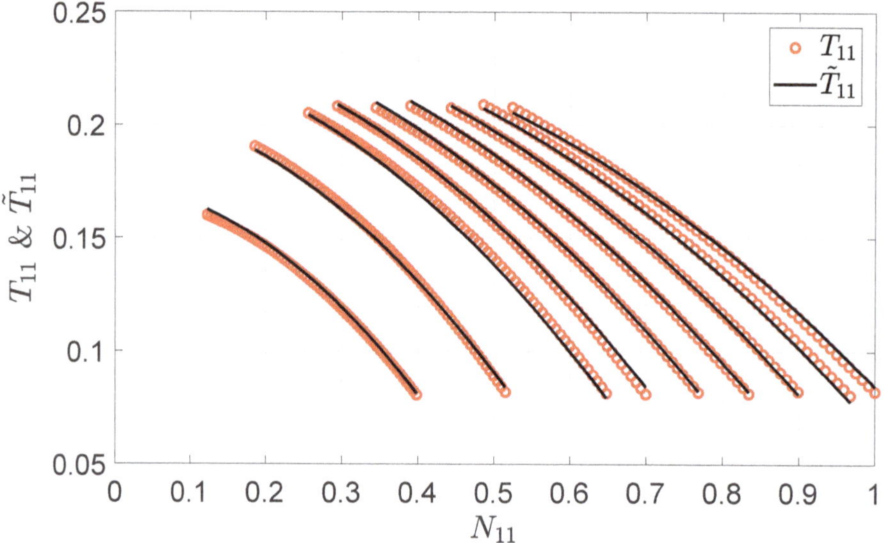

**Figure 4.** $\tilde{T}_{11}$ vs. $N_{11}$ for different values of $\gamma$.

Figures 5 and 6 show the relative regression errors of $Q_{11}$ and $T_{11}$. The relative regression errors are always less than 5% in the range of interest. These optimization models can provide satisfied regression results due to the fact that the main patterns in the 'Hill Charts' have been captured by the chosen functions.

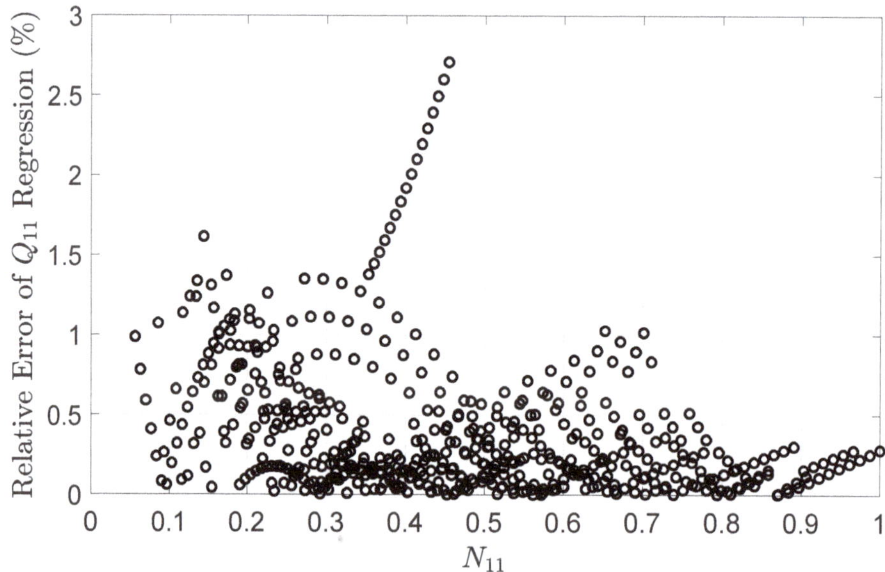

**Figure 5.** Relative error of $Q_{11}$ regression (%).

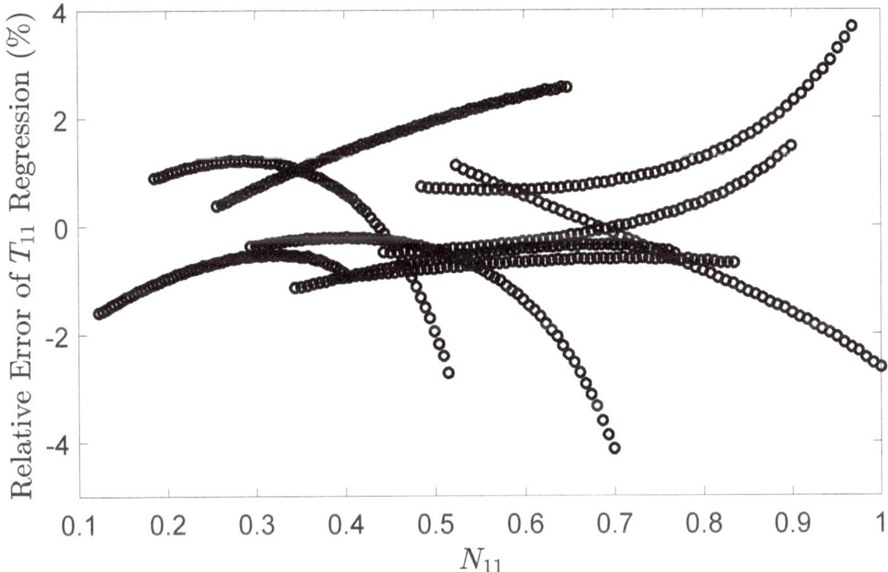

**Figure 6.** Relative error of $T_{11}$ regression (%).

### 2.1.2. Efficiency and Mechanical Power

The mechanical torque and the flow rate can be derived by combining Equations (2) and (3) and the definitions of the unitary variables (1) to get:

$$Q(\gamma, \omega, H, D) = \tilde{Q}_{11} D^2 \sqrt{H} \tag{6}$$

$$T(\gamma, \omega, H, D) = \tilde{T}_{11} D^3 H \tag{7}$$

Once the water head $H$ and the turbine's diameter $D$ are specified, Equations (2)–(7) are then used to compute $Q$ and $T$ for each given $N$ and $\gamma$. The efficiency $\eta$ and the mechanical power $P_m$ (W) are then computed using Equations (8) and (9), for $H = 1$ m and $D = 0.25$ m, which are shown in Figures 7 and 8, respectively:

$$\eta(\gamma, \omega, H, D) = \frac{T(\gamma, \omega, H, D)N}{\rho g H Q(\gamma, \omega, H, D)} \tag{8}$$

$$P_m(\gamma, \omega, H, D) = \eta(\gamma, \omega, H, D)\rho g H Q(\gamma, \omega, H, D) \tag{9}$$

where $\rho$ (kg/m$^3$) is the volume density of water, $g$ (m/s$^2$) is the gravitational acceleration, and $\omega = \frac{2\pi N}{60}$ (rad/s) is the rotational speed.

The water head is supposed to be fixed with respect to the derived static hydraulic features $\eta(\gamma, \omega, H, D)$ and $Q(\gamma, \omega, H, D)$. The reason and its rationality have been analysed in the beginning of Section 2. However, the water head variations are essential to be considered for the hydraulic power computation regarding the variable speed control. The expression (9) is then modified to (10), in which the water head variations $\Delta H$ are taken into account:

$$P_m(\gamma, \omega, H, D) = \eta(\gamma, \omega, H, D)\rho g (H + \Delta H) Q(\gamma, \omega, H, D) \tag{10}$$

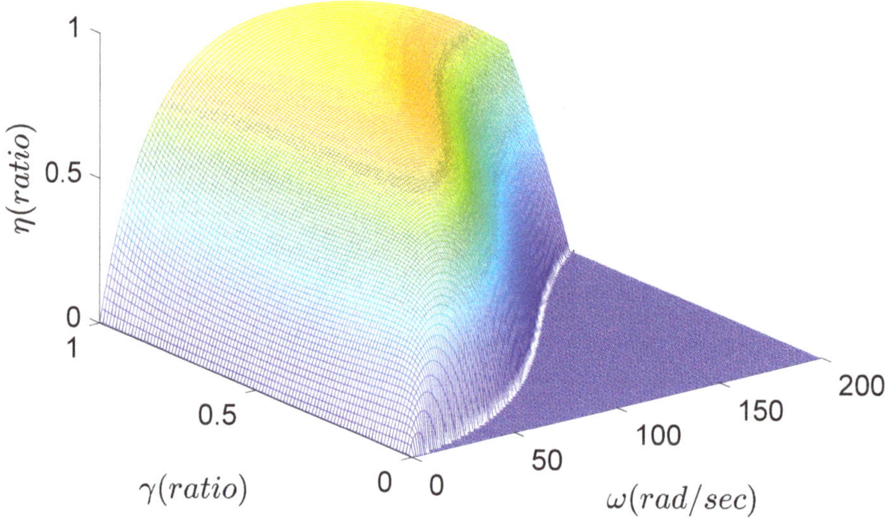

**Figure 7.** $\eta(\gamma, \omega, H, D)$ where negative values are set to zero.

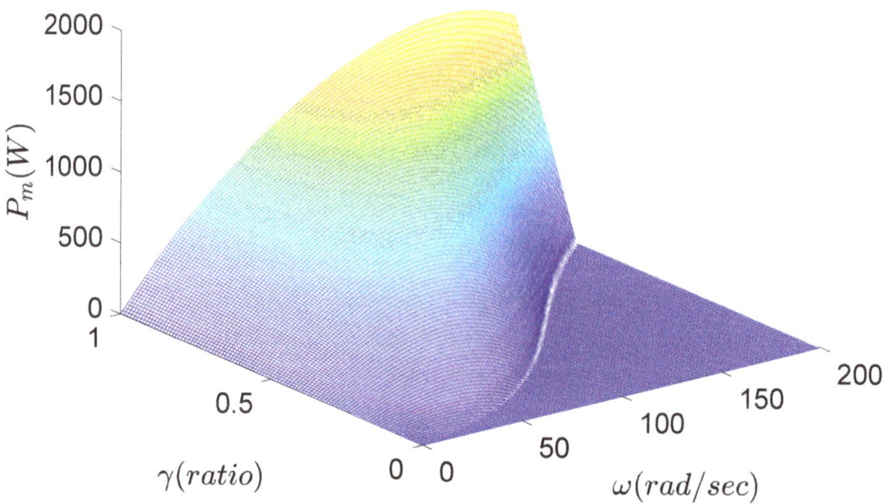

**Figure 8.** $P_m(\gamma, \omega, H, D)$ where negative values are set to zero.

This section provides a flexible philosophy on obtaining mathematical models for a class of hydraulic turbines with 'Hill charts' efficiency characteristics. Different definitions of $\tilde{Q}_{11}$ and $\tilde{T}_{11}$ can be adapted depending on the provided 'Hill Charts' data. The application scope of the proposed modelling approach is thus expanded.

### 2.2. Guide Vane Actuator Dynamic Model

The work in [34] assumes that the guide valve can be opened and closed almost instantaneously. However, it makes sense to consider the guide valve acting process in order to replicate the hydraulic dynamics. The introduced dynamics can help more precisely evaluate how the physical hardware reacts to the dynamics and what would be the impact on the electric control performances [19].

The full nonlinear dynamic model of the guide vane actuator used at GE is schematically presented in Figure 9, which is introduced in this work. The saturation blocks are used to impose limits on the guide vane angle $\gamma$ and the change rate of the guide vane $\dot{\gamma}$, respectively. This model can correctly replicate the nonlinear dynamic behaviours of the guide vane actuator used for a bulb turbine.

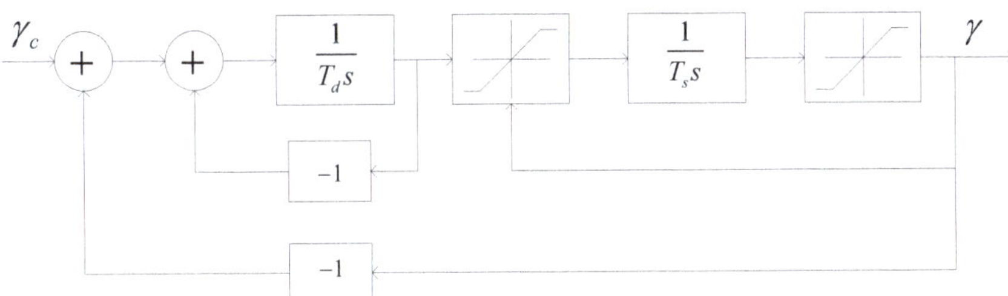

**Figure 9.** Actuator's schematic, $T_s$ and $T_d$ are time constants of the chosen actuator, $\gamma_c$ represents the reference of guide vane opening.

### 2.3. Scaling Operations

#### 2.3.1. Scaling of Hydraulic Turbine

The full size 'Hill charts' data provided by GE have been used to generate the models; these data are related to a 59 MVA bulb turbine rotating at around 8.5 rad/s. The 'Hill charts' can be generated under different working conditions by using the homothetic scaling laws [36,37]. In order to adapt to the laboratory experimental conditions, the diameter of the hydraulic turbine and the water head both can be adjusted, and the resulting regression models can be rescaled as follows:

$$\tilde{Q}_{11}^s(\gamma, \zeta \times N_{11}) = \alpha \times \tilde{Q}_{11} \tag{11}$$

$$\tilde{T}_{11}^s(\gamma, \zeta \times N_{11}) = \beta \times \tilde{T}_{11} \tag{12}$$

where the scaling coefficients $\alpha, \beta, \zeta > 0$.

#### 2.3.2. Scaling of Drive Train

The hydraulic turbine is coupled with the PMSG through the drive train. The dynamic equation between the hydraulic driven torque $T(\gamma, \omega, H, D)$ and the electromagnet torque $T_e$ (Nm) of PMSG can be formulated by

$$J\frac{d\omega}{dt} = T_e - T(\gamma, \omega, H, D) - B\omega \tag{13}$$

where $J$ (kg·m$^2$) is the total inertia, and B (Nm·s) is the friction factor [14].

A Direct Current Motor (DCM) is used to emulate the hydraulic torque in response to the turbine speed (detailed in Section 3). Some similitude concerns must be confirmed to precisely emulate the turbine dynamic model [38]. The original inertia $J$ in Equation (13) needs to be corrected for the emulated system. The scaling rules for a category of turning systems can be found in [38]. The launching time $T_s$ (s) of the turbine emulator is required to be identical to that of the desired turbine as given by

$$T_s = \frac{J\omega_n^2}{P_n} = \frac{J_e\omega_{ne}^2}{P_{ne}} \tag{14}$$

where $J_e$ (kg·m$^2$) is the emulator's inertia, $P_n$ (W), and $P_{ne}$ (W) are the rated power, $\omega_n$ (rad/s), $\omega_{ne}$ (rad/s) are the rated rotational speed, and index '$n$' and '$ne$' indicate rated parameters of the original turbine and its emulator, respectively.

The inertia $J_e$ employed to compute the emulated inertial torque can be corrected as the following:

$$J_e = J \cdot m^2 / n \tag{15}$$

where the scaling factors are respectively given by $m = \omega_n / \omega_{ne}$ and $n = P_n / P_{ne}$ [38].

*2.4. General Comparison with Previous Modellings*

The hydraulic models discussed in Section 1 and the proposed model regarding their implementations on PHIL simulations are generally compared in Table 1, which are discussed from the following aspects: accuracy, computational rapidity, and scalability.

**Table 1.** Comparisons of different modellings for their implementations on PHIL benchmarks.

| Modelling Approaches | Accuracy | Computational Rapidity | Scalability |
|---|---|---|---|
| Linear torque model [10] | Low | High | Low |
| 2D hydraulic model [11,12] | Moderate | Moderate | Low |
| Real-time efficiency model [13] | Moderate | Low | Moderate |
| Look-up table based model [22,23] | Moderate | High | Low |
| Dynamic hydraulic model [25,26] | High | Low | Moderate |
| Previous regression model [27] | Moderate | Moderate | Low |
| Previous 'Hill charts' based model [28] | Moderate | Moderate | Moderate |
| Proposed 'Hill charts' based model | Moderate | Moderate | High |

### 2.4.1. Accuracy Analysis

Compared to the linear torque model [10] and 2D hydraulic models [11,12], more effectors including water flow rate $Q$, guide vane ratio $\gamma$, water head $H$, and turbine speed $\omega$ have been taken into consideration. In addition, we introduce the dynamic model of guide vane actuator, which can help more definitely evaluate the induced effects by the dynamics on the whole control system. Note that the efficiency feature of hydraulic turbine can deteriorate due to the ageing [28]. Then, we can update the models with recent 'Hill Charts' measurements.

### 2.4.2. Computational Rapidity Analysis

Transient dynamics are usually considered for various offline simulation models for hydraulic dynamics study in [25,26]; however, the increased computational burden would be a challenge for real-time simulations. If the execution time devoted to the real-time simulation model exceeds the sampling period, the control system behaviour can be deteriorated. Consequently, the complexity of hydraulic model has to be sacrificed in order to achieve a rapid computational speed. In fact, the hydraulic transient dynamics are much slower compared to that of the electrical control; therefore, there are no great effects induced when the transient dynamics are neglected.

### 2.4.3. Scalability Issues

The scalability refers to the analyticity, the flexible scaling, and the application field. In this work, an analytical modelling approach is proposed by taking use of optimization routines. By following the similarity laws, reduced-scale models are flexibly established for various laboratory operation conditions, the modelling flexibility is thus improved. In addition, this approach is based on 'Hill Charts' data, which can be adapted to various types of hydraulic turbines but not be limited to a particular turbine type. Furthermore, either a continuous model or a discrete model can be chosen, which depends on its application fields.

It can be remarked that the proposed modelling approach enables positive features for their implementations on VS-HEP PHIL real-time simulations systems: decent accuracy, light computational complexity, and high scalability.

## 3. PHIL Real-Time Simulation System

Based on the flexible hybrid PHIL benchmark in the laboratory [19], a VS-HEP test rig is built as shown in Figure 10, which is adapted to the obtained reduced-scale hydraulic model. A systematic PHIL design procedure implemented to wind power systems is described in [19], whose main design steps can be followed. The global design schematic is illustrated in Figure 10a. The experiential benchmark under test is shown in Figure 10b. The primary elements are:

- Real-time Software Simulator (RTSS): the obtained reduced-scale mathematical model of a concerned hydraulic turbine
- Actuator: Direct Current Motor (DCM) and its drive
- Device under test: the controller (dSPACE controller)
- Physical system: the PMSG, back-to-back power electronics converters, the power grid, and measurement sensors

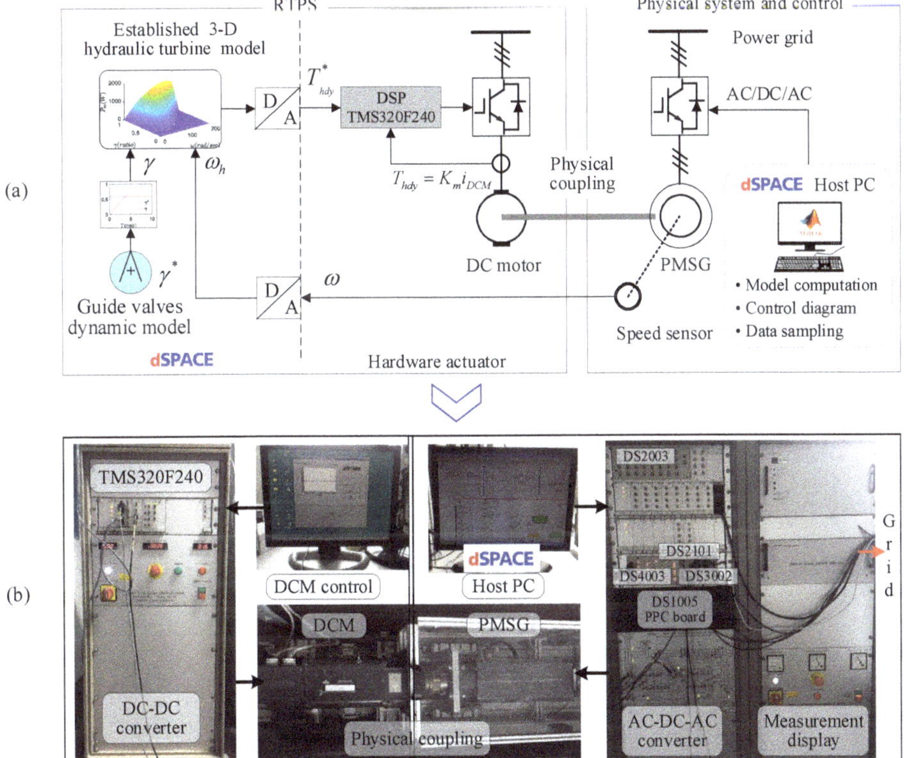


**Figure 10.** PHIL-based hybrid real-time VS-HEP simulation system, (**a**) global design diagram, with $T_{hdy}$ the emulated hydraulic torque, $K_m$ the torque constant, $i_{DCM}$ the direct current of DCM, (**b**) PHIL experiment benchmark

Furthermore, the detailed descriptions of the main subsystems illustrated in Figure 10 will be respectively provided in the following subsections.

### 3.1. dSPACE Modular, dSPACE Controller, and dSPACE Environment

The dSPACE as a rapid-prototyping system allows the implementation of numerical models and any control structures for energetic conversion systems [39,40]. In this work, the model of hydraulic turbine and the control diagram are both achieved with the dSPACE real-time digital simulator based on the MATLAB/Simulink environment. The processor of the dSPACE card is dedicated to compute the model and synchronised on a timer; this allows for running the model in real time [41]. Moreover, some extra hardware interfaces are provided to facilitate the design of hardware prototype, such as Analog to Digital (ADC), Digital to Analog (DAC), Pulse Width Modulation (PWM), Encoder, Resolver, etc. In addition, dSPACE provides add-in toolboxes to Matlab/Simulink software which allows for easy setting of interface cards and easy development of control system in a Simulink environment [39,40].

The dSPACE system under use is composed of a processor card and a number of input/output cards. From the dSPACE host PC, the system can convert the Simulink models into real-time executable codes via the Real-Time Workshop. Then, the codes are executed on one or several processors [41]. The RTI dSPACE 1005 system is currently used in the laboratory, which consists of a DS1005 PowerPC processor card (as the motherboard) that operates at 480 MHz, a DS2003 measurement acquisition card with 32 analog inputs, a DS2101 display card with five analog outputs, a DS3002 speed card with six high resolution inputs for incremental encoders, a DS4003 card with 96 logic inputs/outputs, and finally a DS5101 card with 16 PWM outputs of a resolution 25 ns [41]. The parameters of the dSPACE real-time digital simulator in our work have been configured as follows: the discretization time step Ts = 0.1 ms, the analogue outputs of $\pm 10$ V, and the computation method ode1 (Euler).

More details of the controls development and implementation with the dSPACE software under Matlab/Simulink environment can be found in [39–42].

### 3.2. Real-Time Physical Emulator

Real-Time Physical Emulator (RTPE) is the core module of the PHIL benchmark, which includes two key elements: the RTSS and the actuator (see Figure 10). The RTSS can digitally emulate the hydraulic behaviours by using the obtained mathematical model. In addition, the computed torque and the measured rotational speed information exchange at the interface. The RTSS computes the torque reference and sends it to the DCM control system, and the coupling turbine speed value is then sent back to the RTSS. The hydraulic model is generated in the Matlab/Simulink environment with the dSPACE modular (DS1005). The torque control is implemented in the DSP TMS320F240. The DC/DC power electronics converter is composed of a four-quadrant Pulse Width Module (PWM) IGBT chopper connected with a diode rectifier [19].

The RTPE could correctly emulate the hydraulic physical dynamics and the control algorithms could be executed in real time, and the following conditions must be met:

(1) The computational time devoted to the RTSS needs to be shorter than that of DCM controls in order to replicate hydraulic physical dynamics and static features. The dSPACE modular used in the simulation system has not been renewed for decades due to its high costs. The computational capability is thus even weaker. This is also one of the reasons that some nonlinear transient hydraulic dynamics are neglected for practical considerations.

(2) The torque control dynamics must be faster than the real physical dynamics. The driven torque of DCM is proportional to the induced current; therefore, a direct current controller can be employed to regulate the emulated torque [38]. As the current control relates to the electrical control category, which is sufficiently faster than mechanical dynamics.

(3) The overall execution time of the complete simulation system (the computational time of hydraulic model, the execution time of control laws, the time on ADC and DAC process, etc.) must be shorter than the sampling period of the simulation diagram under the dSPACE environment. Otherwise, if the computational demands of the overall program make the processor take more time than the sampling period, an overrun condition would happen. Then, we could not ensure real-time execution of the control program [40].

### 3.3. Control Laws under Test

The controllers under test are running with the Matlab/Simulink based on the dSPACE DS1005. The control input is then converted to switch commands for the power electronics converters in real time [19]. The digital variables and the analog measurements are exchanged via the I/O interface.

The control diagram of the grid-connected VS-HEP is presented in Figure 11. In the figure, $i_g(a,b,c)$ represents the grid currents, $i_m(a,b,c)$ are the machine currents, $v_g(a,b,c)$ are the grid voltages, $i_{gd}^*$ and $i_{gq}^*$ are the $d$-axis and the $q$-axis grid currents references, respectively, $i_{md}^*$ and $i_{mq}^*$ are the $d$-axis and the $q$-axis machine currents references, respectively, $P_g$ is the grid-injected active power, $\omega$ is the rotational speed, $\omega^*$ is the rotational speed reference, $L_f$ and $R_f$ are the filter inductance and the filter resistance, respectively, $V_{dc}$ is the DC-link voltage, and $C$ is the DC-link capacitance.

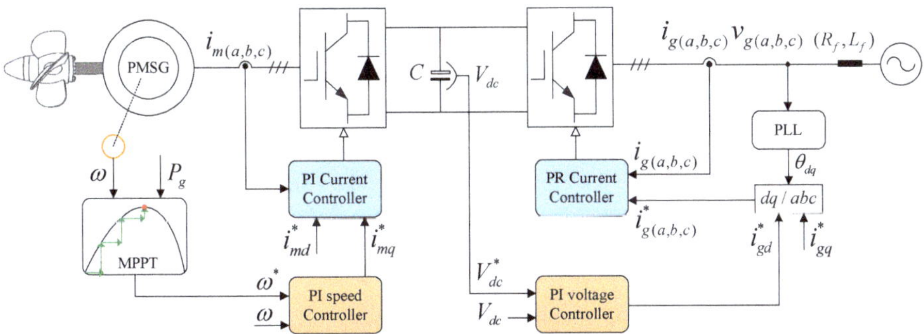


**Figure 11.** Control design diagram of grid-connected VS-HEP.

The VS-HEPs have two mostly common operating configurations: the grid following mode and the grid forming mode [43]. This paper is targeting on the grid following case. Then, the control design is mainly involved in two aspects as follows:

- Machine-side converter control: An adaptive maximum power point tracking method proposed in [5] is employed in this paper. The electrical machine is vector-controlled based on the Park model ($dq$ frame). Proportional Integral (PI) controllers are used in both the outer speed loop and the inner current loop [34].
- Grid-side converter control: A dual-loop control structure is commonly designed for grid-side controls [5,34]. A Phase-Locked Loop (PLL) is employed to obtain the phase of grid voltage [44]. The outer voltage loop maintains a constant DC-link voltage, where the PI controller is employed. The inner current loop is controlled by three Proportional Resonant (PR) controllers. The PR controller is efficient for the grid-side currents control because the grid frequency used to be around the nominal value (50 Hz or 60 Hz).

In fact, the PI controller can ensure a unit gain and zero phase lag for DC components or very low-varying signals [45]. Otherwise, if the reference signal has AC components, it is almost impossible to ensure the zero steady-state error in terms of its amplitude and phase. Moreover, the error would increase with the increasing frequency. The steady-state error can be reduced by increasing the controller gain. However, some limitations have been proved in previous works [45,46]. By introducing a high control gain, the closed-loop bandwidth would be increased. Finally, the system stability is possible to be deteriorated [46].

A PR controller (16) has a proportional term and a resonant integrating term with the frequencies to be corrected. The resonant term introduces theoretically infinite gains at the resonance frequencies, and the static error is hence eliminated:

$$H_{PR}(s) = k_{pr} + \frac{2k_r s}{s^2 + \omega_0^2} \tag{16}$$

where $\omega_0$ is the resonant frequency, $k_r$ is the integration gain, and $k_{pr}$ is the proportional term. To reduce the noise sensitivity [46], the PR controller applied in this work is modified by integrating the parameter $\omega_c$ as follows:

$$H_{PR}(s) = k_{pr} + \frac{2k_r s}{s^2 + 2\omega_c s + \omega_0^2} \tag{17}$$

The controller achieves infinite gains at the resonance frequencies for expression (16). The design corresponding to the (17) can still achieve a relatively high gain at the defined resonant frequency. The steady-state output phase and magnitude error in the closed-loop system would be approximately zero [46]. Furthermore, the multi-resonant controller (18) can be used to correctly control the currents with harmonics injected [47]:

$$H_{MR} = k_{pr} + \sum_{i=1}^{h} \frac{2 \cdot k_{ir} \cdot s}{s^2 + (\omega_i^2)} \tag{18}$$

with $\omega_i$ being the angle frequency of $i$-order harmonic, and $k_{ir}$ the corresponding integration gain.

Note that, due to the variable speed operation, the frequency of machine currents are not constant but varying with time. Consequently, it is difficult to control machine side currents by using a PR controller with a fixed resonant frequency $\omega_0$ [45]. This is why a PI controller upon the $dq$ frame is more appropriate for the machine side converter control.

Lastly, it is worth mentioning that this work is dedicated to the modelling of hydraulic turbines for its implementation on PHIL benchmark. We have not provided many results regarding the comparisons between PI and PR controllers. More comparative experiments between them therefore will be conducted based on the PHIL real-time benchmark discussed in this work.

### 3.4. Physical System Configurations

The emulated hydraulic shaft by DCM is coupled to a PMSG connecting with power grids through back-to-back power electronics converters as shown in Figure 10b. Some ancillary devices such as voltage, current, and position sensors are included as well [34]. The key parameters of the VS-HEP PHIL simulation system are given in Table 2.

**Table 2.** Parameters of the PHIL experiment benchmark.

| RTPE Parameters | | Parameters of Physical System | | | |
|---|---|---|---|---|---|
| Water head $H$ | 1 m | Armature $R_s$ | 0.17 $\Omega$ | Pole pairs | 4 |
| Turbine diameter $D$ | 0.25 m | d-axis $L_q$ | 0.0017 H | PMSG inertia | 0.004 kg·m$^2$ |
| DCM normal power | 6 kW | q-axis $L_d$ | 0.0019 H | DC bus voltage | 400 V |
| DCM normal speed | 3000 rpm | Magnet flux | 0.11 Wb | Power grid | 179 V, 50 Hz |
| DCM inertia | 0.0275 kg·m$^2$ | Friction factor | 0.01 N·m·s | Switch frequency | 10 kHz |

## 4. PHIL Performance Verification and Analysis

Experiments are conducted to assess performance of the obtained reduced-scale model for its implementation on the built PHIL benchmark. The static hydraulic features are firstly verified to be correctly replicated by the actuator (the DCM). Then, the key parameters regarding dynamics and the dynamic response of the real-time physical emulator are verified as well. The control laws employed to machine-side and grid-side controllers are validated in the end.

### 4.1. Replication of Static Features

In order to obtain the static curves, the opening ratio of guide vane and the turbine speed are set manually. The opening ratio of guide vane is firstly set to a specific value. When the actuator reaches the steady state, the rotational speed is then adjusted. Figure 12 illustrates the emulated hydraulic features and the regression models obtained in Section 2: the flow rate in Figure 12a, the torque in Figure 12b, and the output hydraulic power in Figure 12c. Firstly, the results indicate that the hydraulic model emulated by the RTPE can correctly replicate the established regression models. Then, we can observe some features from Figure 12: the water flow rate $Q$ will decrease when the rotation speed increases for a constant valve opening $\gamma$; the water flow rate and the output torque both will increase when we increase the opening ratio of guide vane $\gamma$; there always exists an optimal rotational speed that maximizes the gained hydraulic power for each given guide vane opening $\gamma$.

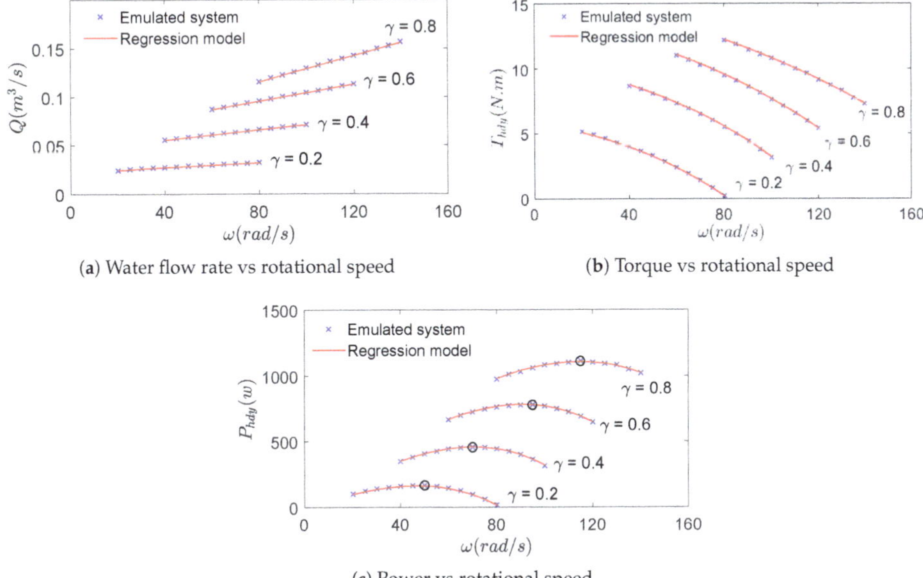


**Figure 12.** Static hydraulic features assessment, (**a**) water flow rate vs rotational speed, (**b**) hydraulic turbine torque vs rotational speed, and (**c**) output hydraulic power vs rotational speed.

### 4.2. Dynamics Verification

In the dSPACE environment, the execution time of each model part can be measured via atomic subsystems, which has been answered in [48]. In a real-time program, there is a variable named 'turnaroundTime' for each task, which can indicate how much time it takes to execute the complete task. The key dynamics parameters are provided in Table 3. Firstly, the execution time taken on the hydraulic model is fast enough compared to the actuator's dynamics and the sampling period. Then, the overall executional time of the complete simulation system (the computational time of hydraulic model, the executional time of employed control laws, the time on digital/analog and analog/digital conversion process, etc.) is shorter than its sampling period. The hydraulic dynamics depend on its capacity, adjusted range, and its device mechanism, which can range from several seconds to minutes [49,50]. The operation time of the guide vane actuator in this work is around 2∼4 s. The torque control of DCM with a settling time of around $4ms$ ensures that the emulated mechanical torque can efficiently replicate the hydraulic dynamics.

**Table 3.** Key parameters regarding dynamics.

| Dynamics Parameters | Time Values |
| --- | --- |
| Executional time of hydraulic model | 4 μs |
| Executional time of whole PHIL system | 0.08 ms |
| Sampling period | 0.1 ms |
| Settling time of DCM current control | 4 ms |
| Dynamics of guide vane actuator | 2∼4 s |

The hydraulic dynamic features in the offline simulation in Matlab/Simulink and the PHIL real-time simulation are presented in Figures 13 and 14, respectively. A soft starting is employed in the beginning; the variable speed control starts at $t = 40$ s, with the guide vane opening of $\gamma = [0.5, 0.6, 0.5]$. The two figures show that the emulated hydraulic torque $T_{hdy}$, the water flow rate $Q$, and the output hydraulic power $P_{hdy}$ in real time could achieve similar dynamic features as that of offline simulations. However, we notice some differences on the dynamic behaviour; the most relevant difference is dealing with the overshoot that affects the torque. Two possible reasons could have caused this difference. Firstly, our PHIL implementation is based on the existed flexible hybrid PHIL benchmark in the laboratory. The reason could be coming from the actuator which emulates the turbine, i.e., the DC motor of Figure 10, as its internal control is not accessible, we cannot change its internal controllers. Another reason could be the aging of the physical system. The offline simulation models (PMSG, converters, line filter, etc.) are defined by applying the parameters originally provided by the manufacturer. However, the hardware system under test has been used for many years. The parameters of the PMSG, the values of line inductance and resistance, etc. could have changed due to the aging factor. The parameters shifts could cause the differences on the control dynamics response. In fact, this is also one of the critical reasons why the control laws must be validated with the real-time PHIL benchmark. Moreover, compared to offline simulations, one of the simulated parts is replaced by the hardware devices in the PHIL system, and the real physical constraints are thus taken into account in the real-time simulation loop. This would help more precisely evaluate the whole electric performances.

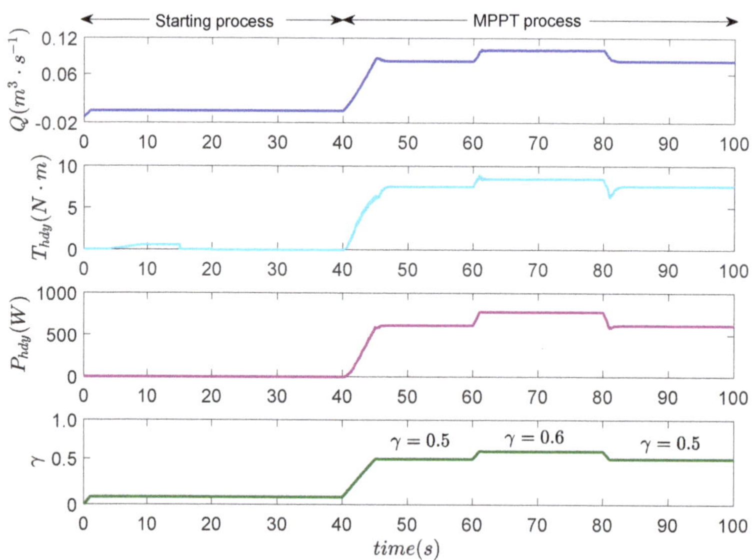



**Figure 13.** Hydraulic dynamics process in the offline simulation.

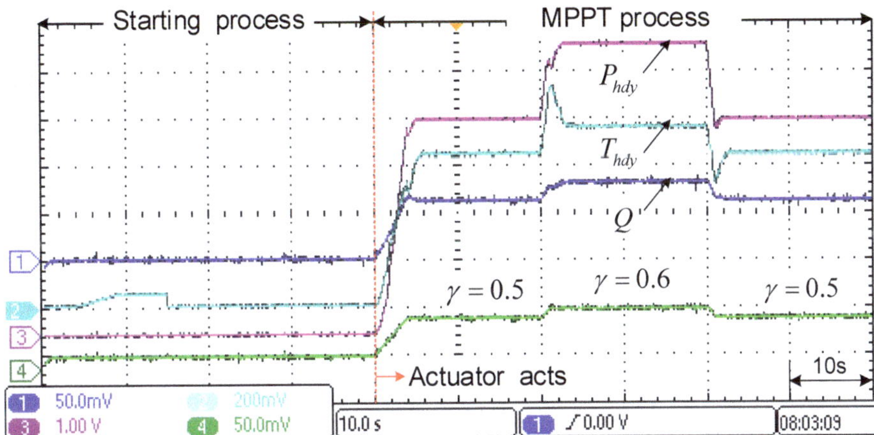


**Figure 14.** Hydraulic dynamics process in the real-time simulation, $P_{hdy}$ the generated hydraulic power (100 w/div), $T_{hdy}$ the turbine torque (2 N·m/div), $Q$ the water flow rate (0.05 m$^3$/s/div).

### 4.3. Control Laws Test

Figure 15 presents the machine-side control performances under the built PHIL benchmark with/without introducing the dynamic model of guide vane opening. In the control design, a soft-starting process is considered in order to smoothly start the machine as follows:

- Firstly, a very small initial torque is increasingly added to overcome its friction torque, and the rotational speed goes up slowly;
- Then, the speed controllers start working under no-load state at $t = 9$ s, and the speed soon reaches the reference value;
- At $t = 40$ s, the variable speed control mode is enabled and the rotational speed changes with guide vane opening adjustments in order to maximize the grid-injected power.

Compared to Figure 15a, the actuator dynamics process could deteriorate both speed and current control performances as shown in Figure 15b. Thus, it makes sense to assess the effect of the induced dynamics on the electrical-side control performance.

The grid-side control results are presented in Figures 16 and 17. In the soft staring process, the DC-link voltage increases until reaching the steady state; no hydraulic power is sent to the grid until the guide vane actuator acts. In the Maximum Power Point Tracking (MPPT) process, the active power $P_g$ and reactive power $Q_g$ (controlled to be zero) injected to the power grid change with the adjustment of guide vane opening; the DC-link voltage $V_{dc}$ almost stays constant. Figure 17 shows the control performance of grid-injected currents. In the beginning, the injected sinusoidal current synchronizes with the grid voltage of phase A well. Then, the reactive current reference steps from 0 A to 2 A, and a phase shift occurs between the grid voltage and the injected current. The grid-injected currents are correctly controlled by using PR controllers.

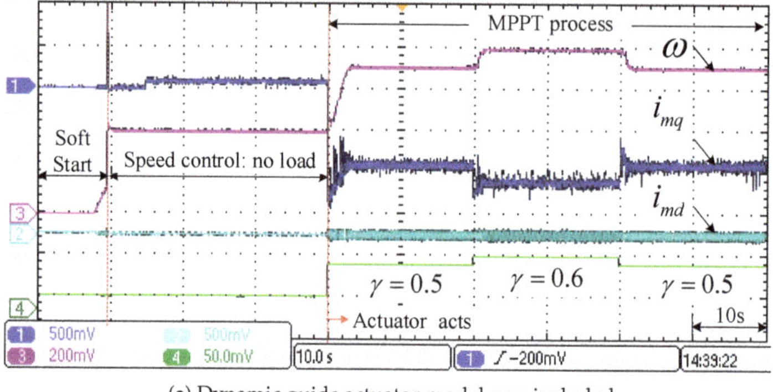


(**a**) Dynamic guide actuator model non-included

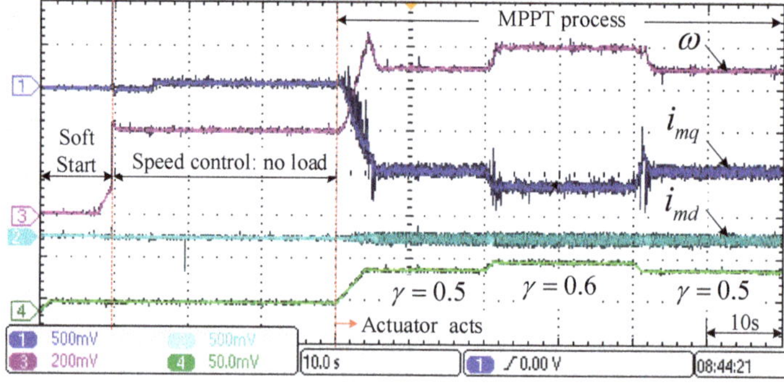


(**b**) Dynamic guide actuator model included

**Figure 15.** Machine-side control performances with/without introducing guide vane opening dynamic model, the $q$-axis current $i_{mq}$ (5 A/div), the $d$-axis current $i_{md}$ (5 A/div), the rotational speed $\omega$ (20 rad/s/div).

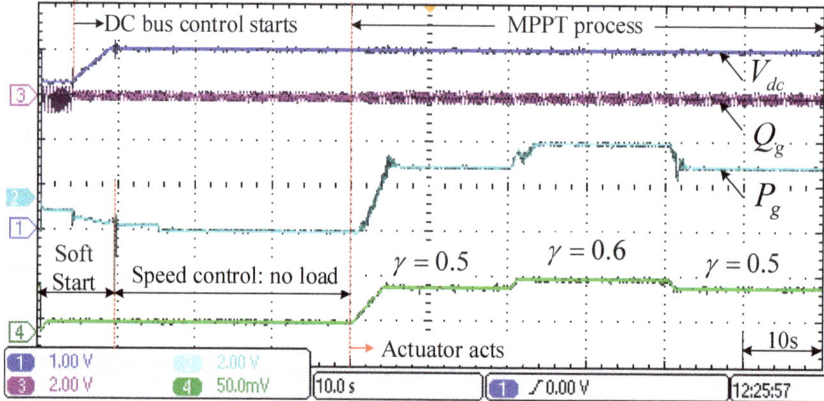


**Figure 16.** Grid-side control performance, the DC-link voltage $V_{dc}$ (100 V/div), the active power $P_g$ (200 W/div), the reactive power $Q_g$ (200 VarA/div).

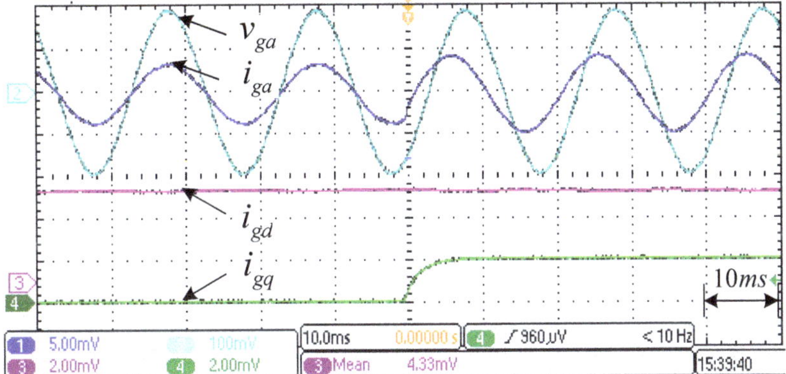

**Figure 17.** Grid-injected current control, $i_{ga}$ the grid current of phase A (5 A/div), $v_{ga}$ the grid voltage of phase A (100 V/div), $i_{gq}$ the reactive current (2 A/div), $i_{gd}$ the active current (2 A/div).

## 5. Conclusions

In this paper, a reduced-scale hydraulic modelling approach is proposed for achieving efficient PHIL simulations. The static features are modelled based on the 'Hill Charts' measurements, which are analytically generated by using optimization routines. Furthermore, the nonlinear dynamics of the actuator model are introduced in order to evaluate the effects of the induced dynamics on the whole electric performance. Moreover, the reduced-scale models can be established for different laboratory conditions. The suggested modelling approach enabled positive features for its implementations on PHIL simulation systems: decent accuracy, light computational complexity, and high scalability.

A bulb turbine coupled to a PMSG with back-to-back PWM converters is chosen as an example for PHIL design and performance assessment. Some conclusions could be highlighted based on the experimental results: Firstly, the RTPE can correctly replicate the hydraulic static features. In addition, the computational time taken on the RTPE is rapid enough to ensure that the emulated turbine torque can efficiently track the hydraulic dynamics. The controls of variable speed operation and the grid-side integration are validated under the established PHIL testing benchmark in the end. The established benchmark can be used for advanced control techniques study of VS-HEPs. Furthermore, the flexible PHIL real-time test benchmark can be adapted to different hydraulic regimes such as pumped-storage hydropower plant for future research.

**Author Contributions:** B.G. wrote this article, designed the PHIL simulation system method, implemented the hardware platform, and conducted the experimental tests and validations. A.M. contributed to the mathematical model of hydraulic turbine. S.B. and M.A. supervised, reviewed, and checked the article manuscript. C.B. helped to confirm the descriptions of dSPACE and the physical system. J.P. helped to acquire the financial support for the project leading to this publication and revised the paper. All authors have read and agreed to the published version of the manuscript.

**Funding:** This research is supported by PSPC Innov'hydro project, which brings together General Electric (GE) Renewable Energy, EDF (Électricité de France), Grenoble INP, and other key players in the hydroelectric sector, in France. The publication is financed by the School of Engineering, HES-SO, Valais, Switzerland.

**Conflicts of Interest:** The authors declare no conflict of interest.

## References

1. Joseph, A.; Chelliah, T.R. A Review of Power Electronic Converters for Variable Speed Pumped Storage Plants: Configurations, Operational Challenges, and Future Scopes. *IEEE J. Emerg. Sel. Top. Power Electron.* **2018**, *6*, 103–119. [CrossRef]
2. Valavi, M.; Nysveen, A. Variable-speed operation of hydropower plants: Past, present, and future. In Proceedings of the 2016 XXII International Conference on Electrical Machines (ICEM), Lausanne, Switzerland, 4–7 September 2016; pp. 640–646, [CrossRef]

3. Guo, B.; Bacha, S.; Alamir, M.; Villamil, A.O. Optimal Management of Variable Speed Pumped-Storage Hydro-Electric Plant: Cases Study. In Proceedings of the 2019 IEEE International Conference on Industrial Technology (ICIT), Melbourne, Australia, 13–15 February 2019; pp. 1119–1125. [CrossRef]

4. Sheibani, M.R.; Yousefi, G.R.; Latify, M.A.; Hacopian Dolatabadi, S. Energy storage system expansion planning in power systems: A review. *IET Renew. Power Gener.* **2018**, *12*, 1203–1221. [CrossRef]

5. Belhadji, L.; Bacha, S.; Munteanu, I.; Rumeau, A.; Roye, D. Adaptive MPPT Applied to Variable-Speed Microhydropower Plant. *IEEE Trans. Energy Convers.* **2013**, *28*, 34–43. [CrossRef]

6. Bortoni, E.; Souza, Z.D.; Viana, A.; Villa-Nova, H.; Rezek, N.; Pinto, L.; Siniscalchi, R.; Bragança, R.; Bernardes, J. The Benefits of Variable Speed Operation in Hydropower Plants Driven by Francis Turbines. *Energies* **2019**, *12*, 3719. [CrossRef]

7. Nababan, S.; Muljadi, E.; Blaabjerg, F. An overview of power topologies for micro-hydro turbines. In Proceedings of the 2012 3rd IEEE International Symposium on Power Electronics for Distributed Generation Systems (PEDG), Aalborg, Denmark, 25–28 June 2012; pp. 737–744. [CrossRef]

8. Singh, R.R.; Chelliah, T.R.; Agarwal, P. Power electronics in hydro electric energy systems—A review. *Renew. Sustain. Energy Rev.* **2014**, *32*, 944–959. [CrossRef]

9. Musa, M.; Ab Razak, J.; Tahir, M.M.; Mohamad, I.S.; Nazri, M. Small scale hydro turbines for sustainable rural electrification program. *J. Adv. Res. Fluid Mech. Termal Sci.* **2018**, *49*, 138–145.

10. Ansel, A.; Robyns, B. Modelling and simulation of an autonomous variable speed micro hydropower station. *Math. Comput. Simul.* **2006**, *71*, 320–332. [CrossRef]

11. Márquez, J.; Molina, M.; Pacas, J. Dynamic modeling, simulation and control design of an advanced micro-hydro power plant for distributed generation applications. *Int. J. Hydrog. Energy* **2010**, *35*, 5772–5777. [CrossRef]

12. de Mesquita, L.M.O.; dos Santos Menas, J.; van Emmenk, E.L.; Aredes, M. Maximum power point tracking applied on small hydroelectric power plants. In Proceedings of the 2011 International Conference on Electrical Machines and Systems, Beijing, China, 20–23 August 2011; pp. 1–6. [CrossRef]

13. Borkowski, D. Maximum Efficiency Point Tracking (MEPT) for Variable Speed Small Hydropower Plant with Neural Network Based Estimation of Turbine Discharge. *IEEE Trans. Energy Convers.* **2017**, *32*, 1090–1098. [CrossRef]

14. Guo, B.; Bacha, S.; Alamir, M.; Mohamed, A.; Boudinet, C. LADRC applied to variable speed micro-hydro plants: Experimental validation. *Control Eng. Pract.* **2019**, *85*, 290–298. [CrossRef]

15. Guo, B.; Bacha, S.; Alamir, M.; Iman-Eini, H. A Robust LESO-based DC-Link Voltage Controller for Variable Speed Hydro-Electric Plants. In Proceedings of the 2019 IEEE International Conference on Industrial Technology (ICIT), Melbourne, Australia, 13–15 February 2019; pp. 361–366.

16. Krishnakumar R., V.; Vigna, K.R.; Gomathi, V.; Ekanayake, J.; Tiong, S. Modelling and simulation of variable speed pico hydel energy storage system for microgrid applications. *J. Energy Storage* **2019**, *24*, 100808. [CrossRef]

17. Reigstad, T.I.; Uhlen, K. Variable Speed Hydropower Conversion and Control. *IEEE Trans. Energy Convers.* **2020**, *35*, 386–393. [CrossRef]

18. Montoya, J.; Brandl, R.; Vishwanath, K.; Johnson, J.; Darbali-Zamora, R.; Summers, A.; Hashimoto, J.; Kikusato, H.; Ustun, T.S.; Ninad, N.; et al. Advanced Laboratory Testing Methods Using Real-Time Simulation and Hardware-in-the-Loop Techniques: A Survey of Smart Grid International Research Facility Network Activities. *Energies* **2020**, *13*, 3267. [CrossRef]

19. Munteanu, I.; Bratcu, A.I.; Bacha, S.; Roye, D.; Guiraud, J. Hardware-in-the-Loop-based Simulator for a Class of Variable-speed Wind Energy Conversion Systems: Design and Performance Assessment. *IEEE Trans. Energy Convers.* **2010**, *25*, 564–576. [CrossRef]

20. Elghali, S.E.B.; Balme, R.; Le Saux, K.; Benbouzid, M.E.H.; Charpentier, J.F.; Hauville, F. A simulation model for the evaluation of the electrical power potential harnessed by a marine current turbine. *IEEE J. Ocean. Eng.* **2007**, *32*, 786–797. [CrossRef]

21. Guo, B.; Bacha, S.; Alamir, M.; Mohamed, A.T. *Variable Speed Micro-Hydro Power Generation System: Review And Experimental Results*; SGE 2018-3ème édition du Symposium de Génie Electrique; Génie Electrique: Nancy, France, 2018.

22. Borghetti, A.; Di Silvestro, M.; Naldi, G.; Paolone, M.; Alberti, M. Maximum efficiency point tracking for adjustable-speed small hydro power plant. In Proceedings of the 16th PSCC, Glasgow, Scotland, 14–18 July 2008.

23. Perez-Diaz, J.I.; Fraile-Ardanuy, J. Neural networks for optimal operation of a run-of-river adjustable speed hydro power plant with axial-flow propeller turbine. Control and Automation. In Proceedings of the 2008 16th Mediterranean Conference on Control and Automation, Las Vegas, NV, USA, 31 March–3 April 2008; pp. 309–314.

24. Shitahun, A.; Ruge, V.; Gebremedhin, M.; Bachmann, B.; Eriksson, L.; Andersson, J.; Diehl, M.; Fritzson, P. Model-Based Dynamic Optimization with OpenModelica and CasADi. In Proceedings of the 7th IFAC Symposium on Advances in Automotive Control, Tokyo, Japan, 4–7 September 2013; Volume 46, pp. 446–451. [CrossRef]

25. Chen, D.; Ding, C.; Ma, X.; Yuan, P.; Ba, D. Nonlinear dynamical analysis of hydro-turbine governing system with a surge tank. *Appl. Math. Model.* **2013**, *37*, 7611–7623. [CrossRef]

26. Mello, C.; KOESSLER, R. Hydraulic turbine and turbine control models for system dynamic studies. *IEEE Trans. Power Syst.* **1992**, *7*, 167–179.

27. Borkowski, D. Laboratory Model of Small Hydropower Plant with Variable Speed Operation. *Masz. Elektr. Zesz. Probl.* **2013**, *Nr 3*, 27–32.

28. Borkowski, D. Analytical Model of Small Hydropower Plant Working at Variable Speed. *IEEE Trans. Energy Convers.* **2018**, *33*, 1886–1894. [CrossRef]

29. Nielsen, T.K. Simulation model for francis and reversible pump turbines. *Int. J. Fluid Mach. Syst.* **2015**, *8*, 169–182. [CrossRef]

30. Yadav, O.; Kishor, N.; Fraile-Ardanuy, J.; Mohanty, S.R.; Pérez, J.I.; Sarasúa, J.I. Pond head level control in a run-of-river hydro power plant using fuzzy controller. In Proceedings of the 2011 16th International Conference on Intelligent System Applications to Power Systems, Hersonissos, Greece, 25–28 September 2011; pp. 1–5. [CrossRef]

31. Kaunda, C.S.; Kimambo, C.Z.; Nielsen, T.K. Hydropower in the Context of Sustainable Energy Supply: A Review of Technologies and Challenges. *ISRN Renew. Energy* **2012**. [CrossRef]

32. Okot, D.K. Review of small hydropower technology. *Renew. Sustain. Energy Rev.* **2013**, *26*, 515–520.[CrossRef]

33. Li, Y.; Song, G.; Yan, Y. Transient hydrodynamic analysis of the transition process of bulb hydraulic turbine. *Adv. Eng. Softw.* **2015**, *90*, 152–158. [CrossRef]

34. Guo, B.; Mohamed, A.; Bacha, S.; Alamir, M. Variable speed micro-hydro power plant: Modelling, losses analysis, and experiment validation. In Proceedings of the 2018 IEEE International Conference on Industrial Technology (ICIT), Lyon, France, 19–22 February 2018; pp. 1079–1084. [CrossRef]

35. Mesnage, H. Modélisation et ContrÔle Avancé pour les Centrales de Turbinage de Moyenne et Haute Chute. Ph.D. Thesis, Université Grenoble Alpes (UGA), Grenoble, France, 2017.

36. Negoiţă, G.C.; Maré, J.C.; Budinger, M.; Vasiliu, N. Scale-laws based hydraulic pumps parameter estimation. *UPB Sci. Bull.* **2012**, *74*, 199–208.

37. Briggs, M.J. *Basics of Physical Modeling in Coastal and Hydraulic Engineering*; Technical Report; US Army Corps of Engineers: Washington, DC, USA, 2013.

38. Munteanu, I.; Bratcu, A.I.; Andreica, M.; Bacha, S.; Roye, D.; Guiraud, J. A new method of real-time physical simulation of prime movers used in energy conversion chains. *Simul. Model. Pract. Theory* **2010**, *18*, 1342–1354. [CrossRef]

39. Quijano, N.; Passino, K.; Jogi, S. *A Tutorial Introduction to Control Systems Development and Implementation with dSPACE*; Tutorial; The Ohio State University: Columbus, OH, USA, 2002.

40. Ghaffari, A. *dSPACE and Real-Time Interface in Simulink*; Department of Electrical and Computer Engineering, San Diego State University: San Diego, CA, USA, 2012.

41. Lam, Q.L. Advanced Control of Microgrids for Frequency and Voltage Stability: Robust Control Co-Design and Real-Time Validation. Ph.D. Dissertation, Université Grenoble Alpes, Grenoble, France, 2018.

42. dSPACE Support. *DS1104 Controller Board, Hardware Installation and Configuration*; dSPACE Support: Cambridge, MA, USA, 2004.

43. Andreica, M.; Bacha, S.; Roye, D.; Exteberria-Otadui, I.; Munteanu, I. Micro-hydro water current turbine control for grid connected or islanding operation. In Proceedings of the 2008 IEEE Power Electronics Specialists Conference, Rhodes, Greece, 15–19 June 2008; pp. 957–962, [CrossRef]

44. Guo, B.; Bacha, S.; Alamir, M.; Hably, A.; Boudinet, C. Generalized Integrator-Extended State Observer With Applications to Grid-Connected Converters in the Presence of Disturbances. *IEEE Trans. Control. Syst. Technol.* **2020**, 1–12. [CrossRef]

45. Bacha, S.; Munteanu, I.; Bratcu, A.I.; others. Power electronic converters modeling and control. *Adv. Textb. Control Signal Process.* **2014**, *454*, 454.

46. Holmes, D.G.; Lipo, T.A.; McGrath, B.P.; Kong, W.Y. Optimized Design of Stationary Frame Three Phase AC Current Regulators. *IEEE Trans. Power Electron.* **2009**, *24*, 2417–2426. [CrossRef]

47. De Heredia, A.L.; Gaztanaga, H.; Etxeberria-Otadui, I.; Bacha, S.; Guillaud, X. Analysis of Multi-Resonant Current Control Structures and Tuning Methods. In Proceedings of the IECON 2006—32nd Annual Conference on IEEE Industrial Electronics, Paris, France, 7–10 November 2006; pp. 2156–2161. [CrossRef]

48. dSPACE FAQ 023. Measuring Execution Times of Blocks and Subsystems. 2018. Available online: https://www.dspace.com/en/pub/home/support/kb/faqs/faq023.cfm#144_32292 (accessed on 10 September 2020).

49. Farell, C.; Gulliver, J. Hydromechanics of variable speed turbines. *J. Energy Eng.* **1987**, *113*, 1–13. [CrossRef]

50. Li, C.; Zhou, J. Parameters identification of hydraulic turbine governing system using improved gravitational search algorithm. *Energy Convers. Manag.* **2011**, *52*, 374–381. [CrossRef]

**Publisher's Note:** MDPI stays neutral with regard to jurisdictional claims in published maps and institutional affiliations.

*Review*

# Advanced Laboratory Testing Methods Using Real-Time Simulation and Hardware-in-the-Loop Techniques: A Survey of Smart Grid International Research Facility Network Activities

Juan Montoya [1,*], Ron Brandl [1,2], Keerthi Vishwanath [2], Jay Johnson [3], Rachid Darbali-Zamora [3], Adam Summers [3], Jun Hashimoto [4], Hiroshi Kikusato [4], Taha Selim Ustun [4], Nayeem Ninad [5], Estefan Apablaza-Arancibia [5], Jean-Philippe Bérard [5,6], Maxime Rivard [5,6], Syed Qaseem Ali [6], Artjoms Obushevs [7], Kai Heussen [8], Rad Stanev [9], Efren Guillo-Sansano [10], Mazheruddin H. Syed [10], Graeme Burt [10], Changhee Cho [11], Hyeong-Jun Yoo [11], Chandra Prakash Awasthi [12], Kumud Wadhwa [12] and Roland Bründlinger [13]

[1]  Fraunhofer IEE, 34119 Kassel, Germany; ron.brandl@der-lab.net
[2]  DERLab, 34119 Kassel, Germany; keerthi.vishwanath@der-lab.net
[3]  Sandia National Laboratories, Albuquerque, NM 87123, USA; jjohns2@sandia.gov (J.J.); rdarbal@sandia.gov (R.D.-Z.); asummer@sandia.gov (A.S.)
[4]  National Institute of Advanced Industrial Science and Technology, Fukushima 963-0298, Japan; j.hashimoto@aist.go.jp (J.H.); hiroshi-kikusato@aist.go.jp (H.K.); selim.ustun@aist.go.jp (T.S.U.)
[5]  CanmetENERGY, Natural Resources Canada (NRCan), Varennes, QC J3X 1S6, Canada; nayeem.ninad@canada.ca (N.N.); estefan.apablaza-arancibia@canada.ca (E.A.-A.); Jean-Philippe.Berard@opal-rt.com (J.-P.B.); maxime.rivard@opal-rt.com (M.R.)
[6]  OPAL-RT Technologies, Montreal, QC H3K 1G6, Canada; Syed.QaseemAli@opal-rt.com
[7]  ZHAW, IEFE, 8401 Winterthur, Switzerland; artjoms.obusevs@zhaw.ch
[8]  DTU, Kgs. 2800 Lyngby, Denmark; kh@elektro.dtu.dk
[9]  TU-Sofia, 1756 Sofia, Bulgaria; rstanev@tu-sofia.bg
[10]  Institute for Energy and Environment, Electronic and Electrical Engineering Department, University of Strathclyde, Glasgow G1 1XQ, UK; efren.guillo-sansano@strath.ac.uk (E.G.-S.); mazheruddin.syed@strath.ac.uk (M.H.S.); graeme.burt@strath.ac.uk (G.B.)
[11]  KERI, Changwon-si 51543, Korea; chcho@keri.re.kr (C.C.); hjyoo@keri.re.kr (H.-J.Y.)
[12]  Power Grid Corporation of India Limited, Gurgaon 122001, India; chandraprakash@powergridindia.com (C.P.A.); kmd@powergridindia.com (K.W.)
[13]  Austrian Institute of Technology, 1210 Vienna, Austria; roland.bruendlinger@ait.ac.at
*  Correspondence: juan.montoya@iee.fraunhofer.de; Tel.: +49-561-7294-211

Received: 20 May 2020; Accepted: 22 June 2020; Published: 24 June 2020

**Abstract:** The integration of smart grid technologies in interconnected power system networks presents multiple challenges for the power industry and the scientific community. To address these challenges, researchers are creating new methods for the validation of: control, interoperability, reliability of Internet of Things systems, distributed energy resources, modern power equipment for applications covering power system stability, operation, control, and cybersecurity. Novel methods for laboratory testing of electrical power systems incorporate novel simulation techniques spanning real-time simulation, Power Hardware-in-the-Loop, Controller Hardware-in-the-Loop, Power System-in-the-Loop, and co-simulation technologies. These methods directly support the acceleration of electrical systems and power electronics component research by validating technological solutions in high-fidelity environments. In this paper, members of the Survey of Smart Grid International Research Facility Network task on Advanced Laboratory Testing Methods present a review of methods, test procedures, studies, and experiences employing advanced laboratory techniques for validation of range of research and development prototypes and novel power system solutions.

**Keywords:** co-simulation; CHIL; geographically distributed simulations; power system protection and control; holistic testing; lab testing; field testing; PHIL; PSIL; pre-certification; smart grids; standards

---

## 1. Introduction

Increasing deployments of distributed energy resources (DER) and smart grid technologies in cyber–physical energy systems, has driven the scientific community and power industries to develop novel technologies. Such technology allows the energy systems to be operated and controlled more optimally, however, its inclusion brings challenges to the power system stability, operation, control, protection schemes, device interoperability, substation automation, wide-area protection mechanisms, cybersecurity, and additional reliance on Internet of Things (IoT) devices, to name a few examples. In order to ensure a smooth transition to a more distributed grid with higher penetration of renewable energy units, new procedures and methods for testing and validating interoperability, reliability, and stability must be developed.

To address this need, researchers have developed advanced laboratory evaluation methods based on real-time simulation (RTS) and Hardware-in-the-Loop (HIL) including: Power Hardware-in-the-Loop (PHIL), Controller Hardware-in-the-Loop (CHIL), Power System-in-the-Loop (PSIL), and co-simulation that improve the fidelity of smart grid simulation tools. Now physical equipment or controllers can be inserted into real-time (RT) or faster than RT simulations to validate equipment performance characteristics. The development of multi-domain simulation environment is also providing new information about the interactions of cyber–physical environments.

To draw a baseline for the reader, a brief explanation of the definitions of RTS and HIL techniques mentioned in this review are provided:

- RTS is a simulation, which is solely digitally executed in a real-time way on an RTSM.
- Co-simulation is a test setup that combines at least two different software tools executed on one or more computational systems.
- HIL is a test setup that combines a real-time simulated system with a physical hardware component or system, where interfaces with physical and simulated systems enabling closed loop interactions.
- Controller HIL or CHIL is a HIL technique where the sensors and actuators of a physical controller are interfaced with a real-time simulation.
- Power HIL or PHIL is a HIL setup, where at least one of the bi-directional interfaces of a setup exchanges power with real, physical power hardware through a Power Amplifier.
- Power System in-the-Loop or PSIL is a novel HIL concept where more than two domains interface each other in order to perform holistic experiments, e.g., a connection between a virtual simulated system (where RTS and co-simulation occur), a controller component (where CHIL occurs), and physical power system (where PHIL occurs).

### 1.1. Motivation of the Review

This review brought together technical experts from multiple investigation areas to identify state-of-the-art testing method trends. The Smart Grid International Research Facility Network (SIRFN) is a worldwide network of smart grid research and test-bed facilities participating under the Annex 5 of the International Smart Grid Action Network (ISGAN). The testing and evaluation capabilities of SIRFN allows the international community to enable: improved design, implementation, and testing of smart grids and their functionalities, including the reliable integration of clean energy technologies.

The SIRFN task on Advanced Laboratory Testing Methods (ALTM) addresses state-of-the-art testing procedures and develops recommendations on future testing techniques of electrical power systems and their domains. SIRFN implements new recommendations by collaborative activities among test infrastructures and identifies potential common activities for future application of advanced methodologies within the network. This community has a good perspective on the quickly evolving landscape of novel simulation techniques such as RTS (PHIL, CHIL, PSIL, etc.) and is in an excellent position to identify promising upcoming advanced laboratory testing methods for other research teams and the power system industry.

### 1.2. Review Structure

In this survey, state-of-the-art capabilities afforded by new simulation tools are enumerated for applications covering power system design, smart grid control, power electronics, communications, and cybersecurity. Specifically, this review surveyed recent activities of the SIRFN ALTM task participants about ongoing and recently finished ALTM, RTS, and HIL research techniques, including the following topics:

- Interfacing methods of PHIL, CHIL, and PSIL simulation;
- HIL testing of power system protection and control;
- HIL testing of smart grid/microgrid controllers, energy management systems, and power electronic converters;
- Co-simulation and RTS integration;
- Geographically distributed HIL and RTS;
- Industrial experiences and HIL in standardized testing.

Each of these topics forms a separate section, which provides a brief introduction and then reports on specific experiences and activities by the SIRFN ALTM members in the additional subsections. Thereafter, a summary of experiments is organized in a table, gathering the hardware used in the reported activities, to give the reader an overview of the equipment used in HIL experiments. In addition, a grouping of the literature and references used for each section is tabulated and presented. Finally, a conclusion and future outlook is included, thereby the authors express their ideas about the review from the perspective of the SIRFN ALTM community.

## 2. Interfacing Methods of PHIL, CHIL, and PSIL Simulation

### 2.1. Introduction

RTS and PHIL technologies are present in activities covering (a) component testing, (b) power system stability studies (as risk-free and close-to-reality alternative to field tests), and (c) development and verification of new control strategies. Establishing a seamless interface between an RTS model and physical devices is a technical challenge because the design of the interfacing method impacts both stability and accuracy of the RT control system [1,2]. For this, an interface algorithm (IA) is implemented between the real-time simulation machine (RTSM) and power amplifier (PA) to ensure a stable and accurate range of operation.

The typical temporal resolution of a PHIL experiment for electro-mechanical transients (EMT) is in the range of 10–100 microseconds, but the interactions between large-scale and local dynamics as well as wide-area and local control systems increases drastically in proportion with the number of nodes of the system. The large-scale dynamics are "slow" (from milliseconds to seconds) and typically represented by root mean squared (RMS) values. One solution to combine these environments, is to create PHIL environments which represent RMS simulations and convert the fast dynamics of the interfaced power equipment to RMS values. It should be noted though, that many PHIL simulations are executed in the fast domain with time steps between 40–120 microseconds. In case of capturing slow dynamics, the Power System-in-the-Loop (PSIL) testing concept [3] can be implemented to exploit the use of quasi-static and quasi-dynamic [4–6] time domain analysis combined with PHIL.

This section classifies the interfacing methods of PHIL, CHIL, and PSIL simulation in two main topics:

1. **Interface Algorithms for Fast Dynamics:** Nowadays, the accuracy and stability of several IAs have been analyzed in detail [1,2,7–12], including: the ideal transformer model (ITM),

partial circuit duplication (PCD), transmission line model (TLM), damping impedance method (DIM), time variant first-order approximation (TFA), and advanced ideal transformer model (AITM); being the ITM and DIM are the most widely used techniques for connecting power equipment to a PHIL RTS.

2. **PHIL Integration at Slow Time Scales:** For application areas where the concerned dynamics are slow (RMS values calculated using tens to hundreds of cycles), the technical requirements of the HIL integration can be simplified, that reduces hardware cost and improves scalability. Examples of system behaviors and relevant functions in these time scales are: power and energy management, active and reactive power balancing, demand side management, voltage control strategies, and determination of the proximity to operating limits. By the use of the PSIL concept, this review present two methods: (a) quasi-dynamic PHIL in Section 2.2.3 and (b) quasi-static PHIL in Section 2.2.4.

*2.2. Reported Experiences and Activities from SIRFN ALTM Members*

2.2.1. Power Amplifier Characterization for RTS

The PA plays a vital role in PHIL simulations. In order to assess the accuracy and stability of PHIL simulations, accurate modeling of the PA or grid simulator must be available. CanmetENERGY and OPAL-RT$^{TM}$ have been collaborating to develop a grid simulator characterization environment. This environment performs frequency sweeps with different operating voltages. The resulting magnitude and phase response data, generates a number of Bode plots and transfer functions which realistically model the amplifiter behavior. This realistic amplifier model can be used to assess the stability of an RT power system simulation before the actual PHIL study is conducted.

As each amplifier is different, amplitude step tests were performed to determine the intrinsic amplifier delays and the communication delays of the system. Once these delays were identified, they were factored out of the recorded data, and included in the transfer function modeling. Since this modeling work has the ultimate goal of providing an amplifier model for all operating conditions, work remains to find a generic methodology to characterize different types of PAs with sufficient accuracy. This activity is an ongoing research and further results might affect the exact implementation of this method.

2.2.2. Stability and Accuracy Comparison for Different Interfacing Methods

In [12], the voltage ideal transformer method (ITM) and the voltage damping impedance method (DIM) were compared during unity and non-unity power factor setting and curtailed real power levels to a baseline, when connecting a physical photovoltaic (PV) inverter to a PHIL simulation. The ITM and DIM methods generally, in simulation, have a low pass filter (LPF) on the output of the RTS to increase stability of the system when the PV inverter connects. The discrete time step of the simulation and LPF introduce a phase shift on the voltage output to the PA. This phase voltage shift injects artificially created or synthetic reactive power into the simulation at the point of connection in the RTSM. A low pass filter lead compensator (LPF LD) can be used to tune the introduced phase shift to minimize the synthetic reactive power injection in the simulation without removing the characteristics of the device under test (DuT). Results from [12] are shown in Table 1. The average error represents the summation of each individual error compared to a baseline, divided by the total number of tests. The execution cycle is the percent of time spent to compute all tasks of the model. The major computation time is expressed in a percentage of time spent by the model to perform block calculations including discrete and continuous state calculations. Each interface method takes computational effort to implement. Knowing the trade-off between computational usage and model accuracy empowers the user to select the appropriate interfacing method for the simulation.

**Table 1.** Real-time simulation (RTS) interface method using one core.

| Method | Average Error (%) | Execution Cycle (%) | Major Computation Time (%) |
|---|---|---|---|
| DIM LPF | 27.755 | 62.21 | 49.19 |
| DIM LPF LD | 3.303 | 65.99 | 52.16 |
| ITM LPF | 14.940 | 49.20 | 42.71 |
| ITM LPF LD | 4.507 | 49.46 | 42.99 |

### 2.2.3. Quasi-Dynamic PHIL

To perform simulations at the continental scale, the electrical dynamics need to be simplified. When dealing with multiple node power systems, the conventional approach solves a full set of differential equations in RT resulting in a significant computational burden. One option to overcome this issue is the so-called "quasi-dynamic" modeling approach for slow dynamics time domain analysis using modified nonlinear algebraic equations [5,6]. The quasi-dynamic PHIL analysis is typically used for long term power system behavior studies covering the time domains from several minutes to several hours or even days where the fast transients are considered to have settled down (to steady state) within the period of each time step.

To represent the power system dynamics depending on the application, a single or several (suitable for power system split and re-synchronization phenomena studies) mass dynamic models could be additionally used at a lower computational cost. This combination provides fast RT computation of the voltage magnitude, phase, and frequency within each individual node of the power system. This approach could be successfully used for large analyses of the interaction between one or several DuTs and the bulk power system.

The test set up at the Power System Stability Laboratory of the Technical University of Sofia (TU-Sofia) for testing Power System Stability support functions of active micro- and nanogrids, is one typical application of this approach. The Figure 1a shows the laboratory setup of a micro/nanogrid model consisting of: (a) programmable grid forming unit, (b) multi-functional laboratory transformer, (c) physical power line emulator, (d) PV emulator, (e) maximum power point charge controller, (f) bidirectional converter, (g) battery storage, (h) smart load controller, (i) physical model of hydro power plant, (k) motor generator sets with controls emulating different types of generation, and (d) a supervisory control and data acquisition (SCADA) system recording the system parameters.

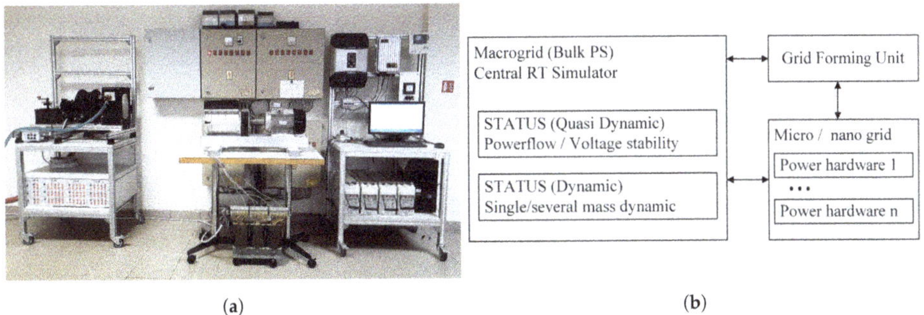


**Figure 1.** Power Hardware-in-the-Loop (PHIL) TU-Sofia Power System Stability Laboratory. (**a**) Micro/ nanogrid lab setup; (**b**) Lab architecture.

While power system stability used to be widely accepted as a Bulk Power System (macrogrid) problem, the micro-, mini-, and nanogrid concepts which are feasible today allow more efficient and reliable system stability support functions based on local and distributed control [13]. To determine the impact on the macrogrid stability and the right settings of this control, a PHIL with the architecture shown in Figure 1b is used. It consists of a central RTSM representing the macrogrid using the STATUS quasi-dynamic and dynamic software. The STATUS quasi-dynamic module provides power flow and voltage stability analysis with a rich set of detailed stability indicators. Combined with the STATUS single/several mass dynamic model the frequency dynamics are given. Through the grid-forming unit the voltage and frequency (U,f) signals are amplified and transferred to the micro/nanogrid power DuT. In each RT time step, the computed power system stability indicators provide feedback on the efficiency of the actions performed by the local control.

Theoretically, the architecture is open and if another "i"-th distinct micro/nanogrid needs to be involved, a (U,f) signal with a time delay corresponding to the phase angle $\delta i$ could be sent from the central RTS to another grid-forming unit.

### 2.2.4. Quasi-Static PHIL

A scale-up on the hardware side of the HIL setup can be achieved, when the requirements of dynamic capabilities of the interfacing converter are reduced. Such a setup is relevant when interactions among several hardware components (e.g., DER) in a physical low-voltage distribution network, need to be investigated when they are coupled to medium and high voltage network dynamics. The PSIL concept [3] can be used for: use cases with slow voltage dynamics and interactions among DER controllers, inverters and upstream controls, and dispatch of coordinated voltage control setpoints, among others. Similar interfacing challenges are considered in remote or geographically distributed PHIL setups, as discussed in Section 6 and [14,15].

An illustration of this setup is found in Figure 2. Here, the RTSM uses as PA a (slow, ~10 Hz update rate) grid-forming converter and similarly infrequent, asynchronous, and non-lockstep measurements; the grid-forming converter is connected to a low-voltage feeder and loads with different responses to voltage and frequency variations. The simulation interface needs therefore to accommodate an asymmetric coupling with update-rates differing by several orders of magnitude.

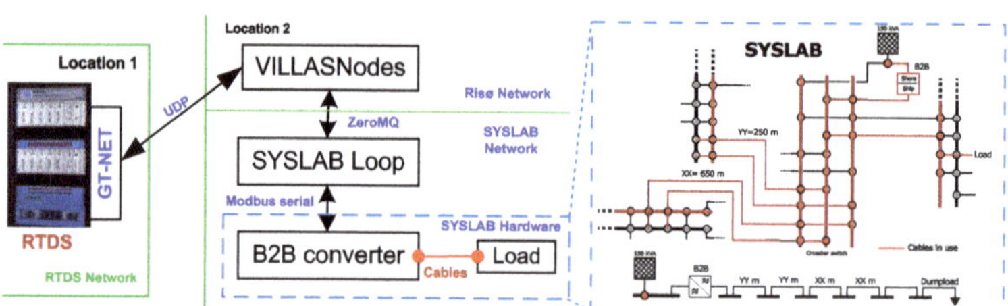


**Figure 2.** Configuration elements of the QsPHIL power system in the loop setup reported in VILLAS4ERIGrid [16].

The technical setup involves:

- a RTSM to simulate the relevant high and/or medium voltage simulation elements, and an interfacing component;
- a physical interfacing device such as a controllable grid-forming converter which can set frequency and voltage;
- a sufficiently fast RMS measurement at the interface device or its local bus.

Mathematically, the coupling is based on the ITM IA. In it, voltage and frequency are set by the RTSM model, while the RMS current and the phase ($I, cos\varphi$) are the feedback set by the physical branch. The exchanged signals are as follows:

- the RTSM system sends voltage and frequency RMS setpoints;
- the Interface device (grid-forming component of the lab) realizes these setpoint with some dynamic delay;
- the RMS P and Q values (or I, $cos\varphi$) are measured at the interface component, and transmitted to the RTSM component.

In this setup, the physical interface voltage and the current feedback are out-of-step during transients due to the hardware and software physical delays. Experiment designs account for this delay and consider only quasi-static phenomena: Relevant dynamics are simulated on the hardware power system components at slower time scales of 100 milliseconds or longer, relevant to energy management considerations, and voltage control coordination.

Due to the difference of update rates between RTSM at 20 kHz and the physical system operating at about 2 Hz, a multi-rate interface technique is necessary. Further, to ensure rapid convergence the long and non-deterministic delays between hardware (measurement, 10 Hz, 10 ms jitter; actuation 0.5–2 Hz) and RTSM elements are accounted for by suitable interface algorithms. Compensation methods enable accelerated convergence to steady state values. Two interfacing methods are described and evaluated in [16]. Due to the suitability for time delays, quasi-static PHIL methods are also applicable to geographically distributed setups, as further discussed in Section 6.1.2.

## 3. HIL Testing of Power System Protection and Control

### 3.1. Introduction

With growing installations of DER and the inclusion of the microgrid (MG) concept to enhance power system reliability, challenges appear in protection and control of transmission and distribution power systems. These problems are associated with additional fault current sources, increase of the total short-circuit level in the grid, and the presence of microprocessor based relays or Intelligent Electronic Devices (IEDs) [17]. This increases the threat to protection schemes due to: blinding of feeder protection, sympathetic tripping, failed reclosing, unintentional islanding, and recloser-fuse miscoordination [17–19], driving the protection and control engineers to re-evaluate long-established practices, as well as the way the relays are tested and certified. Adaptive protection is a protection strategy that is widely studied by the scientific community which tries to tackle some of these protection issues by adjusting protection equipment set points based on system operations [17,20–23]. The modeling of faults and protection strategies also needs to be studied and verified with transients (EMT simulations), which makes PHIL a good option to validate fault behaviors [24,25]. Finally, the requirements to certify such IEDs for the new power-system needs, requires frameworks suitable for close-to-reality testing, bringing an opportunity for HIL techniques to be implemented as a solution, like in [26]. Other alternatives for control and management of the power grid, such as wide area control by the use of synchrophasors and phasor measurement units (PMUs), is providing new reliability and controllability benefits to the power system [27]. It is important to have the capability to control and monitor the system in order to perform actions on the power system, such as the re-synchronization of an islanded MG [28], ensuring the conditions and requirements of local/international codes like the IEEE Std. 1547 [29]; or to dampen power systems oscillation by the inclusion of power system stabilizer strategies with PMU measurements [30].

### 3.2. Reported Experiences and Activities from SIRFN ALTM Members

#### 3.2.1. HIL Validation of Fault Locator Accuracy without Line Reactor Current in Distance Protection Scheme

In this HIL study, a transmission line with actual parameters connected between two 400 kV substations was simulated in a RTDS™ RTSM as shown in Figure 3 to study the impact of the non-subtraction of a line reactor current on a fault locator function accuracy in a distance relay, while using the *IEC 61850 Process Bus based Full Digital substation* implementation in POWERGRID India.

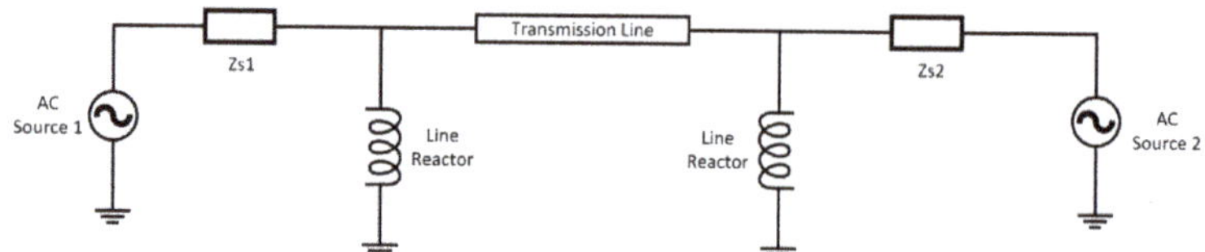

**Figure 3.** Model in RTDS for Hardware-in-the-Loop (HIL) study of line reactor in distance protection scheme.

To carry out the HIL study, a protection IED used at POWERGRID substations was interfaced with the RTSM and associated with one end of the modeled line. In order to visualize the impact of reactor current, two cases were studied. Figure 4 shows the position of the current transformer (CT) connection for both cases. In the first case, the current to the IED was fed from the CT connected between the source and reactor as shown in Figure 4a (reactor current included) and in the second case the IED was fed through the current from the CT connected towards the line, after the line reactor (reactor's current excluded) as shown in Figure 4b.

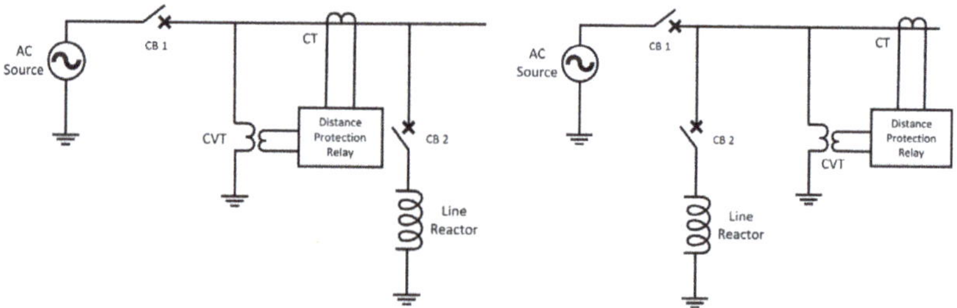

(a) Relay fed through CT connected before reactor.  (b) Relay fed through CT connected after reactor
**Figure 4.** Connection point of current transformers (CTs) for Case 1 (a) and 2 (b).

Both cases were studied for different types of faults and fault locations along the line and compared to determine the impact of reactor current on fault location determination. Table 2 shows the results of the fault locator performance for both scenarios in case of a line-to-ground fault at different fault positions.

**Table 2.** Comparison of fault locator performance.

| Fault Position | | Case 1—Bus Side CT | | Case 2—Line Side CT | |
|---|---|---|---|---|---|
| % Length | km | Fault Locator | % error | Fault Locator | % error |
| | | Reading (km) | | Reading (km) | |
| 5 | 15 | 14.90 | 0.67 | 14.92 | 0.53 |
| 15 | 45 | 44.62 | 0.84 | 44.64 | 0.80 |
| 25 | 75 | 74.17 | 1.11 | 74.68 | 0.43 |
| 35 | 105 | 104.03 | 0.92 | 104.82 | 0.17 |
| 45 | 135 | 133.61 | 1.03 | 135.05 | 0.04 |
| 55 | 165 | 163.06 | 1.18 | 165.64 | 0.39 |
| 65 | 195 | 193.77 | 0.63 | 196.90 | 0.97 |
| 75 | 225 | 223.78 | 0.54 | 228.32 | 1.48 |
| 85 | 255 | 254.02 | 0.38 | 259.89 | 1.92 |
| 95 | 285 | 285.04 | 0.01 | 292.46 | 2.62 |

### 3.2.2. Distance Protection Relay Type Testing Framework

The IEC 60255-121 standard specifies the minimum requirements for functional and performance evaluation of distance protection functions. It defines the tests to be performed in order to assess the relay characteristics such as operate time, transient overreach, phase selection, etc. In contrast to standard relay test sets, using a RTSM allows closed-loop testing of relays by simulating the power system, generating the test signals, and recording the resulting waveforms. The authors in [26] presented this concept and developed a framework to automate these tests. Figure 5 presents the workflow of the proposed HIL testbed. First, test scenarios and their respective parameters were defined and fixed. A power system RTS is created and interfaced with the relay under test. Low-level signals representing voltage and current are sent to the relay and resulting trip signals and any required simulated signal are recorded for post-processing and analysis. These test results are exported, relay performance is assessed, and a report is generated. This process is repeated automatically for every test defined in the spreadsheet.

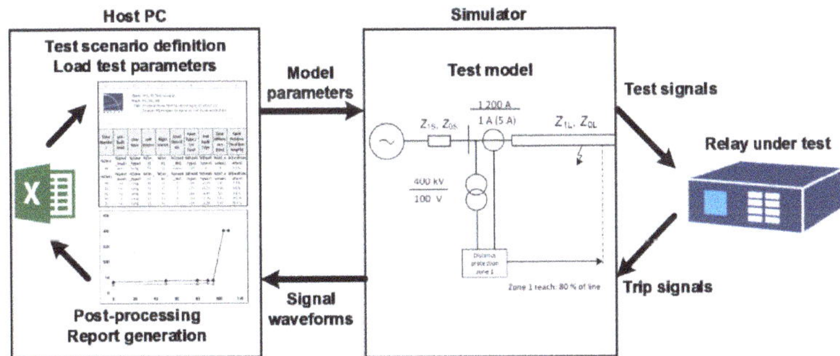


**Figure 5.** Workflow of the distance protection relay testing.

### 3.2.3. Adaptive Protection with HIL

The use of RTS to analyze and simulate power systems in both the steady state and faulted states provides valuable insight to the protection coordination of the power system. With the use of Analog and Digital I/Os, communication protocols (C37.118, GOOSE, and IEC 61850 Sample Values), and protection relays, an adaptive protection platform can be realized. Adaptive protection is needed when the power system has bi-directional power flows, usually from some type of DER source, that conventional protection cannot coordinate. The RTS adaptive protection platform allows designers and test engineers to evaluate and validate new equipment and methodologies in a realistic scenario [20]. In [21], an adaptive protection scheme that overcomes the challenges of the dual overcurrent relays (DOCR) in a power system with PV is proposed. Data are communicated about the PV plant to update the trip settings of the DOCR improving the protection selectivity and reliability.

### 3.2.4. Fault Modeling and Validation Between Simulation Tools

When simulating distribution systems with high penetrations of DER, protection studies with short-circuit analysis and phasor analysis in the frequency domain are possible using quasi-static time-series simulation tools [24]. The computational time is fast, but no transient dynamics are included in the simulation. Therefore, some researchers have turned to MATLAB/Simulink time-domain simulations of the fault analysis, so protection studies can be performed at a much higher resolution and include transients. In certain situations the DER dynamics, switching transients, and output filter capacitor discharge behaviors are important to capture. In those cases PHIL simulations with physical DER can further improve the fidelity of the analysis.

In [25], PHIL tests were performed by connecting the physical PV inverter to a 15-bus distribution system with 2 other simulated PV systems, emulated using a three-phase *dq* design. Faults were simulated at locations closest to the substation, furthest from the substation, and at the PHIL-interfaced PV inverter). This work demonstrated the difference in fault currents using OpenDSS quasi-static time-series simulations, RT MATLAB/Simulink simulations, and a PHIL approach. The cases with

PV that show the largest effects on the system are selected. OpenDSS provides steady-state results, whereas MATLAB/Simulink and the RT PHIL Simulation capture fault dynamics. The improvement in fidelity from the PHIL PV inverter is because it includes inductive and capacitive values that are not emulated in the *dq* PV inverter model and that are nonexistent in the OpenDSS PV inverters. Figure 6 shows the RMS PV inverter current when a 3 phase fault is applied at near the substation. Based on the modeling approach, the PV inverter phase A current varies. For OpenDSS, there is no instantaneous response available for the PV inverter. MATLAB/Simulink captures those transients, but interfacing the physical PV inverter provides an additional improvement in the model fidelity because it includes the output filter of the DER.

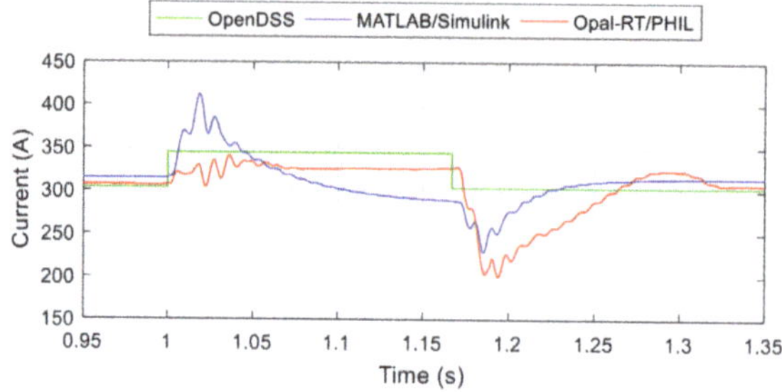


**Figure 6.** Comparison between physical photovoltaic (PV) inverter currents during a 3Ø Fault at Bus 4 [25]. Reproduced with permission from Rachid Darbali-Zamora, Feeder Fault Comparison Utilizing a Real-Time Power Hardware-in-the-Loop Approach for Photovoltaic System Applications; published by IEEE, 2019.

### 3.2.5. Wide Area Controller HIL Testing for Power Systems Oscillation Damping

In this work, different algorithms were investigated for dampening oscillations on a well-known benchmark Kundur system [31]. Controller behavior was studied by closing the loop between PMUs and an OPAL-RT RTSM. The setup is as shown in Figure 7. The Simscape Power Systems toolbox of MATLAB was used to model the two-area Kundur model in the OPAL-RT RTSM, with 4-th order model generator equipped with a simple exciter model for transient stability studies. The analog output channels were assigned to the model voltage measurements of generators 1 and 3 and connected to two NI$^{TM}$ Open PMUs. The Raspberry Pi$^{TM}$ devices communicate with the PMUs with a Phasor Data Concentrator (PDC), to reproduce a realistic communication loop. In the Raspberry Pi, a conventional PI controller was implemented as the complement of the PSS of Generators 1 and 3. This control was translated in voltage and assigned to the analog output of the Raspberry Pi. This controller signal is added to the PSS output and directly connected to the input of the automatic voltage regulator. The results obtained following a 3 phase short circuit are depicted in Figure 8. The implemented wide area controller improves the power system behavior as expected during the test. A detailed paper is expected to be published with application of more advanced control algorithm applying ZHAW dynamic power system technique [30] soon.

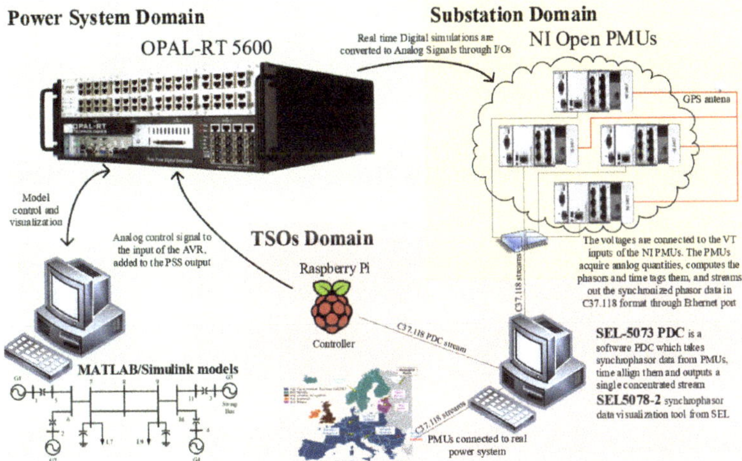


**Figure 7.** Controller Hardware-in-the-Loop (CHIL) testbed setup for controller validation.

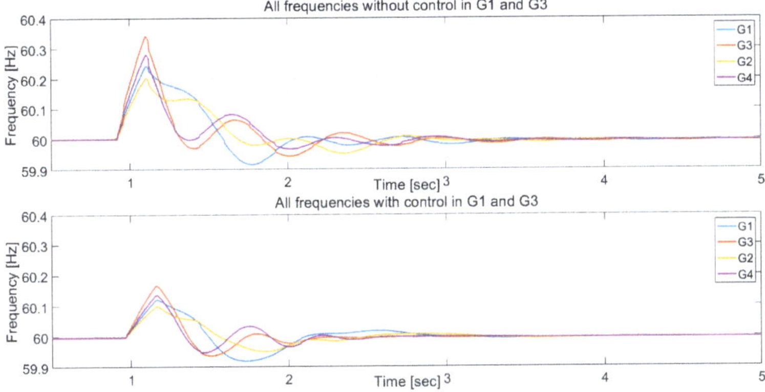


**Figure 8.** Frequency response without and with controller.

## 4. HIL Testing of Smart Grid/Microgrid Controllers, Energy Management Systems, and Power Electronic Converters

### 4.1. Introduction

Microgrids, defined as small-scale power systems consisting of load and distributed generators, are considered as an essential part of the future grid. Testing and development of MGs require detailed modeling of various components that operate over a wide range of time constants from hours (for distributed generation dispatch) to milliseconds (for accurate representation of stability and protection related phenomena). HIL testing provides these capabilities while greatly reducing the complexities of traditional test methods. By reducing the risk, cost, and total time required to test complex embedded systems, HIL systems have been used to study many MGs around the world. This section condenses the SIRFN ALTM experiences:

1. **Microgrid prototyping and validation:** Testing AC, DC, and Hybrid MGs requires a wide variety of techniques and methodologies including: benchmarks and prototyping platforms [32–35], testing chains and procedures for centralized and decentralized controllers [36–41], and rapid

control prototyping (RCP) for a quick development of new control strategies in a real hardware environments [42–44].

2. **Microgrid control strategies:** Advanced testing methods are helpful in the establishment of the control strategies for loads and generators, reconfiguration equipment, and isolation and re-synchronization actions [28].

3. **Development of inverter and power electronics functions:** DER functions and device interoperability support large, traditional power systems and microgrids alike. These technologies can be evaluated prior to implementation using CHIL and RT simulation techniques [45–52]. The same testing approach can be used for large-scale power electronics integration studies [53,54].

*4.2. Reported Experiences and Activities from SIRFN ALTM Members*

4.2.1. Integrated PHIL and Laboratory Testing for Microgrid Controller

A test setup with a physical microgrid controller (MGC), multiple microgrid components, a Data Aquisition System (DAS), and the RTSM shown in Figure 9 can be constructed by integrating PHIL technologies with a lab testing approach [33,34]. This is useful for evaluating several functions included in the MGC, which manage DER and support the operations typically reserved for diesel generators. The controller was tested in the PHIL configuration with a remote island MG. The RTSM and PA were used to emulate the diesel genset, which was not present in the lab. Device power ratings were scaled with the feedback current and voltage values to the RTSM to represent the physical system. In this PHIL setup, the MGC was interconnected to the communication systems and tested as a blackbox. In other words, the lab tests were performed without the need to know the internal design of the controller. Furthermore, testing with real hardware equipment increases the fidelity of the test results while RTS adds microgrid asset diversity. Such validation is crucial to successful on-site deployments on commissioning and site acceptance tests.

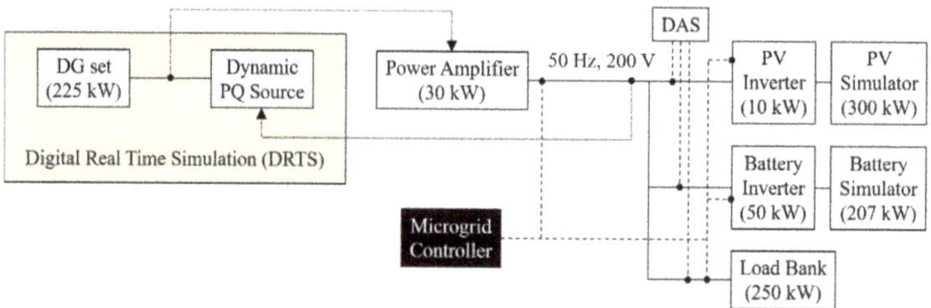


**Figure 9.** PHIL lab setup for testing microgrid (MG) capabilities.

The test demonstrated that, after including in the testbed the MGC as a blackbox, it was able to stabilize the MG frequency providing set-points to the PV and Battery inverters, proving the capability of the laboratory setup to perform PHIL and laboratory testing for MGC.

4.2.2. Development of a Droop Frequency Control of Stand-Alone Multi-Microgrid System with HIL

To operate multiple MGs with different frequencies, a framework of the stand-alone multi-microgrid (MMG) system was proposed in [43]. In the proposed MMG system, each MG connects to the common DC line through an AC/DC interlinking converter as shown in Figure 10.

An HIL system was used to validate the performance of the proposed controller. The RT digital simulator (OP5600) was used to emulate the MMG system instead of the physical plant. The proposed control algorithm implemented in a digital signal processor (DSP) can be tested easily with OP5600.

The signal exchange between DSP controller and OP5600 for $MG_1$ of MMG system is shown in Figure 11a. DSP controller receives the analog signal (voltage, current, and dc link voltage) from OP5600. Then, the pulse width modulated (PWM) signals generated by a DSP controller are sent to the IC1 in OP5600. The overall HIL system for the MMG system is shown in Figure 11b.

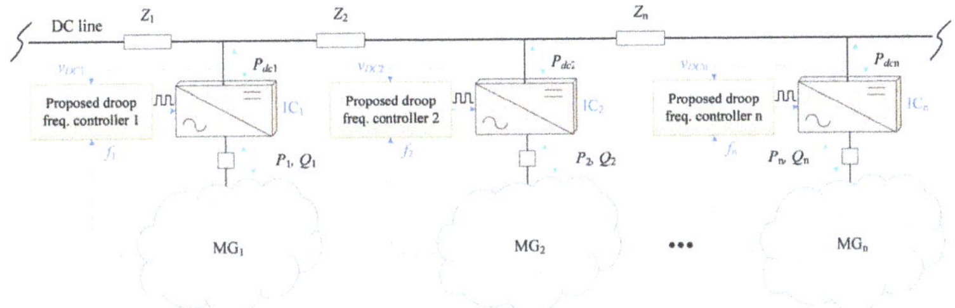

**Figure 10.** Proposed stand-alone multi-microgrid system [43]. Reproduced with permission from Hyeong-Jun Yoo, A Droop Frequency Control for Maintaining Different Frequency Qualities in a Stand-Alone Multimicrogrid System; published by IEEE, 2018.

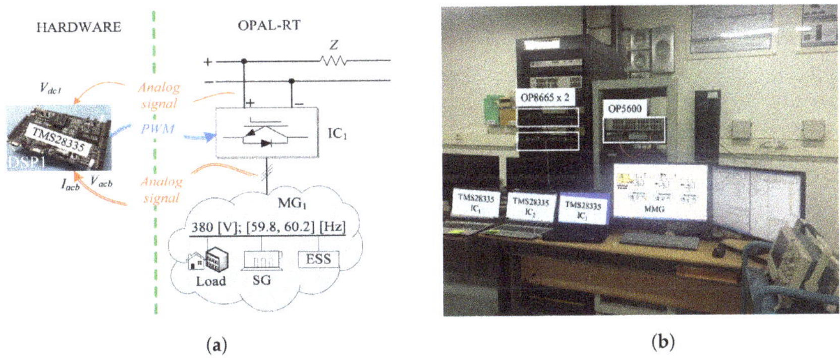

(a)          (b)

**Figure 11.** HIL setups [43]. Reproduced with permission from Hyeong-Jun Yoo, A Droop Frequency Control for Maintaining Different Frequency Qualities in a Stand-Alone Multimicrogrid System; published by IEEE, 2018. (**a**) I/O signals between DSP and OP5600 for the $MG_1$; (**b**) HIL platform for proposed MMG system.

### 4.2.3. Microgrid Re-Synchronization with PMU Measurements

In this HIL simulation, PMU measurements were used to re-synchronize an MG with the utility grid. Use of HIL for this test case is beneficial since it involves potentially dangerous events leading to MG islanding and re-connection to an utility grid that may cause large power swings. Furthermore, co-simulation of power and communication exchanges with RTSM enables testing of the developed resynchronization techniques within the same test setup. The simulation of PMUs is carried out in the RTDS RTSM using a dedicated processing card, known as GTNET PMU card, which interfaces the RTSM with external equipment over a LAN connection using various standard protocols. For the test system used in this experiment, GTNET-PMU8 component protocol is selected in the RSCAD library. This component can emulate up to 8 PMUs that communicate with IEEE Std C37.118.2 protocol. The GTNET-PMU8 component is shown in Figure 12. For synchronizing, GTNETx2 uses a Global Positioning system (GPS) clock to provide the absolute time signal reference to the RTSM through a GTSYNC card where the phase of the signals computed in the RTSM will shift relative to the signals of external equipment. Each PMU takes 6 inputs of voltages and currents values for each phase of a 3-phase bus. The MG synchronization can occur either through active and passive MG synchronization. In active synchronization, integrated control mechanism is used to match voltage, phase angle, and frequency of the islanded system to the grid. This approach requires special infrastructure like dedicated communication facilities which involve complex controlling techniques mostly applied for inverter-based systems with high capital costs. In passive synchronization, both sides of the connection are monitored to ensure that the voltages, frequencies, and phases are within tolerance before closing the contactor.

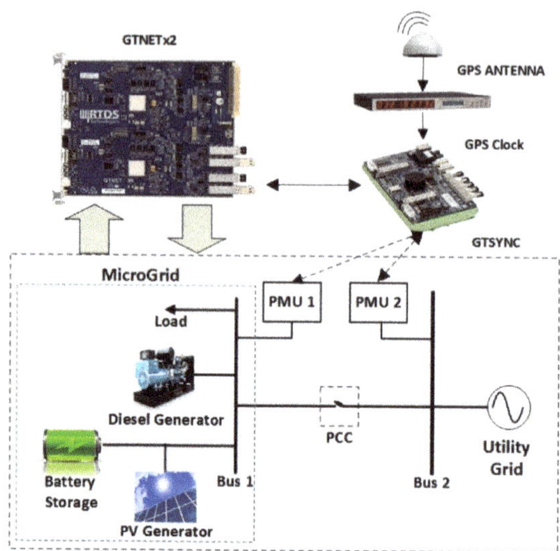

**Figure 12.** HIL test set-up for model validation.

### 4.2.4. Distributed Coordination Control in Hybrid AC/DC Microgrid with RCP

The study in [44] presents the MG shown in the Figure 13, consisting of an AC and DC microgrid, connected with an interlinking converter. The interlinking converter in a hybrid AC/DC MG system plays an important role of maintaining power sharing between AC and DC MG systems. The proposed coordination control strategy not only regulates accurate reactive power and DC current sharing among distributed generation in AC and DC MGs but also maintains power sharing among two MGs and restores the AC frequency and DC voltage to their nominal values. To verify the feasibility and the effectiveness of the proposed control strategies, an RCP system was designed as depicted in Figure 13.

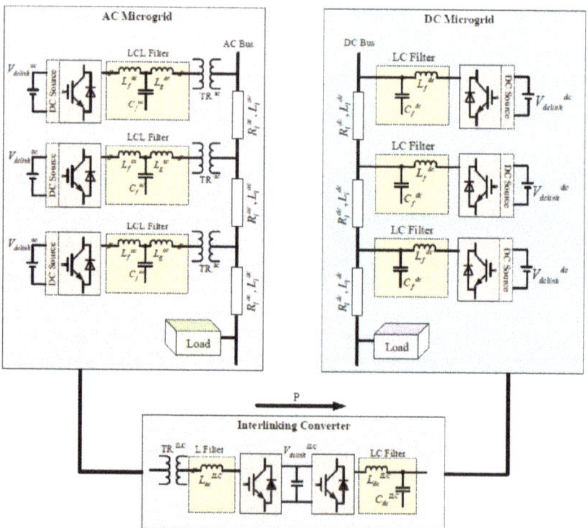

**Figure 13.** Circuit diagram of hybrid AC/DC MG [44]. Reproduced with permission from Hyeong-Jun Yoo, Consensus-Based Distributed Coordination Control of Hybrid AC/DC Microgrids; published by IEEE, 2020.

Two OP4510 RTSMs were used to control distributed generation units and interlinking converter. The measured voltages and currents from PT and CT were sent to OP4510s. The pulse width modulated signals generated by the controllers in OP4510s were sent to distributed generation units and interlinking converter. Detailed structure of the configuration of the tested system and exchanged signals between OP4510s and real hardware system can be found in [44].

### 4.2.5. Design and Validation of a Rule-Based Microgrid Controller

Since most MGs contain a limited number of assets, simple rule-based dispatch is sufficient for operation. The authors in [37] present the design and validation of a rule-based MGC. The controller topology is centralized in nature, i.e., all the measurements, instrumentation, and any requests from the utility (such as islanding) is communicated to the MGC which sends out the dispatch set-points and other control signals to all the assets. Validation of the controller was performed using an HIL testbed as shown in Figure 14. The communication between the controller and the assets was achieved using IEC 61850 GOOSE messages.

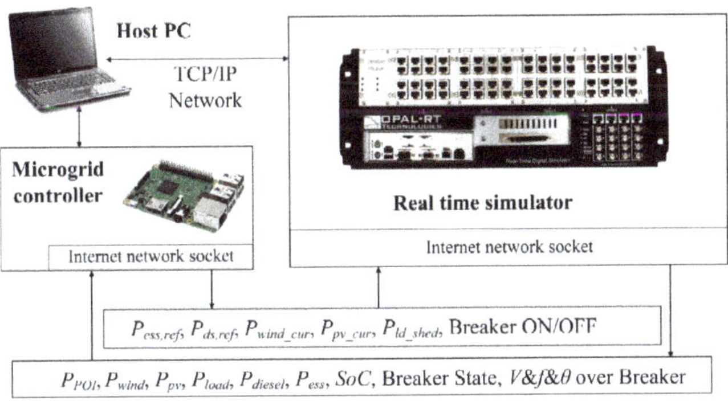


$P_{ess,ref}$, $P_{ds,ref}$, $P_{wind\_cur}$, $P_{pv\_cur}$, $P_{ld\_shed}$, Breaker ON/OFF

$P_{POI}$, $P_{wind}$, $P_{pv}$, $P_{load}$, $P_{diesel}$, $P_{ess}$, SoC, Breaker State, $V\&f\&\theta$ over Breaker

**Figure 14.** Rule based microgrid controller design and validation testbed [37].

### 4.2.6. Decentralized Microgrid Control Systems

Several decentralized MG control systems were proposed and tested using HIL testbeds. A multi-agent based control system with simulated agents in an RTSM and implemented agents on Raspberry Pi[TM] as edge devices was presented [38]. The work presented focuses on the handling of unplanned islanding events by the use of consensus-based algorithms between several agents or a multi-agent system. A similar approach was adopted in [39,40] with the proposed resilient information architecture platform for the smart grid (RIAPS). The control system uses Beaglebone Black Boards as edge devices to implement the decentralized control functionality.

### 4.2.7. Microgrid Controller Development with an Advanced Testing Chain Methodology

The development of a centralized MGC with the advanced testing chain methodology [55] was based on the work in [36]. The MGC takes on coordinating tasks in the microgrid to ensure the safe operation of the components and the longest possible supply to the connected loads. MGC functions include (a) initiation, (b) coordination and monitoring of the reconstruction of the MG after major disruptions, (c) energy management in stand-alone/island operation, (d) re-synchronization, and (e) connection to the higher-level network or to another MG. In addition, all relevant measurement values and operating parameters are saved in the MGC in a second resolution for the purpose of subsequent system monitoring. For operation, a graphical user interface was created, which shows an overview of the most important operating parameters and enables the input of user commands. The operation of the MGC has been tested in pure simulation to validate the general behavior and functionality. After that, the MGC was implemented in a real-time target, in this case, a Programmable Logic Controller

(PLC) Bachmann MX220. The PLC was connected to the MG model running in a OPAL-RT OP5600 RTSM through a communication interface with a MODBUS protocol. To perform a PHIL experiment, simulated components where exchanged for real devices. The PLC manages parts of the simulated MG and sends set-points to a real battery inverter, which is connected to the RTSM via an AMETEK™ RS90 amplifier. The development process of the MGC can be seen in the Figure 15.

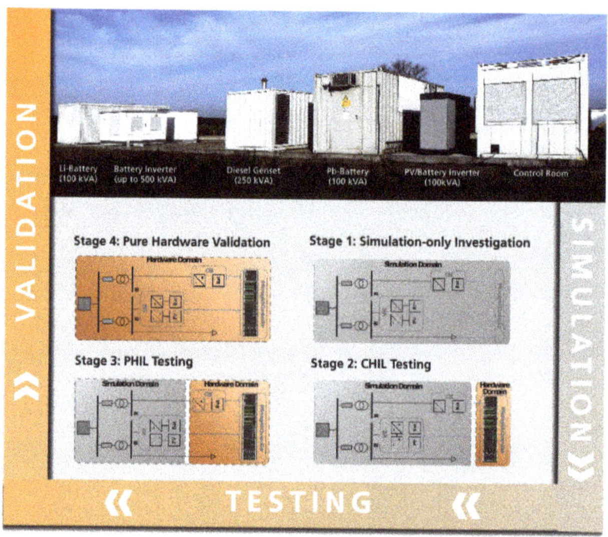


**Figure 15.** Testing process for a controller development at Fraunhofer IEE, Germany.

### 4.2.8. Generic Microgrid Controller Development, Testing, and Validation

In order to facilitate the deployment of MGs, substantial efforts have been made in recent years to standardize MGC specifications. IEEE Std. 2030.7 and 2030.8 [56] cover various functions common to MGs, regardless of their topology and configuration, as well as test procedures. As a co-chair of the IEEE 2030.7 working group, the University of California Irvine (UCI) Advanced Power and Energy Program defined specifications and developed a generic MGC that can be adapted to various MG topologies [41].

A CHIL testbed, as shown in Figure 16, was used in order to test and validate the MGC for two MG systems: the 20 MW-class UCI MG and the 10 MW-class UCI Medical Center MG. Detailed EMT models of these two MG systems, including their loads, lines, transformers, circuit breakers, and DER, were developed using Simulink and simulated on an OPAL-RT RTSM. The MGC was implemented on a platform provided by ETAP™ and time critical MGC functions were transferred to a Schweitzer Engineering Laboratories RT Automation Controllers (SEL RTACs) in order to emulate a field-deployed load controller. Monitored values from the simulated MG models are sent to the MGC using IEC 61850 GOOSE/DNP3 and resulting commands and setpoints from the dispatch functions and load shedding schemes were sent back to the simulation. Testing the MGC using HIL allowed identification of the MGs operational limits for a continuous operation and safe transition to islanded mode. An islanding of the MG was then conducted to further validate the MGC specifications and corroborate the CHIL test results. The UCI MG was successfully islanded for 75 minutes during which various load changes were applied and seamlessly reconnected to the utility grid.

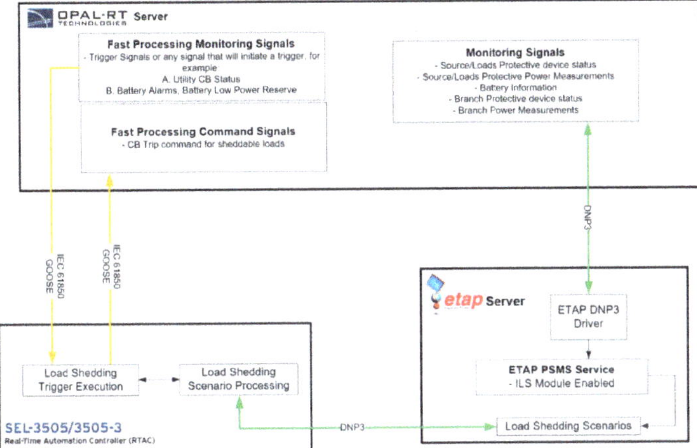

**Figure 16.** Generic microgrid controller CHIL testbed [41]. Reproduced with permission from Elsevier, A generic microgrid controller: Concept, testing, and insights; published by Elsevier, 2018.

### 4.2.9. CHIL for Grid-Support Functions of Inverters

Advancements in CHIL technologies are enabling power electronics control systems to be accurately modeled with low-power benchtop equipment. This accelerates the design cycle by providing quick results with pre-deployment control and communications code. For instance, the Austrian Institute of Technology (AIT), in collaboration with Typhoon HIL Inc. and Sandia National Laboratories, evaluated a prototype DER controller and interoperability SunSpec Modbus interface before implementing the technology in a physical converter design [45]. By evaluating the equipment in the CHIL, software bugs could be quickly corrected in a low-consequence environment before moving to the next stage of the converter design. The AIT Smart Grid Converter (ASGC) was also evaluated against multiple North American interconnection standards, including Underwriters Laboratories (UL) 1741 and IEEE 1547-2018 [46,47], before having physical hardware to evaluate. This gave the team confidence that the equipment would perform appropriately with costly hardware. Later, the fully-built physical converter was found to have similar characteristics to the CHIL simulations for active power curtailment, frequency-watt, volt-watt, and under-voltage ride-through experiments [48,49].

### 4.2.10. PHIL Smart Inverter Testing with Megawatt Scale Grid Simulator

Impact assessment testing by using PHIL technologies are introduced to confirm capabilities of smart inverter functions. National Institute of Advanced Industrial Science and Technology (AIST) in collaboration with a Japanese utility and manufacturer evaluated a 500 kVA PV smart inverter at global scale smart system research facility at Fukushima Renewable Energy Institute, AIST (FREA-G). FREA-G is one of the largest power electronics testing facility with 5 MVA grid simulator and 3.3 MVA bi-directional DC source. AIST also have a one tenth scale DER testing laboratory and a design of FREA-G is based on same set up of DER testing laboratory [35]. A testing configuration is shown in Figure 17.

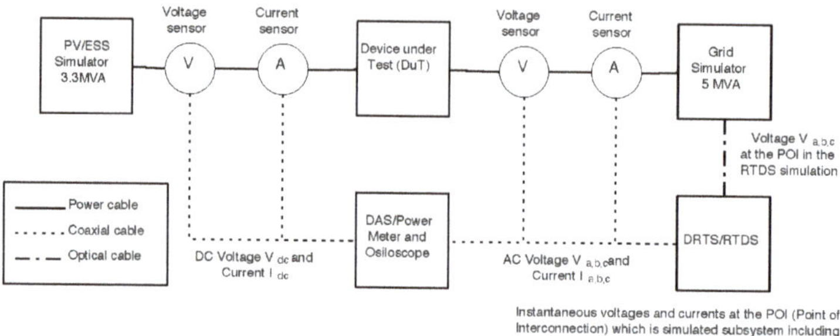

**Figure 17.** Megawatt scale PHIL setup for smart inverter assesment test.

In this case, a RTDS/NovaCore RTSM simulates a testing power system which is shown in Figure 18a. A simulated instantaneous voltage at the point of interconnection of the PV system is sent to grid simulator and amplified to emulate a voltage of DuT. Measured current and/or voltage between the grid simulator and DuT is fed back to the RTSM to introduce behavior of DuT for power system simulation. Two test cases have been conducted with this setup. The first test case is volt-var function with low voltage ride through condition shown in Figure 18a. The second test case is a frequency-watt function with load shedding event shown in Figure 18b. The result was compared with Controller HIL (CHIL) testing to conclude the PHIL setup has the potential for assessing the impact of smart inverter functions at a large-scale. Preliminary results from a battery energy storage system can be found in [57].

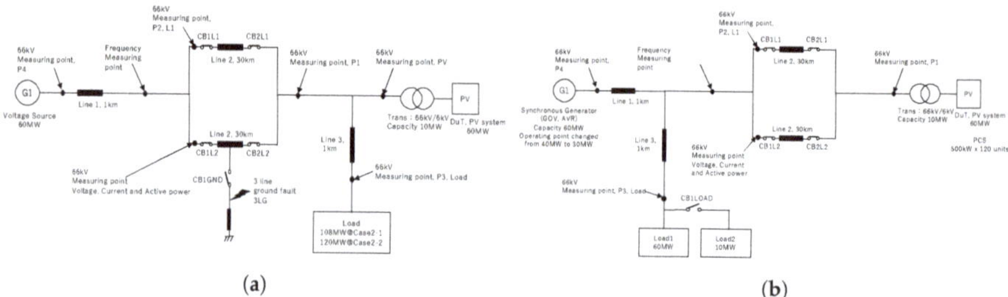

(a)  (b)

**Figure 18.** Power system for combination test for (**a**) volt-var and low voltage ride through and (**b**) frequency-watt and other functions.

### 4.2.11. CHIL for Validation of Unintentional Islanding

AIT developed a CHIL based testbed for unintentional islanding (UI) schemes to assess the risk-of-islanding with large shares of inverter-based DER in a project by EPRI collaborating with public utilities, inverter industry, the NY State, and Sandia [50,51].

For the CHIL UI validation testbed, a set of generic, non-proprietary UI schemes, developed in an earlier project sponsored by EPRI [52], were implemented into its Smart Grid Converter controller (SGC): (a) Sandia Frequency Shift (SFS), including Q-F-function, (b) Quasi SFS, (c) Negative sequence injection, and (d) Rate-of-Change-of-Frequency (RoCoF).

The characteristic parameters of each UI scheme can be adjusted during run-time, also allowing to enable or disable, each of the UI schemes. To enable the operation of multiple UI schemes at the same time, a special "decision-tree" concept allows a quasi-parallelization of the UI schemes. This allows to utilize the advantages of the individual schemes while reducing the need to apply "aggressive" settings to the individual parameters. Through this concept, the reliability of the UI detection can be increased and at the same time, negative effects of UI schemes on the voltage quality, grid stability, etc. can be mitigated. The CHIL setup consists of the control board taken from the AIT SGC which was connected to the RTSM (Typhoon HIL 602). Using this setup, the AIT SGC controller can be operated under the same full range of operating conditions as for the physical laboratory tests, as shown in Figure 19.

The model running on the RTSM includes SGC's power train, a PV array, and a power grid model, represented by an adjustable RLC circuit for the simulation of the local system load. Using the model, a wide range of tests were performed to investigate and validate the response of the UI schemes, the sensitivity of their parameter settings as well as the impact of advanced grid support functions and ride-through capabilities. For each condition, the impact of key control parameters on the effectiveness of the UI schemes, i.e., accurately and reliably detect the islands and trip the inverter, was assessed.

The work successfully demonstrated the benefits of the CHIL testbed to assess the UI schemes implemented in inverter-based DER. As part of the project [50], studies were performed to investigate the sensitivity of control parameters on the overall performance of studied UI schemes, the resulting run-on-time and other characteristics was assessed. In addition, the impact of advanced grid support functions on the capability to reliably detect islanded conditions was assessed.

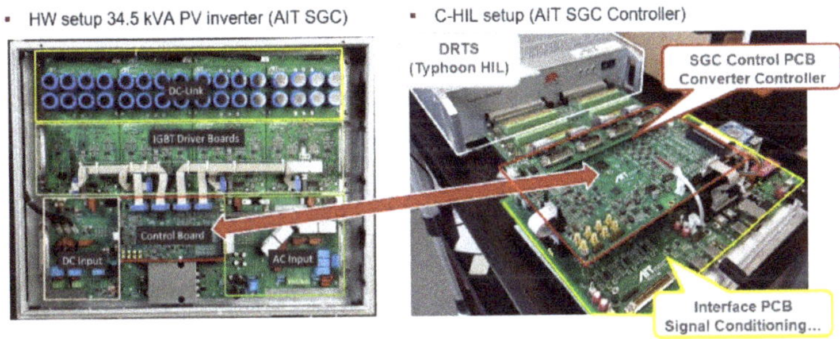


**Figure 19.** CHIL islanding testbed with AIT SGC HIL controller connected to Typhoon-HIL RTSM.

## 5. Co-Simulation and RTS Integration

### *5.1. Introduction*

Smart grid co-simulations are growing in popularity. The need for integration of new technologies in the power systems is challenging and has driven the industry and scientific community to develop their own tools and software. Co-simulation plays a big role in these assessment approaches as showed in [58]. However, one specific problem with co-simulation is coupling multiple simulation platforms. The majority of them involve a large number of distributed physical devices controlled by various software components and advanced algorithms, with a common interface based in communication technologies.

This section contains SIRFN ALTM experiences including integration of RTSM to co-simulation platforms [59,60], network emulators, and cyber–physical systems in co-simulation and HIL [61–63], development of tools with enhanced cybersecurity features for co-simulation and HIL experiments [64,65], and integration via co-simulation of electrical vehicles charging station interfaces [66].

### *5.2. Reported Experiences and Activities from SIRFN ALTM Members*

### 5.2.1. Asynchronous Integration of RTSM with Co-Simulation Platforms

In [59], possibilities for testing and validating advanced control strategies and smart grid applications with RTSM co-simulations are studied. It presents two approaches based on a Message Bus architecture, depicted in Figure 20, for co-simulation and rapid prototyping of networked systems:

- Lablink: Simulation Message Bus (SMB) based Implementation
- OpSim: Representational State Transfer (REST) based Implementation

Depending on the used approach, coupling software simulators to an RTSM will introduce latency in which improvements below the 1 ms threshold cannot be expected, and this will limit the spectrum of applicability of the co-simulation approaches for system-based testing.

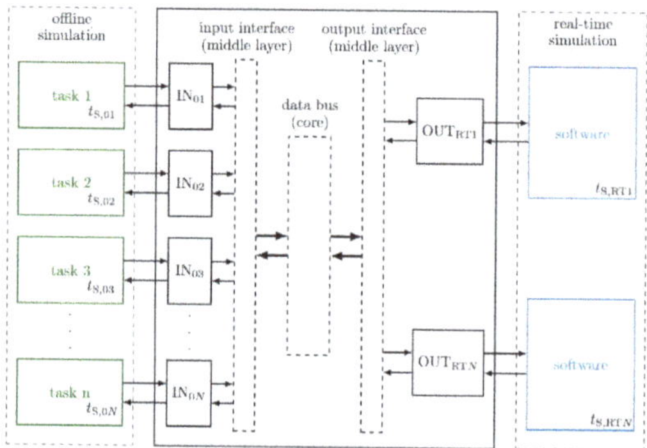


**Figure 20.** Simulation message bus architecture for co-simulation of RT and non-RT systems [59]. Reproduced with permission from Juan Montoya, Asynchronous Integration of Real-Time Simulators for HIL-based Validation of Smart Grids; published by IEEE, 2019

A co-simulation environment developed by Fraunhofer IEE and the University of Kassel, called OpSim, was created for both RT and non-RT co-simulations of complex power systems [60]. The objective of this co-simulation platform is to make research on smart grids affected or operated by several stakeholders by interconnecting different simulation models form different software tools. The core of its platform is a REST based message bus that manages the exchange of information between clients by means of a client/proxy architecture.

5.2.2. Co-Simulation of Cyber–Physical Systems

Co-simulation of the communication network on the network emulation software (ExataCPS) along with the electric power network simulated with a RT EMT type software (HYPERSIM and RT-LAB) was created in [61]. The differentiating factor between the two proposed configurations of the co-simulation was the implementation of the communication link and the number of RTSMs involved. In one configuration the RTSM was used to run the power network simulation and the controller was connected to the simulator via a computer running the communication network emulation software, as shown in Figure 21b. The second configuration involved a single RTSM where the power network simulation and the communication network simulation were run on different cores with a virtual link between them as shown in Figure 21a. In both cases, the authors specifically showed co-simulation of a MGC communicating with the assets on a MG. The specific cyber phenomena studied was communication delay due to: cyber attack, data manipulation, and man-in-the middle attacks. Similar testbeds with physical communication connections were proposed by the authors in [62,63], with the exception that the communication system emulator used was OPNET™. In both cases, the testbed involved the interfacing of the RTS with the monitoring and control node via a host computer running OPNET exchanging information by its System-in-the-Loop (SITL) interface. The distinct feature of such testbeds is the synchronization of the two kinds of simulations. The RTS usually runs a time-stepping simulation that is either EMT or phasor domain, whereas the network emulators are discrete event based simulators.

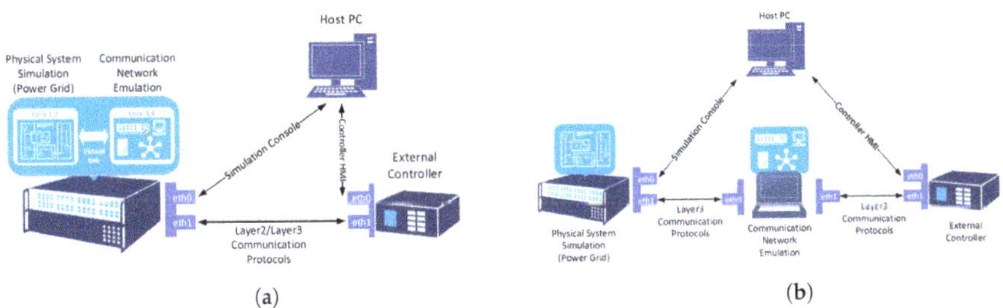

**Figure 21.** Cyber–physical co-simulation setups. (**a**) Single simulators and virtual communication link; (**b**) Two simulators and physical communication link.

### 5.2.3. SCEPTRE: Suite of Tools Providing an ICS Co-Simulation Environment

There is an increasing desire to establish co-simulation environments that capture the characteristics of multiple domains. For smart-grid applications it is often important to model both the power system dynamics as well as communication network dynamics. To this end, many researchers have been constructing cyber–physical simulation tools. Sandia National Laboratories has developed a suite of tools called SCEPTRE (capitalized but not an acronym) that provide an industrial control system (ICS) co-simulation environment. SCEPTRE has been used in a number of projects to capture the dynamics of communications and power systems. One good example is a project that established a distribution power system simulation with large PV plant at the end of the feeder. Simulated PV inverters were created that interacted with the power system so that any changes to the controllers would be reflected in the power simulation and vice versa. An adversary-based (red team) assessment was performed for different cybersecurity defenses (segmentation, encryption, and moving target defense) in this environment and given a cybersecurity score [64,65].

### 5.2.4. Electrical Vehicle/Charging Station Integration Testing

AIST developed a smart electric vehicle (EV) management communication solution in accordance with IEC 61850 for the communication link between EV and charging station (CS) [66]. First, a new information model was developed to emulate CSs. Second, a model based on IEC 61850-90-8 was altered to accommodate reverse power flow by the CS model. Steps of the EV management scheme were transferred via IEC 61850 messages based on their transmission purpose and nature. Simple measurements were sent with Sampled Values (SV), event-based instructions were sent as Generic Object-Oriented Substation Event (GOOSE), and other ad-hoc information exchange is down with Manufacturing Messaging Specification (MMS)

In order to validate the standardized models and messages, an HIL test was performed with the topology shown in Figure 22. An IEC 61850 emulator was utilized to model an EV while a digital RTSM modeled the CS (and connection to the grid). In the future, the RTSM will be integrated with an optimization software to run power system simulations and centralized control algorithms.

**Figure 22.** HIL test set-up for model validation.

## 6. Geographically Distributed HIL and RTS

### 6.1. Introduction

The operation and planning of power systems is continuously increasing in its complexity with the integration of innovative components, controls, and novel architecture paradigms. Accordingly, aspects of scalability and stable system dynamics must be evaluated with advanced laboratory testing methods to provide confidence in the technologies prior to deployment. The requirements for a single test bed to cope with such a diverse and large scale system are extensive; however, the RT coupling of geographically distributed test beds can overcome any such limitations that a single test site might face. This expands the realms of validation not only by incorporation of additional equipment available but more importantly brings wider experienced individuals together.

The classification of geographically distributed (GD) simulation setups is currently being undertaken within the IEEE PES Task Force on Interfacing Techniques for Simulation Tools, with the broadly identified classification reported here as:

1. **Geographically distributed RTS:** in this group only RTSMs at both ends are coupled together. Typically, this is used to overcome large simulation complexity that would require large RT resources that might not be available in a single test site. This has been the most common type of GD simulations reported in the literature [67–71].
2. **Geographically distributed HIL experiments:** this group involves configurations where hardware is coupled remotely to a RTS. By using this configuration, the testing of hardware equipment (controllers or power components) in a system environment when no RTS is available at the premises is made possible. The involvement of hardware in this group requires careful consideration based on the interfacing method and communications [72–76], as discussed in Section 2.

The main challenges arising from the implementation of GD simulation have been identified as the interfacing method and the communication medium utilized for the realization of the interface, and these are briefly discussed next.

### 6.1.1. Interfacing GD Simulations

The accuracy and stability of an interface algorithm utilized for the coupling of subsystems presents a challenge. The interface algorithms used for GD simulations are in practice similar to the ones used for monolithic PHIL or CHIL simulations but with significant increase in time delay (due to the geographic separation of the infrastructures), which impacts the stability of these configurations in comparison to single-site PHIL and CHIL.

The representation of the signals exchanged between the subsystems, such as instantaneous, RMS, or any transformation thereof, plays an important role in a successful implementation of a GD simulation setup. Studies to establish the appropriateness of signal representations have been widely reported, where the selection of signal representation can be concluded as a trade-off between accuracy and stability for the given application [69]. The implemented representation can also be important for allowing the compensation of time delays as the ideal place for this compensation is at the interface and the format selected can decide on the feasibility of the time delay compensation [71,77].

### 6.1.2. Communications

The communication over long distances between geographically separated assets is the key characteristic that differentiates monolithic HIL simulations from GD simulation. For feasibility purposes, communication generally takes place over the Internet with time delays depending on the available network infrastructure between the sites and in contrast with monolithic setups present non-deterministic time delays [78]. Latencies over the Internet between different cities have been evaluated in [69,72]. Most of the publications in the field use the User Datagram Protocol (UDP) for transmission of data due to its RT capabilities [71,74,76,79], although the use of other protocols such as Transmission Control Protocol (TCP) or RT Protocol (RTP) have also been reported [14,67,69,75]. Additionally, an orchestrator can be utilized to coordinate the protocols, time steps and also for visualization and data logging aspects.

### 6.2. Reported Experiences and Activities from SIRFN ALTM Members

#### 6.2.1. Geographically Distributed CHIL for Advanced Validation of a Distributed Control Algorithm

To increase the realism of the distributed control algorithm validation environment, GD simulation can be configured with the controller units geographically separated from the simulated power system (emulating the environment in which they could be deployed). An example of such configuration is presented in Figure 23, in which a distributed control for a MG is evaluated [72]. By using the GD architecture, the data exchange between the controller units as well as between the power assets and controller units can be thoroughly evaluated under a real communication network with uncertainties such as latency or packet-losses that might occur during the validation procedure. In this case, the RTS of the power system is hosted at Nanyang Technological University, Singapore, while the clusters of distributed controllers are located at University of Strathclyde, UK and University Grenoble Alpes, France. The local communication between the controllers is established by TCP/IP, for the communication with the power system UDP protocol through a cloud server service is used (Redis).

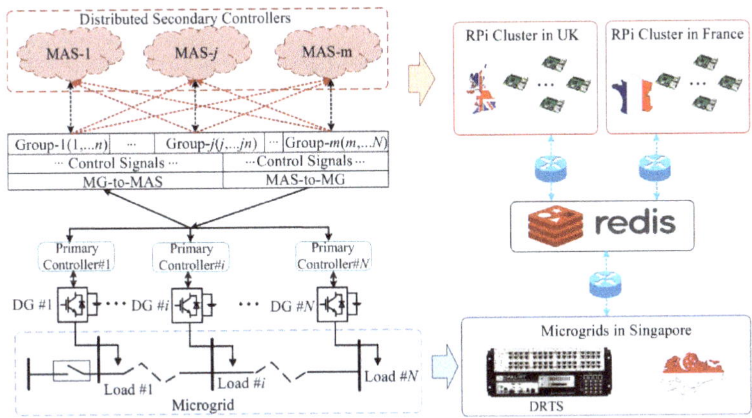


**Figure 23.** Geographically distributed CHIL for advanced validation of a distributed control algorithm [72]. Reproduced with permission from Efren Guillo-Sansano, A Distributed Control Scheme of Microgrids in Energy Internet and Its Multi-Site Implementation; published by IEEE, 2020.

#### 6.2.2. Delay Assessment for Geographically Distributed CHIL Experiment

The author in [75] presents a delay assessment for two GD devices (a PC with a controller and a RTSM) interfaced via a co-simulation platform. Both devices were interfaced between Germany and Greece (Kassel-Athens) through an asynchronous TCP/IP connection. The so-called co-simulation platform OpSim was used as an orchestrator as shown in Figure 24. The objective of this work was to share resources of GD research infrastructures, to assess delays and latencies of GD simulators, and to define the limits and RT capability of the OpSim platform.

This particular experiment presented the limits in latencies between OpSim and each GD simulator by measuring the Round-Trip Time (RTT). In addition, the validation of a coordinated voltage control (CVC) was made by benchmarking the GD simulation with pure simulation results from Matlab.

The results indicated that OpSim has RT capabilities to interconnect GD simulators from different research laboratories by means of asynchronous communication and to perform slow (low-bandwith) grid voltage control. Based on the user-defined values, variables, and publishing rates, estimation tools provide a way to analyze if a RT co-simulation experiment can be performed *a priori*.

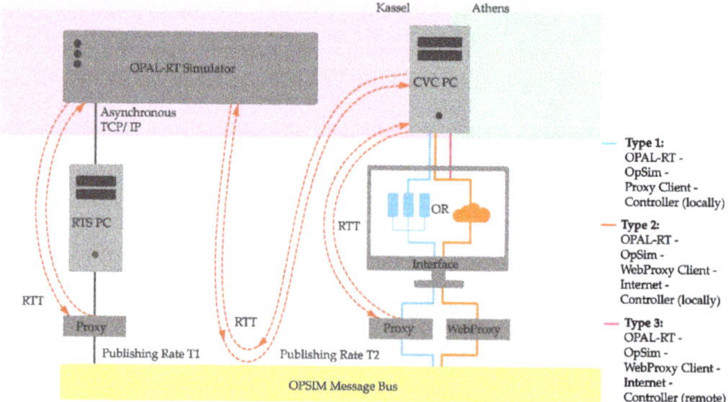

**Figure 24.** Test setup for geographically distributed co-simulation with OpSim [59]. Reproduced with permission from Juan Montoya, Asynchronous Integration of Real-Time Simulators for HIL-based Validation of Smart Grids; published by IEEE, 2019.

### 6.2.3. Geographically Distributed PHIL for Testing of a Voltage Controller

This case study investigated the feasibility of geographically remote power equipment to be incorporated for the validation of a control algorithm to enable a systems testing environment.

The GD-PHIL configuration is shown in Figure 25, where a CIGRE low voltage test network is simulated in RT at University of Strathclyde with the control algorithm under validation in a CHIL implementation. To incorporate the behaviour of a real hardware device, and to characterize its consequent impact on the performance of the control algorithm, a battery energy storage system at the premises of RSE (Italy) was interconnected at PCC4. Further details on the implementation can be found in [14,80].

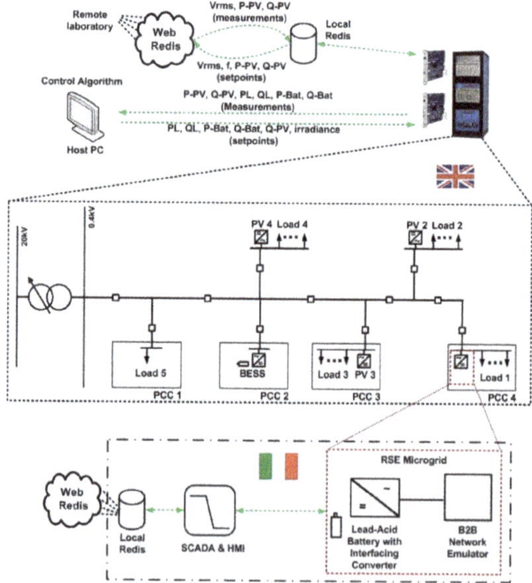

**Figure 25.** Geographically distributed PHIL for testing of a voltage controller [14].

### 6.2.4. Global RT Super Lab Demonstration

In certain situations, connecting multiple RT power simulations provides greater fidelity or realism than a monolithic simulation. For instance in the RT Super Lab demonstration, a geographically distributed RTS (GD-RTS) representing transmission-distribution co-simulation was created with different PHIL- and CHIL-interfaced equipment connected at each of the facilities. Idaho National Laboratory ran the transmission simulation and seven other laboratories located in North America and Europe ran distribution simulations that interfaced with the transmission simulation. The goal of the simulation was to see if transporting energy intercontinentally would be possible to avoid blackouts, assuming there was transmission capacity to do so [76]. The co-simulation was possible using VILLASframework, a decentralized software suite that enabled gateways to communicate using UDP, Message Queuing Telemetry Transport (MQTT), Advanced Message Queuing Protocol (AMQP), or IEC 61850.

### 6.3. Future Outlook

Although the concept of GD simulation has been around for over a decade, more recent advancements in the past couple of years have led to a renewed interest in their utilization. With a wide array of implementations being reported, a few research gaps still remain to be addressed, such as:

- With a number of different interface algorithms, signal transformations, and communication protocols being reported, there is a lack of formalization or guidance available for selection in regards to application. A set of combinations need to be appraised for applications such as transient studies, dynamic studies and steady state evaluations.
- Recognizing that the communications delays are dominated by the non-deterministic characteristic of the Internet, options such as use of dedicated bandwidth should be explored.
- In distinction to a monolithic PHIL simulation where a DuT is connected to a RTSM, the GD simulation presents a significant challenge in determining system partitioning. The optimal approach to split a system for simulation over the GD simulation requires further assessment.
- With the number of subsystems within one GD simulation expected to increase, where more than two research infrastructures are expected to be interconnected, a streamlined facilitation of initialization is required. A lot of work for co-simulation setups has been reported and their applicability for GD simulation needs to be explored.

## 7. Industrial Experiences and HIL in Standardized Testing

### 7.1. Introduction

Recent standards revisions have acknowledged the use of HIL as a way to test compliance. The IEEE Std. 2030.8-2018 [56] accepts testing environments ranging from fully simulated testbeds to field installed equipment. HIL testing with RTS for full or partial testing of the MG control systems is included within the accepted simulation environments. Similarly, the forthcoming IEEE Std. 1547.1 revision [81] also accepts HIL as a way to test compliance, and also demonstrates an example of performing unintentional anti-islanding test with a PHIL setup. However, the research community keeps putting efforts on development of platforms for validation according to grid codes and pre-certification of units [47,82–84], definition of procedures for compliance testing and acceptance tests [85–89], and validation of marine- and aero-electrical power systems [90–93].

Furthermore, the ongoing IEEE Standards Association Project *P2004-HIL Simulation Based Testing of Electric Power Apparatus and Controls* aims to provide recommended practices for using HIL as a method of testing electric apparatus and controls for standards that accept it as a testing method. In addition, the ERIGRID project outcomes include a book to be published under the name *European Guide to Power System Testing: The ERIGrid Holistic Approach for Evaluating Complex Smart Grid Configurations* [14] and includes the experiences and suggestions for standardization of testing and validation of cyber–physical energy systems (CPES) by the use of simulation, co-simulation, and HIL laboratory testing, which can act as a reference for researchers interested in developing standardization of HIL techniques in the future.

### 7.2. Reported Experiences and Activities from SIRFN ALTM Members

#### 7.2.1. Compliance Testing of a Hybrid UPS According to JEC2433-2016

The authors in [85] used HIL testing for design and compliance testing of a hybrid UPS based on an emergency diesel generator and battery energy storage system. The hybrid UPS was tested to meet the class 2 UPS requirement from the Japanese standard on UPS, JEC2433-2016. The authors adopted the CHIL approach to evaluate the compliance of the designed controller. The CHIL testbed and all the results of the compliance can be further seen in [85].

#### 7.2.2. Development of a PMU Pre-Certification Platform

OPAL-RT and Vizimax collaborated on the development of a HIL phasor measurement unit (PMU) pre-certification platform [82]. This collaboration allowed Vizimax to successfully validate their PMU compliance with IEEE Std. C37.118.1 prior to testing at the National Institute of Standards and Technology (NIST). Furthermore, it allowed OPAL-RT to validate their test equipment performance and develop a standard compliant automated HIL PMU test platform. It was demonstrated that using HIL is a valuable method for pre-certification of monitoring, control, and protection devices with an accuracy comparable to that of calibration lab equipment. Furthermore, it provides the capability to go beyond the standard requirements. A diagram of the testbed is shown in Figure 26.

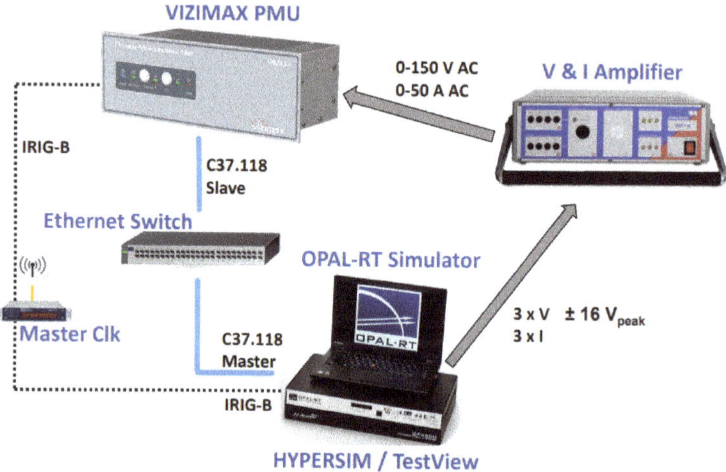


**Figure 26.** HIL testbed for pre-certification of a phasor measurement unit (PMU).

#### 7.2.3. Factory Acceptance Tests Based on HIL Testing

Hydro Quebec has been involved in HIL testing of their AC/DC interconnected network with hybrid (analog and digital) and fully digital simulators [86,87]. The authors perform CHIL studies with a one-to-one replica of the HVDC line commutated converters controller of both terminals of the high voltage DC (HVDC) lines. The power system components including the power electronic switches, DC filters, DC transmission lines, etc. were all simulated while the controllers were actual controllers interfaced with the simulation link using I/O. Réseau de Transport d'Électricité (RTE) in France applies a similar approach for testing of static var compensators (SVC) connected to their network [88]. All the power components were simulated on the RTSM while the controllers were interfaced with the simulation via I/Os. Factory acceptance tests (FAT) based on HIL testing of a replica controller of a five terminal modular multilevel converter based HVDC project was presented in [89]. The power components were simulated on a multi-rate simulation platform using both the CPU processor and the FPGA for high fidelity simulation requirement of the application. Many other utilities adopt a similar approach for various other Flexible AC transmission and HVDC devices as well including series compensators, synchronous condensers etc.

### 7.2.4. System Validation Platform HIL Based Grid Code Testing Aspects of DER Inverter

Several international laboratories have collaborated for years to evaluate the grid support functions of DER devices [47,84]. For this purpose, the group uses a versatile open-source DER testing and certification platform known as Sunspec System Validation Platform (SVP) [83]. SVP automates the test procedures by executing sequences of testing logic that change the settings on the DuT, grid simulator, PV simulator, and data acquisition system using Python test scripts. The group has developed drivers for a wide range of communication interfaces, power simulation and data acquisition equipment. For CHIL tests, drivers for Typhoon HIL RTSM were implemented. The group has assessed the test procedures from different standards, i.e., UL 1741 SA [94], forthcoming IEEE 1547.1 [81], etc. and provided feedback to the standards development organizations for corrections and enhancements of the test procedures. The implementation of the test procedures from IEEE P1547.1 standard required tests with unbalanced voltage magnitudes and phases, and maintaining specific rate of change of frequency (RoCoF), which are not achievable directly in certain grid simulators when they are controlled directly through communication link from SVP. To address these issues, a RTS based signal generator was developed in OPAL-RT eMegasim$^{TM}$ environment. The signal generator implemented all the test requirements of AC source for IEEE P1547.1 and was easily controlled through the Python API of RT-LAB (ver. 2019.2.3). In this case the Python API supports test automation and is directly controlled from SVP. This approach has been used to collect the test results in [47] and this methodology is proven to be reliable. The UL 1741 SA and IEEE 1547.1 test scripts used in this SVP based activities are also available at Github.

### 7.2.5. HIL for Marine Electrical Power Systems (MEPS)

The advancements brought to power systems validation with the use of RTS and HIL methods have prompted its use in other sectors such as MEPS and aero-electrical power systems. These sectors have recently begun the integration of smart grids concepts to leverage on their advantages. However, these novel concepts need to be thoroughly assessed by the use of laboratory testing methods before their deployment [90]. Such testing methodologies have been utilized for a number of applications, such as the assessment of the impact of incorporating novel components like high power dense direct current loads and their power electronic interfaces [90,91], smart coordinated control strategies [92], and MEPS architectures [93].

### 8. Summary of Testing Methods and Configurations

This review takes into account various experiments and experiences in different fields of research, using different research infrastructures and devices. Below, the methods described in each subsection are cited. Table 3 condenses the hardware used in the reported activities from the SIRFN ALTM members subsections to give the reader an overview of the equipment used in HIL experiments.

**Table 3.** Summary of experiments and hardware used by SIRFN ALTM Members.

| Section | Topic | RTS/HIL Type | Interfaces and Protocols | Hardware and Equipment |
|---|---|---|---|---|
| Section 2.2.1 | Power amplifier characterization for RTS | PHIL | Analog I/O | RTSM: OPAL OP5700 PA: AMETEK |
| Section 2.2.2 | Stability and accuracy comparison for different interfacing methods | PHIL [12] | Analog I/O | RTSM: OPAL OP5600 PA: AMETEK RS90 |
| Section 2.2.3 | Quasi-dynamic PHIL | PHIL/ PSIL* | IA: ITM Digital/Soft I/O | RTSM: Workstation PA: Studer XTM 4000, Electroinvent ELDI |
| Section 2.2.4 | Quasi-static PHIL | PHIL/ PSIL* [16] | IA: ITM Digital/Soft I/O | RTSM: RTDS PA: ABB PCS100 SFC |
| Section 3.2.1 | HIL validation of fault locator accuracy in distance protection scheme | CHIL | Analog I/O IEC 61850 9-2 SV | RTSM: RTDS Signal Amp: Omicron |
| Section 3.2.2 | Distance protection relay type testing framework | CHIL [26] | Analog I/O Digital I/O IEC 61850 /60255-121 | RTSM: OPAL OP5600 |
| Section 3.2.3 | Adaptive protection with HIL | PHIL [20,21] | Analog I/O IEC 61850 SV C37.118 GOOSE | RTSM: OPAL OP5600 PA: AMETEK RS90 |
| Section 3.2.4 | Fault modeling and validation between simulation tools | PHIL [25] | IA: ITM, DIM Analog I/O OpenDSS | RTSM: OPAL OP5600 PA: AMETEK RS90 |
| Section 3.2.5 | Wide Area Controller HIL testing for Power Systems Oscillation Damping | CHIL [30] | C37.118 GPS Analog I/O | WAC: Raspberry Pi PMU: NI cRIO Open PMU, NI9467 PDC: SEL-5073 PDC |
| Section 4.2.1 | Integrated PHIL and laboratory testing for microgrid controller | PHIL [33,34] | IA: ITM Analog I/O | RTSM: RTDS NovaCore PA: SanRex 500kVA DAS: Yokogawa WT3000E/WT1800 DuT: NK-EMS Load: SanRex RLC bank |
| Section 4.2.2 | Droop frequency control of stand-alone multi-microgrid system with HIL | CHIL [43] | Analog I/O Digital I/O | RTSM: OPAL OP5600 Control Unit: OP8665 |
| Section 4.2.3 | Microgrid re-synch with PMU measurements | CHIL | C37.118 GPS | RTSM: RTDS/GTNET |
| Section 4.2.4 | Distributed coordination control in hybrid AC/DC MG with RCP | PHIL [44] | Analog I/O Digital I/O Modbus TCP/IP | RTSM: OPAL OP4510 |
| Section 4.2.5 | Design and validation of a rule-based microgrid controller | CHIL [37] | IEC 61850 | RTSM: OPAL OP4510 MGC: SEL 3360 |
| Section 4.2.6 | Decentralized microgrid control systems | CHIL [38] | IEC 61850 GOOSE | RTSM: OPAL OP5600 Raspberry PI |
|  |  | CHIL [39,40] | MODBUS C37-118 | RTSM: OPAL OP031 + OPAL OP5607 TI F28377S, Beaglebone Black Boards |

**Table 3.** *Cont.*

| Section | Topic | RTS/HIL Type | Interfaces & Protocols | Hardware & Equipment |
|---|---|---|---|---|
| Section 4.2.7 | Microgrid controller development with an Advanced Testing Chain methodology | CHIL/ PHIL/ PSIL* [55] | Analog I/O Modbus TCP/IP IA: DIM | RTSM: OPAL OP5600 PA: AMETEK RS90 DAS: DEWETRON 800 Current Source Inverter: SMA SCS500 |
| Section 4.2.8 | Generic microgrid controller development, testing, and validation | CHIL [41] | IEC 61850 GOOSE DNP3 | RTSM: OPAL OP5600 Load control: SEL 3505 RTAC MGC: ETAP |
| Section 4.2.9 | CHIL for Grid-support functions of inverters | CHIL [45,48,49] | Analog I/O Modbus TCP/IP | RTSM: Typhoon HIL602 DuT: AIT SGC |
| Section 4.2.10 | PHIL smart inverter testing with megawatt scale grid simulator | PHIL | IA: ITM Analog I/O | RTSM: RTDS NovaCore PA: SanRex 5MVA DAS: HIOKI PW6001, MR8827 |
| Section 4.2.11 | CHIL for validation of unintentional islanding | CHIL [50,51] | Analog I/O Modbus TCP/IP | RTSM: Typhoon HIL602 DuT: AIT SGC |
| Section 5.2.1 | Asynchronous integration of RTSM with co-simulation platforms | CHIL [59] | Co-Sim: Lablink UDP | RTSM: OPAL OP5600 Gateway: Raspberry Pi |
| | | CHIL [60] | Co-Sim: OpSim Async. TCP/IP | RTSM: OPAL OP5600 |
| Section 5.2.2 | Co-simulation of cyber-physical systems | CHIL [61] | IEC 61850 Co-Sim: TCP/IP Virtual Link with Exata CPS | RTSM: OPAL OP4510 |
| | | CHIL [62,63] | IEC 61850 Co-Sim: Ethernet with Opnet | RTSM: OPAL OP4510 |
| Section 5.2.3 | SCEPTRE: suite of tools providing an ICS co-simulation environment | CHIL/ PHIL [64,65] | Real TCP/IP packets running over simulated network. Physical interfaces to the network can be presented to users/equipment | RTSM: Custom Power Simulation runing in PowerWorld Dynamics Studio PA: AMETEK RS180 |
| Section 5.2.4 | Electrical vehicle/ charging station integration testing | CHIL [66] | Co-Sim: MATLAB and IEC61850 SV Sender (Commercial software) | RTSM: RTDS/GTNET |
| Section 6.2.1 | GD-CHIL for advanced validation of a distributed control algorithm | GD-CHIL [72] | UDP TCP/IP | RTSM: OPAL OP5600 Controller: Raspberry Pi |
| Section 6.2.2 | Delay assessment for geographically distributed CHIL experiment | GD-CHIL [75] | Co-Sim: OpSim Message Bus architecture TCP/IP | RTSM: OPAL OP5600 Controller: Coordinated Voltage Control in Matlab |
| Section 6.2.3 | GD-PHIL for testing of a voltage controller | GD-PHIL PSIL* [80] | IA: ITM UDP | RTSM: RTDS DuT: Lead-Acid Battery |
| Section 6.2.4 | Global RT SuperLab | GD-PHIL/ GD-CHIL/ PSIL* [76] | IA: Multiple Comm. protocol: VILLASnode | RTSM: OPAL OP5600, Typhoon HIL, RTDS RS: Multiple |

**Table 3.** *Cont.*

| Section | Topic | RTS/HIL Type | Interfaces and Protocols | Hardware and Equipment |
|---|---|---|---|---|
| Section 7.2.1 | Compliance testing of a hybrid UPS according to JEC2433-2016 | CHIL [85] | Analog I/O Digital I/O JEC2433-2016 | RTSM: OPAL OP5600 RCP: Custom |
| Section 7.2.2 | Development of a PMU pre-certification platform | CHIL [82] | Analog I/O Digital I/O | PMU: Vizimax PMU Signal Amp: Omicron |
| Section 7.2.3 | Factory Acceptance Tests based on HIL testing | CHIL [86,87] | Analog I/O Digital I/O | RTSM: SGI Altix UV300s Parallel Computers Controller: ABB HVDC Controller |
| | | CHIL [88] | Analog I/O Digital I/O | RTSM: SGI UV100 Parallel Computers Controller: Static Var Compensator Controllers |
| | | CHIL [89] | Analog I/O Digital I/O | RTSM: OPAL OP5600 + OP7020 Controller: HVDC MMC controller |
| Section 7.2.4 | System Validation Platform HIL Based Grid Code Testing | PHIL [47] | IA: ITM, DIM Analog I/O | RTSM: OPAL OP5700 PA: Ametek |
| | Aspects of DER Inverter | CHIL [47] | Analog I/O Digital I/O | RTSM: Typhoon HIL602 DuT: AIT SGC |
| Section 7.2.5 | HIL for marine electrical power systems (MEPS) | PHIL [90] | IA: ITM, DIM Analog I/O | RTSM: RTDS PA: Triphase PM90 |

* recommended HIL concept to adopt in the future.

Additionally, this review provides a summary of the reported literature used for each section as a fast-track to references, and is listed in the Table 4.

**Table 4.** Database of literature used for this review.

| Section | Topic | Reports in Literature |
|---|---|---|
| Section 2 | Interfacing methods of PHIL, CHIL, and PSIL simulation | [1–15] |
| Section 3 | HIL testing of power system protection and control | [17–31] |
| Section 4 | HIL testing of smart grid/microgrid controllers, energy management systems, and power electronic converters | [28,32–56,95] |
| Section 5 | HIL co-simulation and CPES | [58–66] |
| Section 6 | Geographically distributed HIL and RTS | [14,59,67–80] |
| Section 7 | Industrial experiences and HIL in standardized testing | [14,47,56,81–94] |

## 9. Conclusions and Future Outlook

A key to advancing the deployment of smarter, cleaner electric grids is the development and validation of technologies, protocols, standards, and systems that can function effectively in a variety of geographies and grid environments. Laboratories within the SIRFN community are actively expanding laboratory testing methods to evaluate and demonstrate new power system control, optimization, and design technologies. ALTM capabilities are increasingly critical to develop modern power system solutions to enable the grid to accept greater penetrations of renewable and distributed energy resources. This paper surveyed a range of applications for RTS and HIL technologies in order to provide an overview to the reader of current research trends and approaches within international practices. This paper also reviewed new testing concepts like PSIL that, although it is not widely used, provides a framework to future researchers in HIL to exploit simulations with slow dynamics and characterize specific HIL setups besides CHIL and PHIL. It was found that RTS and HIL can be applied to a wide range of R&D areas including power system protection, microgrid controllers, energy management systems, and power electronic converters, as well as standardization and industry developments. Additionally, a trend to co-simulation and geographically distributed testing was recognized, in cases that research institutes lack from specific equipment or computational power and system-based holistic validation is an alternative.

In the future, the SIRFN community of research laboratories will be collaborating in these applied topic areas to further develop power system ALTM technologies and solutions for the grid as it continues to evolve. The ALTM group works in collaboration with specific and challenging tasks on development of interoperable DER certification protocols, microgrid testing and power system testing from SIRFN. The authors encourage the reader to make part of the ISGAN Annex 5: SIRFN and the European Distributed Energy Resources Laboratories (DERLab), in order to actively contribute further to the development of novel power systems technologies. More information and contact details can be found in http://www.iea-isgan.org/our-work/annex-5/.

**Author Contributions:** Conceptualization, formal analysis, project administration, supervision, and visualization: J.M.; funding acquisition: J.M., K.V.; investigation, resources, and software: J.M., R.B. (Ron Brandl), J.J., R.D.-Z., A.S., J.H., H.K., T.S.U., N.N., E.A.-A., J.-P.B., M.R., S.Q.A., A.O., K.H., R.S., E.G.-S., M.H.S., G.B., C.C., H.-J.Y., C.P.A., K.W., R.B. (Roland Bründlinger); methodology: J.M., J.J., J.H., N.N., K.H., E.G.-S., C.C.; validation: J.M., J.J., K.H., R.S., E.G.-S.; writing—original draft: J.M., J.J., R.D.-Z., A.S., J.H., H.K., T.S.U., N.N., E.A.-A., J.-P.B., M.R., S.Q.A., A.O., K.H., R.S., E.G.-S., M.H.S., C.C., H.-J.Y., C.P.A., K.W., R.B. (Roland Bründlinger); writing—review and editing: J.M., K.V., J.J. All authors have read and agreed to the published version of the manuscript.

**Funding:** This research is funded by the German Ministry for Economic Affairs and Energy (BMWi) and the Projektträger Jülich (PTJ) within the project *"Netzregelung 2.0 - Regelung und Stabilität im stromrichter-dominierten Verbundnetz"* (FKZ0350023A). This paper does not necessarily reflect the consolidated opinion of the project consortium *"Netzregelung 2.0"*.

**Acknowledgments:** This review has been performed by a group of researchers, members of the ALTM Task of ISGAN Annex 5: SIRFN, as part of their efforts to enhance the close collaboration among test facilities and identifies potential activities for future application and standardization of ALTM. In this context, the contributions from the members are acknowledged as follows:

- Fraunhofer IEE contributions are supported by the European Community's Horizon 2020 Program (H2020/2014–2020) under the project "ERIGrid" (Grant Agreement No.654113), and by the German Ministry for Economic Affairs and Energy (BMWi) and the Projektträger Jülich (PTJ) within the project "Netzregelung 2.0 - Regelung und Stabilität im stromrichter-dominierten Verbundnetz" (FKZ0350023A).
- Sandia National Laboratories is a multimission laboratory managed and operated by National Technology and Engineering Solutions of Sandia, LLC, a wholly owned subsidiary of Honeywell International, Inc., for the U.S. Department of Energy's National Nuclear Security Administration under contract DE-NA0003525.Its contributions to this work are supported by the U.S. Department of Energy Office of International Affairs.
- National Institute of Advanced Industrial Science and Technology (AIST) contributions to this work are supported by the Ministry of Economy, Trade and Industry (METI).
- CanmetENERGY is a federal research laboratory in Canada; financial support for this research work was provided by Natural Resources Canada (NRCan) through the Program on Energy Research and Development (PERD) in the framework of REN-2 Smart Grid and Microgrid Control for Resilient Power Systems Project.
- Korea Electrotechnology Research Institute (KERI) participation was supported by the Korea Institute of Energy Technology Evaluation and Planning (KETEP) and the Ministry of Trade, Industry & Energy (MOTIE) of Republic of Korea (No.20178530000210).
- University of Strathclyde and Technical University of Denmark (DTU) contributions are supported by the European Community's Horizon 2020 Program (H2020/2014–2020) under the project "ERIGrid" (Grant Agreement No. 654113).
- Zurich University of Applied Science (ZHAW), Institute of Energy Systems and Fluid Engineering (IEFE) contributions are supported by the Swiss Federal Office of Energy (SFOE) and activities of the Swiss Centre for Competence in Energy Research on the Future Swiss Electrical Infrastructure (SCCER-FURIES), which is financially supported by the Swiss Innovation Agency (Innosuisse - SCCER program).
- The participation of AIT within ISGAN-SIRFN is funded in the frame of the IEA Research Cooperation program by the Austrian Federal Ministry for Climate Action, Environment, Energy, Mobility, Innovation and Technology (FFG no. 870646). The development of the AIT SGC was supported by the Austrian Ministry

for Transport, Innovation and Technology (bmvit) and the Austrian Research Promotion Agency (FFG) under the "Energy Research Program 2015" in the SPONGE project (FFG no.848915).
- Contributions from Power Grid Corporation of India Limited are the part of research work carried out at POWERGRID Advanced Research & Technology Centre (PARTeC) located at Manesar. PARTeC is the R&D establishment of Power Grid Corporation of India Ltd.

**Conflicts of Interest:** The authors declare no conflict of interest.

## Abbreviations

The following abbreviations are used in this manuscript:

| | |
|---|---|
| AC | Alternate Current |
| ALTM | Advanced Laboratory Testing Methods |
| CHIL | Controller Hardware-in-the-Loop |
| DAS | Data Acquisition System |
| DC | Direct Current |
| DER | Distributed Energy Resources |
| DIM | Damping Impedance Method |
| DuT | Device Under Test |
| EMT | Electro-magnetic transients |
| GD | Geographically Distributed |
| HIL | Hardware-in-the-Loop |
| IA | Interface Algorithm |
| ITM | Ideal Transformer Method |
| MG | Microgrid |
| MGC | Microgrid Controller |
| PA | Power Amplifier |
| PDC | Phasor Data Concentrator |
| PHIL | Power Hardware-in-the-Loop |
| PMU | Phasor Measurement Unit |
| PSIL | Power System-in-the-Loop |
| RCP | Rapid Control Prototyping |
| RMS | Root Mean Square |
| RT | Real-time |
| RTS | Real-Time Simulation |
| RTSM | Real-Time Simulation Machine |
| WAC | Wide Area Control |

## References

1. Brandl, R. Operational range of several interface algorithms for different power hardware-in-the-loop setups. *Energies* **2017**, *10*, 1946. [CrossRef]
2. Lauss, G.F. Interfacing challenges in PHIL simulations for investigations on P-Q controls of grid connected generation units in electric power systems. *IFAC-PapersOnLine* **2017**, *50*, 10964–10970. [CrossRef]
3. Brandl, R.; Montoya, J.; Strauss-Mincu, D.; Calin, M. Power System-in-the-Loop testing concept for holistic system investigations. In Proceedings of the 2018 IEEE International Conference on Industrial Electronics for Sustainable Energy Systems (IESES), Hamilton, New Zealand, 31 January–2 February 2018; pp. 560–565. [CrossRef]
4. Ebe, F.; Idlbi, B.; Stakic, D.E.; Chen, S.; Kondzialka, C.; Casel, M.; Heilscher, G.; Seitl, C.; Bründlinger, R.; Strasser, T.I. Comparison of power hardware-in-the-loop approaches for the testing of smart grid controls. *Energies* **2018**, *11*, 3381. [CrossRef]
5. Stanev, R.; Krusteva, A.; Tornelli, C.; Sandroni, C. A quasi-dynamic approach for slow dynamics time domain analysis of electrical networks with distributed energy ressources. *Proc. Tech. Univ. Sofia* **2013**, *63*, 273–281.

6.  Andrén, F.; Lehfuss, F.; Jonke, P.; Strasser, T.; Rikos, E.; Kotsampopoulos, P.; Moutis, P.; Belloni, F.; Tornelli, C.; Sandroni, C.; et al. DERri Common Reference Model for Distributed Energy Resources—modeling scheme, reference implementations and validation of results. *Elektrotechnik Inf.* **2014**, *121*, 378–385. [CrossRef]

7.  Langston, J.; Schoder, K.; Steurer, M.; Edrington, C.; Roberts, R.G. Analysis of Linear Interface Algorithms for Power Hardware- in - the- Loop Simulation. In Proceedings of the IECON 2018—44th Annual Conference of the IEEE Industrial Electronics Society, Washington, DC, USA, 21–23 October 2018; pp. 4005–4012.

8.  Paran, S.; Edrington, C.S. Improved power hardware in the loop interface methods via impedance matching. In Proceedings of the 2013 IEEE Electric Ship Technologies Symposium (ESTS), Arlington, VA, USA, 22–24 April 2013; pp. 342–346.

9.  Ren, W.; Steurer, M.; Baldwin, T.L. Improve the Stability and the Accuracy of Power Hardware-in-the-Loop Simulation by Selecting Appropriate Interface Algorithms. *IEEE Trans. Ind. Appl.* **2007**, *44*, 1286–1294. [CrossRef]

10.  Paran, S. Utilization of Impedance Matching to Improve Damping Impedance Method-Based Phil Interface. Master's Thesis, Florida State University, Tallahassee, FL, USA, 2013.

11.  Crăciun, B.; Kerekes, T.; Séra, D.; Teodorescu, R.; Brandl, R.; Degner, T.; Geibel, D.; Hernandez, H. Grid integration of PV power based on PHIL testing using different interface algorithms. In Proceedings of the IECON 2013—39th Annual Conference of the IEEE Industrial Electronics Society, Vienna, Austria, 10–13 November 2013; pp. 5380–5385.

12.  Summers, A.; Hernandez-Alvidrez, J.; Darbali-Zamora, R.; Reno, M.J.; Johnson, J.; Gurule, N.S. Comparison of Ideal Transformer Method and Damping Impedance Method for PV Power-Hardware-In-The-Loop Experiments. In Proceedings of the 2019 IEEE 46th Photovoltaic Specialists Conference (PVSC), Chicago, IL, USA, 16–21 June 2019; pp. 2989–2996.

13.  Stanev, R. A control strategy and operation paradigm for electrical power systems with electric vehicles and distributed energy ressources. In Proceedings of the 2016 19th International Symposium on Electrical Apparatus and Technologies (SIELA), Bourgas, Bulgaria, 29 May–1 June 2016; pp. 1–4.

14.  Strasser, T.; de Jong, E.C.W.; Sosnina, M. (Eds). *European Guide to Power System Testing: The ERIGrid Holistic Approach for Evaluating Complex Smart Grid Configurations*, 1st ed.; Springer International Publishing AG: Basel, Switzerland, 2020. [CrossRef]

15.  Strasser, T.I.; Babazadeh, D.; Heussen, K.; Pelegrino, L.; Arnold, G.; Nguyen, V.H.; Palensky, P.; Kotsampopoulos, P.; Kontou, A.; Hatziargyriou, N.; et al. Virtual ERIGrid Final Conference. 2020. Available online: https://zenodo.org/record/3769631#.XvBJEXERXIU (accessed on 27 April 2020).

16.  Vogel, S.; Stevic, M.; Nguyen, H.T.; Jensen, T.V.; Heussen, K.; Rajkumar, V.S.; Monti, A. Distributed Power Hardware-in-the-Loop Testing using a Grid-forming Converter as Power Interface. *Energies* **2020**, 1–22, submitted.

17.  Papaspiliotopoulos, V.A.; Korres, G.N.; Kleftakis, V.A.; Hatziargyriou, N.D. Hardware-In-the-Loop Design and Optimal Setting of Adaptive Protection Schemes for Distribution Systems With Distributed Generation. *IEEE Trans. Power Deliv.* **2017**, *32*, 393–400. [CrossRef]

18.  Jennett, K.I.; Booth, C.D.; Coffele, F.; Roscoe, A.J. Investigation of the sympathetic tripping problem in power systems with large penetrations of distributed generation. *IET Gener. Transm. Distrib.* **2015**, *9*, 379–385. [CrossRef]

19.  Coffele, F.; Booth, C.; Dyśko, A.; Burt, G. Quantitative analysis of network protection blinding for systems incorporating distributed generation. *IET Gener. Transm. Distrib.* **2012**, *6*, 1218–1224. [CrossRef]

20.  Montoya, L.A.; Montenegro, D.; Ramos, G. Adaptive protection testbed using real time and hardware-in-the-loop simulation. In Proceedings of the 2013 IEEE Grenoble Conference, Grenoble, France, 16–20 June 2013; pp. 1–4.

21.  Mishra, P.; Pradhan, A.K.; Bajpai, P. Adaptive Relay Setting for Protection of Distribution System with Solar PV. In Proceedings of the 2018 20th National Power Systems Conference (NPSC), Tiruchirappalli, India, 14–16 December 2018; pp. 1–5.

22.  Jain, R.; Lubkeman, D.L.; Lukic, S.M. Dynamic Adaptive Protection for Distribution Systems in Grid-Connected and Islanded Modes. *IEEE Trans. Power Deliv.* **2019**, *34*, 281–289. [CrossRef]

23.  Safari-Shad, N.; Franklin, R.; Negahdari, A.; Toliyat, H.A. Adaptive 100% Injection-Based Generator Stator Ground Fault Protection with Real-Time Fault Location Capability. *IEEE Trans. Power Deliv.* **2018**, *33*, 2364–2372. [CrossRef]

24. Darbali-Zamora, R.; Quiroz, J.E.; Hernández-Alvidrez, J.; Johnson, J.; Ortiz-Rivera, E.I. Validation of a Real-Time Power Hardware-in-the-Loop Distribution Circuit Simulation with Renewable Energy Sources. In Proceedings of the 2018 IEEE 7th World Conference on Photovoltaic Energy Conversion (WCPEC) (A Joint Conference of 45th IEEE PVSC, 28th PVSEC 34th EU PVSEC), Waikoloa Village, HI, USA, 10–15 June 2018; pp. 1380–1385. [CrossRef]

25. Darbali-Zamora, R.; Hernandez-Alvidrez, J.; Summers, A.; Gurule, N.S.; Reno, M.J.; Johnson, J. Distribution Feeder Fault Comparison Utilizing a Real-Time Power Hardware-in-the-Loop Approach for Photovoltaic System Applications. In Proceedings of the 2019 IEEE 46th Photovoltaic Specialists Conference (PVSC), Chicago, IL, USA, 16–21 June 2019; pp. 2916–2922. [CrossRef]

26. Li, S.; Hilbrich, D.; Bonetti, A.; Paquin, J.N. Evolution to model-based testing of protection systems and the publication of the IEC 60255-121 standard. In Proceedings of the 8th PAC World Conference, Wroclaw, Poland, 26–29 June 2017.

27. Stifter, M.; Cordova, J.; Kazmi, J.; Arghandeh, R. Real-time simulation and hardware-in-the-loop testbed for distribution synchrophasor applications. *Energies* **2018**, *11*, 876. [CrossRef]

28. Aleem, S.K.A.; Aftab, M.A.; Hussain, S.M.S.; Ali, I.; Ganesh, V.; Ustun, T.S. Real-Time Microgrid Synchronization using Phasor Measurement Units. In Proceedings of the 2019 IEEE International Conference on Intelligent Systems and Green Technology (ICISGT), Visakhapatnam, India, 29–30 June 2019; pp. 19–193.

29. IEEE. *IEEE Standard for Interconnecting Distributed Resources with Electric Power Systems*; IEEE Std 1547-2003; IEEE: Piscataway, NJ, USA, 28 July 2003; pp. 1–28.. [CrossRef]

30. Baltensperger, D.; Dobrowolski, J.; Obushevs, A.; Segundo Sevilla, F.; Korba, P. Scaling Version of Kundur's Two-Areas System for Electromechanical Oscillations Representation. In Proceedings of the 2020 International Symposium on Power Electronics, Electrical Drives, Automation and Motions (SPEEDAM), Sorrento, Italy, 24–26 June 2020; pp. 1–7.

31. Kundur, P.; Balu, N.J.; Lauby, M.G. *Power System Stability and Control*; McGraw-hill: New York, NY, USA, 1994; Volume 7.

32. Salcedo, R.; Corbett, E.; Smith, C.; Limpaecher, E.; Rekha, R.; Nowocin, J.; Lauss, G.; Fonkwe, E.; Almeida, M.; Gartner, P.; et al. Banshee distribution network benchmark and prototyping platform for hardware-in-the-loop integration of microgrid and device controllers. *J. Eng.* **2019**, *2019*, 5365–5373. [CrossRef]

33. Kikusato, H.; Ustun, T.S.; Suzuki, M.; Sugahara, S.; Hashimoto, J.; Otani, K.; Shirakawa, K.; Yabuki, R.; Watanabe, K.; Shimizu, T. Microgrid Controller Testing Using Power Hardware-in-the-Loop. *Energies* **2020**, *13*, 2044. [CrossRef]

34. Kikusato, H.; Ustun, T.S.; Suzuki, M.; Sugahara, S.; Hashimoto, J.; Otani, K.; Shirakawa, K.; Yabuki, R.; Watanabe, K.; Shimizu, T. Integrated Power Hardware-in-the-Loop and Lab Testing for Microgrid Controller. In Proceedings of the 2019 IEEE Innovative Smart Grid Technologies—Asia (ISGT Asia), Chengdu, China, 21–24 May 2019; pp. 2743–2747. [CrossRef]

35. Ustun, T.S.; Konishi, H.; Hashimoto, J.; Otani, K. Hardware-in-the-loop simulation based testing of power conditioning systems. In Proceedings of the 2018 IEEE International Conference on Industrial Electronics for Sustainable Energy Systems (IESES), Hamilton, New Zealand, 31 January–2 February 2018; pp. 546–551.

36. Brandl, R.; Kotsampopoulos, P.; Lauss, G.; Maniatopoulos, M.; Nuschke, M.; Montoya, J.; Strasser, T.I.; Strauss-Mincu, D. Advanced Testing Chain Supporting the Validation of Smart Grid Systems and Technologies. In Proceedings of the 2018 IEEE Workshop on Complexity in Engineering (COMPENG), Florence, Italy, 10–12 October 2018; pp. 1–6. [CrossRef]

37. Sun, C.; Joos, G.; Ali, S.Q.; Paquin, J.N.; Rangel, C.M.; Jajeh, F.A.; Novickij, I.; Bouffard, F. Design and Real-time Implementation of a Centralized Microgrid Control System with Rule-based Dispatch and Seamless Transition Function. *IEEE Trans. Ind. Appl.* **2020**, *56*, 3168–3177. [CrossRef]

38. Al Jajeh, M.F.; Qaseem Ali, S.; Joos, G.; Novickij, I. Islanding of a Microgrid Using a Distributed Multi-agent Control System. In Proceedings of the 2019 IEEE Energy Conversion Congress and Exposition (ECCE), Baltimore, MD, USA, 29 September–3 October 2019; IEEE: Baltimore, MD, USA, 2019; pp. 6286–6293. [CrossRef]

39. Du, Y.; Tu, H.; Lukic, S.; Lubkeman, D.; Dubey, A.; Karsai, G. Development of a Controller Hardware-in-the-Loop Platform for Microgrid Distributed Control Applications. In Proceedings of the 2018 IEEE Electronic Power Grid (eGrid), Charleston, SC, USA, 12–14 November 2018; IEEE: Charleston, SC, USA, 2018; pp. 1–6. [CrossRef]

40. Tu, H.; Du, Y.; Yu, H.; Dubey, A.; Lukic, S.; Karsai, G. Resilient Information Architecture Platform for the Smart Grid (RIAPS): A Novel Open-Source Platform for Microgrid Control. *IEEE Trans. on Ind. Electron.* **2019**. [CrossRef]

41. Razeghi, G.; Gu, F.; Neal, R.; Samuelsen, S. A generic microgrid controller: Concept, testing, and insights. *Appl. Energy* **2018**, *229*, 660–671. [CrossRef]

42. Bagudai, S.K.; Ray, O.; Samantaray, S.R. Evaluation of Control Strategies within Hybrid DC/AC Microgrids using Typhoon HIL. In Proceedings of the 2019 8th International Conference on Power Systems (ICPS), Jaipur, India, 20–22 December 2019; pp. 1–6.

43. Nguyen, T.T.; Yoo, H.J.; Kim, H.M. A Droop Frequency Control for Maintaining Different Frequency Qualities in a Stand-Alone Multimicrogrid System. *IEEE Trans. Sustain. Energy* **2018**, *9*, 599–609. [CrossRef]

44. Yoo, H.J.; Nguyen, T.T.; Kim, H.M. Consensus-Based Distributed Coordination Control of Hybrid AC/DC Microgrids. *IEEE Trans. Sustain. Energy* **2020**, *11*, 629–639. [CrossRef]

45. Johnson, J.; Ablinger, R.; Bründlinger, R.; Fox, B.; Flicker, J. Design and Evaluation of SunSpec-Compliant Smart Grid Controller with an Automated Hardware-in-the-Loop Testbed. *Technol. Econ. Smart Grids Sustain. Energy* **2017**, *2*, 16. [CrossRef]

46. Johnson, J.; Ablinger, R.; Bruendlinger, R.; Fox, B.; Flicker, J. Interconnection Standard Grid-Support Function Evaluations Using an Automated Hardware-in-the-Loop Testbed. *IEEE J. Photovolt.* **2018**, *8*, 565–571. [CrossRef]

47. Ninad, N.; Apablaza-Arancibia, E.; Bui, M.; Johnson, J.; Gonzalez, S.; Moore, T. Development and Evaluation of Open-Source IEEE 1547.1 Test Scripts for Improved Solar Integration. In Proceedings of the EU PVSEC 2019: 36th European Photovoltaic Solar Energy Conference and Exhibition, Marseille, France, 9–13 September 2019.

48. Bründlinger, R.; Stöckl, J.; Miletic, Z.; Ablinger, R.; Leimgruber, F.; Johnson, J.; Shi, J. Pre-certification of Grid-Code Compliance for Solar Inverters with an Automated Controller-Hardware-In-The-Loop Test Environment. In Proceedings of the 8th Solar Integration Workshop, Stockholm, Sweden, 16–17 October 2018.

49. Stöckl, J.; Miletic, Z.; Bründlinger, R.; Schulz, J.; Ablinger, R.; Tremmel, W.; Johnson, J. Pre-Evaluation of Grid Code Compliance for Power Electronics Inverter Systems in Low-Voltage Smart Grids. In Proceedings of the 20th European Conference on Power Electronics and Applications (EPE'18 ECCE Europe), Riga, Latvia, 17–21 September 2018.

50. EPRI. *Performance Assessment of Inverter on-Board Islanding Detection with Multiple Testing Platforms*; Final Project Report, Product Id: 3002014051; Technical Report of Electric Power Research Institute: Palo Alto, CA, USA, 2020.

51. EPRI. *Risk of Islanding Study Utilizing Hardware in the Loop*; Technical Update Report, Product Id: 3002017225; Technical Report of Electric Power Research Institute: Palo Alto, CA, USA, 2020.

52. EPRI. *Inverter On-board Detection Methods to Prevent Unintended Islanding: Generic Response Models*; Technical Update Report, Product Id: 3002014049; Technical Report of Electric Power Research Institute: Palo Alto, CA, USA, 2018.

53. Steurer, M.M.; Schoder, K.; Faruque, O.; Soto, D.; Bosworth, M.; Sloderbeck, M.; Bogdan, F.; Hauer, J.; Winkelnkemper, M.; Schwager, L.; et al. Multifunctional megawatt-scale medium voltage DC test bed based on modular multilevel converter technology. *IEEE Trans. Transp. Electrif.* **2016**, *2*, 597–606. [CrossRef]

54. Langston, J.; Schoder, K.; Steurer, M.; Faruque, O.; Hauer, J.; Bogdan, F.; Bravo, R.; Mather, B.; Katiraei, F. Power hardware-in-the-loop testing of a 500 kW photovoltaic array inverter. In Proceedings of the IECON 2012—38th Annual Conference on IEEE Industrial Electronics Society, Montreal, QC, Canada, 25–28 October 2012; pp. 4797–4802.

55. Nuschke, M.; Brandl, R.; Montoya, J. Entwicklung und Test eines Microgrid Controllers. RET.Con 2018: 1. Regenerative Energietechnik-Konferenz, 2018. pp. 172–180. Available online: https://www.hs-nordhausen. de/fileadmin/daten/aktuelles/veranstaltungen/ret.con/tagungsband_retcon_2018_web.pdf (accessed on 20 April 2020).

56. *IEEE Std 2030.8-2018 IEEE Standard for the Testing of Microgrid Controllers*; IEEE: Piscataway, NJ, USA, 24 August 2018; pp. 1–42. [CrossRef]

57. Hashimoto, J.; Ustun, T.S.; Otani, K. Smart Inverter Functionality Testing for Battery Energy Storage Systems. *Smart Grid Renew. Energy* **2017**, *8*, 337–350. [CrossRef]

58. Vogt, M.; Marten, F.; Braun, M. A survey and statistical analysis of smart grid co-simulations. *Appl. Energy* **2018**, *222*, 67–78. [CrossRef]

59. Gavriluta, C.; Lauss, G.; Strasser, T.I.; Montoya, J.; Brandl, R.; Kotsampopoulos, P. Asynchronous Integration of Real-Time Simulators for HIL-based Validation of Smart Grids. In Proceedings of the IECON 2019—45th Annual Conference of the IEEE Industrial Electronics Society, Lisbon, Portugal, 14–17 October 2019; Volume 1, pp. 6425–6431. [CrossRef]

60. Vogt, M.; Marten, F.; Montoya, J.; Töbermann, C.; Braun, M. A REST based co-simulation interface for distributed simulations. In Proceedings of the 2019 IEEE Milan PowerTech, Milan, Italy, 23–27 June 2019; pp. 1–6. [CrossRef]

61. Zhang, L.; Li, S.; Wihl, L.; Kazemtabrizi, M.; Ali, S.Q.; Paquin, J.N.; Labbé, S. Cybersecurity Study of Power System Utilizing Advanced CPS Simulation Tools. In Proceedings of the 2019 PAC World Americas Conference, Raleigh, NC, USA, 19–22 August 2019; p. 13.

62. Bian, D.; Kuzlu, M.; Pipattanasomporn, M.; Rahman, S.; Wu, Y. Real-time co-simulation platform using OPAL-RT and OPNET for analyzing smart grid performance. In Proceedings of the 2015 IEEE Power & Energy Society General Meeting, Denver, CO, USA, 26–30 July 2015; IEEE: Denver, CO, USA, 2015; pp. 1–5. [CrossRef]

63. Armendariz, M.; Chenine, M.; Nordstrom, L.; Al-Hammouri, A. A co-simulation platform for medium/low voltage monitoring and control applications. In Proceedings of the ISGT 2014, Washington, DC, USA, 19–22 February 2014; IEEE: Washington, DC, USA, 2014; pp. 1–5. [CrossRef]

64. Johnson, J.; Onunkwo, I.; Cordeiro, P.; Wright, B.; Jacobs, N.; Lai, C. Assessing DER Network Cybersecurity Defenses in a Power-Communication Co-Simulation Environment. *IET Cyber-Phys. Syst. Theory Appl.* **2020**. [CrossRef]

65. Onunkwo, I.; Wright, B.; Cordeiro, P.; Jacobs, N.; Lai, C.; Johnson, J.; Hutchins, T.; Stout, W.; Chavez, A.; Richardson, B.T.; et al. *Cybersecurity Assessments on Emulated DER Communication Networks*; Technical Report SAND2019-2406; Sandia National Laboratories, Albuquerque, NM, USA, 2019.

66. Ustun, T.S.; Hussain, S.M.S. Implementation of IEC 61850 Based Integrated EV Charging Management in Smart Grids. In Proceedings of the 2019 IEEE Vehicle Power and Propulsion Conference (VPPC), Hanoi, Vietnam, 14–17 October 2019; pp. 1–5. [CrossRef]

67. Faruque, M.O.; Dinavahi, V.; Sloderbeck, M.; Steurer, M. Geographically distributed thermo-electric co-simulation of all-electric ship. In Proceedings of the 2009 IEEE Electric Ship Technologies Symposium, Baltimore, MD, USA, 20–22 April 2009; pp. 36–43.

68. Vogel, S.; Stevic, M.; Kadavil, R.; Mohanpurkar, M.; Koralewicz, P.; Gevorgian, V.; Hovsapian, R.; Monti, A. Distributed Real-Time Simulation and its Applications to Wind Energy Research. In Proceedings of the 2018 IEEE International Conference on Probabilistic Methods Applied to Power Systems (PMAPS), Boise, ID, USA, 24–28 June 2018; pp. 1–6.

69. Vogel, S.; Rajkumar, V.S.; Nguyen, H.T.; Stevic, M.; Bhandia, R.; Heussen, K.; Palensky, P.; Monti, A. Improvements to the Co-simulation Interface for Geographically Distributed Real-time Simulation. In Proceedings of the IECON 2019—45th Annual Conference of the IEEE Industrial Electronics Society, Lisbon, Portugal, 14–17 October 2019; Volume 1, pp. 6655–6662.

70. Ravikumar, K.G.; Schulz, N.N.; Srivastava, A.K. Distributed simulation of power systems using real-time digital simulator. In Proceedings of the 2009 IEEE/PES Power Systems Conference and Exposition, Seattle, WA, USA, 15–18 March 2009; pp. 1–6.

71. Stevic, M.; Estebsari, A.; Vogel, S.; Pons, E.; Bompard, E.; Masera, M.; Monti, A. Multi-site European framework for real-time co-simulation of power systems. *IET Gener. Transm. Distrib.* **2017**, *11*, 4126–4135. [CrossRef]

72. Wang, Y.; Nguyen, T.L.; Syed, M.H.; Xu, Y.; Nguyen, V.H.; Guillo-Sansano, E.; Burt, G.; Tran, Q.T.; Caire, R. A Distributed Control Scheme of Microgrids in Energy Internet and Its Multi-Site Implementation. *IEEE Trans. Ind. Inform.* **2020**. [CrossRef]

73. Lundstrom, B.; Palmintier, B.; Rowe, D.; Ward, J.; Moore, T. Trans-oceanic remote power hardware-in-the-loop: Multi-site hardware, integrated controller, and electric network co-simulation. *IET Gener. Transm. Distrib.* **2017**, *11*, 4688–4701. [CrossRef]

74. Palmintier, B.; Lundstrom, B.; Chakraborty, S.; Williams, T.; Schneider, K.; Chassin, D. A Power Hardware-in-the-Loop Platform with Remote Distribution Circuit Cosimulation. *IEEE Trans. Ind. Electron.* **2015**, *62*, 2236–2245. [CrossRef]

75. Montoya, J.; Brandl, R.; Vogt, M.; Marten, F.; Maniatopoulos, M.; Fabian, A. Asynchronous Integration of a Real-Time Simulator to a Geographically Distributed Controller Through a Co-Simulation Environment. In Proceedings of the IECON 2018—44th Annual Conference of the IEEE Industrial Electronics Society, Washington, DC, USA, 21–23 October 2018; pp. 4013–4018. [CrossRef]

76. Monti, A.; Stevic, M.; Vogel, S.; De Doncker, R.W.; Bompard, E.; Estebsari, A.; Profumo, F.; Hovsapian, R.; Mohanpurkar, M.; Flicker, J.D.; et al. A Global Real-Time Superlab: Enabling High Penetration of Power Electronics in the Electric Grid. *IEEE Power Electron. Mag.* **2018**, *5*, 35–44. [CrossRef]

77. Guillo-Sansano, E.; Roscoe, A.J.; Burt, G.M. Harmonic-by-harmonic time delay compensation method for PHIL simulation of low impedance power systems. In Proceedings of the 2015 International Symposium on Smart Electric Distribution Systems and Technologies (EDST), Vienna, Austria, 8–11 September 2015; pp. 560–565.

78. Guillo-Sansano, E.; Syed, M.; Roscoe, A.J.; Burt, G.; Coffele, F. Characterization of Time Delay in Power Hardware in the Loop Setups. *IEEE Trans. Ind. Electron.* **2020**. [CrossRef]

79. Wiezorek, C.; Parisio, A.; Kyntäjä, T.; Elo, J.; Gronau, M.; Johannson, K.H.; Strunz, K. Multi-location virtual smart grid laboratory with testbed for analysis of secure communication and remote co-simulation: Concept and application to integration of Berlin, Stockholm, Helsinki. *IET Gener. Transm. Distrib.* **2017**, *11*, 3134–3143. [CrossRef]

80. Strasser, T.I.; Pellegrino, L.; Degefa, M.Z.; Lagos, D.; Syed, M. Demonstration of Multi Research Infrastructure Integration Tests. 2019. ERIGrid Webinar. Available online: https://doi.org/10.5281/zenodo.3553979 (accessed on 22 April 2020).

81. IEEE. *IEEE Approved Draft Standard Conformance Test Procedures for Equipment Interconnecting Distributed Energy Resources with Electric Power Systems and Associated Interfaces*; IEEE P1547.1/D9.9; IEEE: Piscataway, NJ, USA, 2020; pp. 1–283.

82. Paquin, J.N. Synchrophasor Application Studies using Real-Time Simulators. In Proceedings of the 1st International Synchrophasor Symposium, Atlanta, GA, USA, 22–24 March 2016.

83. Sunspec System Validation Platform (SVP). 2020. Available online: https://sunspec.org/svp/ (accessed on 1 May 2020).

84. Johnson, J.; Apablaza-Arancibia, E.; Ninad, N.; Turcotte, D.; Prieur, A.; Ablinger, R.; Bründlinger, R.; Moore, T.; Heidari, R.; Hashimoto, J.; et al. International Development of a Distributed Energy Resource Test Platform for Electrical and Interoperability Certification. In Proceedings of the 2018 IEEE 7th World Conference on Photovoltaic Energy Conversion (WCPEC) (A Joint Conference of 45th IEEE PVSC, 28th PVSEC 34th EU PVSEC), Waikoloa Village, HI, USA, 10–15 June 2018; pp. 2492–2497. [CrossRef]

85. Qiu, D.; Paquin, J.N.; Uda, S.; Kashihara, H.; Fukuda, Y.; Okazaki, N.; Nishimura, S.; Kawasaki, Y.; Gao, F.; Ali, S.Q. Design and HIL Testing of a Hybrid UPS with Seamless Transition. In Proceedings of the 2019 IEEE Power & Energy Society General Meeting (PESGM), Atlanta, GA, USA, 4–8 August 2019; IEEE: Atlanta, GA, USA, 2019; pp. 1–5. [CrossRef]

86. Paré, D.; Turmel, G.; Marcoux, B.; McNabb, D. Validation Tests of The Hypersim Digital Real Time Simulator with a Large AC-DC Network. In Proceedings of the International Conference on Power Systems Transients (IPST2003), New Orleans, LA, USA, 28 September–2 October 2003; p. 6.

87. Guay, F.; Chiasson, P.A.; Verville, N.; Tremblay, S.; Askvid, P. New Hydro-Québec Real-Time Simulation Interface for HVDC Commissioning Studies. In Proceedings of the International Conference on Power Systems Transients (IPST2017), Seoul, Korea, 26–29 June 2017; p. 8.

88. Vernay, Y.; Martin, C.; Petesch, D.; Dennetière, S. Hardware in the loop simulations to test SVC performances on the French Grid. In Proceedings of the International Conference on Power Systems Transients (IPST2015), Cavtat, Croatia, 15–18June, 2015; p. 8.

89. Li, G.; Dong, Y.; Tian, J.; Wang, W.; Li, W.; Belanger, J. Factory acceptance test of a five-terminal MMC control and protection system using hardware-in-the-loop method. In *2015 IEEE Power & Energy Society General Meeting*; IEEE: Denver, CO, USA, 2015; pp. 1–5. [CrossRef]

90. Syed, M.H.; Guillo-Sansano, E.; Avras, A.; Downie, A.; Jennett, K.; Burt, G.M.; Coffele, F.; Rudd, A.; Bright, C. The Role of Experimental Test Beds for the Systems Testing of Future Marine Electrical Power Systems. In Proceedings of the 2019 IEEE Electric Ship Technologies Symposium (ESTS), Washington, DC, USA, 14–16 August 2019; pp. 141–148.
91. Langston, J.; Steurer, M.; Schoder, K.; Borraccini, J.; Dalessandro, D.; Rumney, T.; Fikse, T. Power hardware-in-the-loop simulation testing of a flywheel energy storage system for shipboard applications. In Proceedings of the 2017 IEEE Electric Ship Technologies Symposium (ESTS), Arlington, VA, USA, 14–17 August 2017; pp. 305–311. [CrossRef]
92. Sanchez, J.; Wetz, D.; Dong, Q.; Heinzel, J. Integration and study of hardware in the loop diesel generator with a hybrid energy storage module for naval applications. In Proceedings of the 2017 IEEE Electric Ship Technologies Symposium (ESTS), Arlington, VA, USA, 14–17 August 2017; pp. 580–585. [CrossRef]
93. Strank, S.; Feng, X.; Gattozzi, A.; Wardell, D.; Pish, S.; Herbst, J.; Hebner, R. Experimental test bed to de-risk the navy advanced development model. In Proceedings of the 2017 IEEE Electric Ship Technologies Symposium (ESTS), Arlington, VA, USA, 14–17 August 2017; pp. 352–358. [CrossRef]
94. UL 1741. *Inverters, Converters, Controllers and Interconnection System Equipment for Use with Distributed Energy Resources*, 2nd ed.; UL Standard, Northbrook, IL, USA, 2018.
95. Nuschke, M. Development of a microgrid controller for black start procedure and islanding operation. In Proceedings of the 2017 IEEE 15th International Conference on Industrial Informatics (INDIN), Emden, Germany, 24–26 July 2017; pp. 439–444.

MDPI

St. Alban-Anlage 66

4052 Basel

Switzerland

Tel. +41 61 683 77 34

Fax +41 61 302 89 18

www.mdpi.com

*Energies* Editorial Office

E-mail: energies@mdpi.com

www.mdpi.com/journal/energies